Food Science and Technology

nova
science publishers
www.novapublishers.com

Food Science and Technology

Food Byproducts: Valorization Through Nutraceutical Production
Sandra Teresita Martín-del-Campo, PhD (Editor)
Anaberta Cardador-Martínez, PhD (Editor)
Jessica del Pilar Ramírez-Anaya, PhD (Editor)
2023. ISBN: 979-8-88697-612-0 (Hardcover)
2023. ISBN: 979-8-88697-876-6 (eBook)

Bioprospecting of Natural Compounds in Food, Pharmaceutical and Biomedical Science
Siddhartha Pati, PhD (Editor)
Tanmay Sarkar, PhD (Editor)
Dibyait Lahiri, PhD (Editor)
2022. ISBN: 979-8-88697-361-7 (Hardcover)
2022. ISBN: 979-8-88697-431-7 (eBook)

Proceedings of BIOSPECTRUM: The International Conference on Biotechnology and Biological Sciences: Biotechnological Intervention Towards Enhancing Food Value
Sanket Joshi, PhD (Editor)
Susmita Mukherjee (Editor)
Moupriya Nag, PhD (Editor)
2022. ISBN: 978-1-68507-985-7 (Hardcover)
2022. ISBN: 979-8-88697-069-2 (eBook)

Nutraceuticals: Food Applications and Health Benefits
Anita Kumari, PhD (Editor)
Gulab Singh, PhD (Editor)
2022. ISBN: 978-1-68507-488-3 (Hardcover)
2022. ISBN: 978-1-68507-512-5 (eBook)

More information about this series can be found at
https://novapublishers.com/product-category/series/food-science-and-technology-series/

Gulcin Yildiz
Editor

A Closer Look at Homogenization

nova
science publishers
www.novapublishers.com

https://doi.org/10.52305/CIBQ6632

NOTICE TO THE READER

Library of Congress Cataloging-in-Publication Data

ISBN: 979-8-89113-656-4 (Hardcover)
ISBN: 979-8-89113-669-4 (eBook)

Published by Nova Science Publishers, Inc. † New York

Contents

Preface

This book includes seven chapters about homogenization fundamentals. Chapter one describes the basic homogenization principles. Chapter two provides data related to the asymptotic expansion methods in homogenization. Chapter three summarizes the ultrasonic homogenization in the food industry. Chapter four presents the selected studies that illustrate the evolution and the state of the art in mathematical and computational modeling applied to the homogenization of grain storage conditions. Chapter five highlights the effect of high-pressure homogenization on the quality characteristics of meat and meat products. Chapter six details the impact of high-pressure homogenization on milk and dairy products. Chapter seven discusses the importance of homogenization on plant-based oils and fats.

Chapter 1

Principle of Homogenization

Gulcin Yildiz*

Department of Food Engineering, Igdir University, Iğdır, Turkey

Abstract

The process of homogenization involves changing the particle size of a product to achieve homogeneity throughout [1]. It can be divided into three main categories:

- Mechanical homogenization
- Pressure homogenization
- Ultrasonic pressure homogenization [2].

There are numerous uses of homogenization in the food, chemical, pharmaceutical, and chemical industries. Milk homogenization is the most significant use of the homogenization process in the food sector. The homogenizer selection criteria depend on the specific application.

Keywords: homogenization, mechanical, milk, pressure, ultrasonic

Introduction

The terms *homogeneity* and *heterogeneity* refer to a material's uniformity or lack thereof. A substance that is heterogeneous lacks regularity in one of these characteristics while being homogeneous in composition or character. Every combination, from atoms to populations of people or animals to galaxies, can

* Corresponding Author's Email: gulcn86@gmail.com.

In: A Closer Look at Homogenization
Editor: Gulcin Yildiz
ISBN: 979-8-89113-656-4

benefit from this notion. A substance may therefore be homogenous on a greater scale as opposed to heterogeneous on a smaller scale within the same substance [3].

Auguste Gaulin received a patent for his homogenizer in 1899. The invention was for a three-piston pump that pushed liquid through one or more hair-like tubes while under pressure. In 1900, Auguste Gaulin displayed his invention—a method for "treating" milk—at the Paris World's Fair. According to literature and periodicals of the time, milk processed by the Gaulin machine is where the term "homogenized" first appeared. Therefore, the processes and tools created by Gaulin are what the phrases "homogenization" and "homogenizer" historically refer to. The homogenizer is essentially a positive displacement pump to which a homogenizing valve assembly is coupled. Under pressure, the pump pushes fluids through the homogenizing valve. The action that takes place inside the homogenizing valve assembly is referred to as "homogenization" in this context [2].

History of High Pressure Homogenization Applications

High pressure homegenization (HPH) is a technology that has been used commercially for many years in ceramic materials, synthetic quartz, and many other industrial applications [4-6]. In the 1970s and 1980s, great advances were made in the use of high pressure in the ceramics and metallurgy industry [5]. However, the applications of high pressure in the food industry are very new and are still in the research phase. Particularly in Japan, intensive studies are being carried out on this subject and this technique is being developed for the food industry [4].

As an alternative to high-temperature treatment of foods, the effectiveness of HPH applications was first demonstrated by chemist Bert Hite in 1899 [4, 7]. At the beginning of the century, Hite succeeded in designing a machine that could apply pressures of over 600 MPa. Hite and those who later worked on this subject conducted research on the usability of HPH application in many food production processes.

Hite (1899) first examined the numerical change in the microorganism population in milk under a pressure of 680 MPa for 10 minutes at room temperature. As a result of the research, he detected a decrease of five to six decimal values in the number of microorganisms in milk. In the same years, many researchers did experiments on the subject. Bridgman is one of these researchers [6].

Hite and Bridgman pioneered the study of the use of HPH in the late 1800s and early 1900s by detecting the inactivation of bacteria in milk and the denaturation of egg albumin proteins with HPH. In the last century, HPH applications in foods have been extensively studied [8]. The lack of availability of appropriate equipment prevented the early implementation of HPH applications [5].

Until the 1990s, many studies were conducted examining the relationship between HPH and food processes. The most important of these are:

- In 1914, Bridgman's study of inactivation of egg white proteins by pressure.
- Tomson and Short's study on the inactivation of microorganisms in raw milk by pressure in 1965.
- In 1974, Wilson's study on the sterilization of low-acid foods using pressure and pasteurization temperature.
- Charm's study in 1977 showing that pressure around 200 MPa increased the storage life of foods at refrigerator temperatures.
- The study of Elgasim and Kennick in 1980 on the effect of pressure on steak protein quality.
- Hoover et al. [9] can be summarized as follows: Molecular and cellular changes in systems subjected to hydrostatic pressure.

Nowadays, this technique has found a wide application area in the food industry [6]. It is stated that this technique has been successfully applied to all foods, from meat to fruit juice, at room temperature, in the pressure range of 100-900 MPa [10]. The first commercial products treated with HPH appeared on the Japanese market in 1991 [5].

Nowadays, jams, sauces, ready-made desserts and fruit juices with HPH applied are offered for consumption primarily in Japan, but also in Europe and the USA [11]. HPH-treated oysters and guacamole (a kind of sauce) are on the shelves in the USA [12] while sliced cooked ham and orange juice are on the shelves in Europe [5].

High Pressure Homogenization Systems

Industrial high pressure homogenization systems consist of four main parts [10]:

1. High pressure vessel where pressure is applied.
2. High pressure generation mechanism.
3. Temperature control mechanism.
4. Material placement system.

The high pressure vessel is the heart of the system. This container is made of high-tensile stainless steel in a cylindrical shape. The wall thickness of the pressure vessel varies depending on the high pressure and working capacity used [4, 10, 11]. Different 'lid' designs are used depending on the features of the system; lids that can be opened and closed quickly are preferred for applications in the food industry. The design of the high pressure vessel is very important in the economic evaluation of high pressure applications [10].

During application, the substance must be in a pressure-conducting environment [11]. In cold and warm isostatic pressure applications, water is mostly used as the pressure transmitting medium. Water contains a small amount of oil to ensure slipping and prevent corrosion during the process [12, 13]. After all the air in the system is evacuated, high pressure is provided by one of the direct, indirect or ambient heating methods [4, 11].

Methods of Obtaining High Pressure

Direct Compression

Piston type compression is applied in direct compression method. Pressure of the transmitting medium is provided directly by a piston [4]. This method provides very fast compression by providing a dynamic relationship between the piston and the inner surface of the container. The limitations created by high pressure make it mandatory to use this method in pilot scale systems [4, 11, 12].

Indirect Compression

In this method, the pressure transmitter uses a booster used to pump the media from the reservoir into the closed pressure vessel, raising the media to the desired pressure. In this method, the required connections are static. This method is used in industrial isostatic pressure applications [4, 11, 12].

Heating the Pressure Medium

In this method, high pressure is provided by the expansion that occurs by increasing the temperature of the pressure transmitting medium [4]. The internal volume and temperature of the pressure vessel must be controlled very well [11]. This method is especially used in applications where high pressure and high temperature are used in combination [4, 11].

Advantages of High-Pressure Homogenization

1. It increases the shelf life of food products by destroying saprophyte microorganisms that cause spoilage in foods.
2. The high pressure method does not change or slightly changes the quality characteristics of foods such as color, flavor and nutrients at room temperature.
3. It inactivates enzymes that affect the quality of food.
4. Since the high pressure process is carried out at room temperatures, it reduces the amount of energy consumed after reaching the desired pressure.
5. Small molecules such as vitamins, amino acids, simple sugars and flavor compounds are not affected by high pressure
6. Since the pressure is distributed homogeneously throughout the food, it ensures homogeneous preservation of the food [13-24].

Disadvantages of High-Pressure Homogenization

1. High pressure technique is carried out as a batch process and requires high investment and application costs.
2. This method may cause some undesirable textural changes in the fruits and vegetables to be sterilized in whole pieces or slices.
3. Some enzymes that affect food quality show great resistance to high pressure application.
4. High pressure application is only effective in preserving acidic foods [21, 24, 25].

Inactivation Mechanisms of High Pressure

The inactivation mechanism of high pressure can be explained physically by Le Chatelier's principle. According to the law known as escape from force, when an external effect is applied to a physical system that is in balance, the system changes itself to minimize this effect. If the pressure increases, the volume decreases and the density increases, so the equilibrium shifts to the side with fewer moles and the chemical balance of the system changes.

Changes in the Cell Membrane

Pressurized cell membranes often show altered permeability. With a reduction in the cross-section of each phospholipid molecule, a decrease in volume occurs. In general, it has been understood that the first area that pressure damages in microorganisms is the cell membrane. Malfunctions (malfunctions) in the cell membrane caused by pressure cause inhibition of amino acid uptake, probably due to denaturation of membrane proteins. Many studies have shown the loss of intracellular elements of microorganisms after pressure application. Leakage of these components from cells indicates damage to the cell membrane, and as the rate of loss from cells increases, the degree of death and damage increases accordingly [26, 27].

Changes Caused by Pressure in the Cell

In aqueous systems, volume decrease occurs as a result of the increase in the concentration of ionized groups. With the effect of pressure, electrostatic ion bonds are separated and as a result the pH value decreases (While the pH of pure water is 7 at 25°C and 0.1 MPa, it is 6.25 at 100 MPa). Cell morphology changes with the application of pressure and cell division slows down with the application of high pressure. Cells elongate and protoplasm viscosity changes. Various morphological changes that occur as a result of these changes can be counted as separation of the cell wall and cell membrane, followed by thickening and curling of the cell wall.

Effects on Metabolism

Important protein-containing regions in pressure inactivation of microbial cells are enzymes and especially membrane-bound ATPases [28, 29]. It is assumed that pressure plays an important role in the denaturation of key enzymes in the death and damage caused by pressure in some organisms. Factors that promote enzyme inactivation are changes in intramolecular structures and changes in the active sites of the enzyme. Factors affecting enzyme inactivation under pressure include pH, substrate concentration, and subunit structure of the enzyme. Energy-releasing reactions can be prevented by the effect of pressure, slowing down the vital activities of the cell. Pressure affects reaction systems by (a) decreasing the available molecular space required for the reaction, (b) increasing the interchain reaction.

Effects on the Genetic Mechanism

Stabilization occurs in the natural structure of DNA. DNA is more tolerant to pressure than proteins due to the large number of intermolecular hydrogen bridge bonds present in its structure. On the other hand, DNA replication and transcription catalyzed by enzymes are also prevented.

The Homogenizing Industry

The homogenizing field covers a huge range of topics. The definition of homogenous is "having the same composition, structure, or character throughout," whereas the definition of homogenize is "to make or render homogeneous" [30].

Homogenizing is an example of an umbrella term, or a word that refers to a very broad concept. When someone says they are homogenizing, they may be disrupting, emulsifying, dispersing, stirring, blending, mixing, or any other number of processes. There are three main types of homogenizing procedures currently in use:

- mechanical,
- pressure-based, and
- ultrasonic [31].

Pressure Homogenization

For a long time, microbial cells have been broken apart using high-pressure homogenizers. The technique has been found to be generally acceptable for a variety of bacteria, yeast, and mycelia, with the exception of very filamentous microorganisms [32].

This kind of homogenizer functions by applying pressure to a very small channel or aperture and driving cell suspensions through it. They may or may not thereafter impinge at high velocity on a hard-impact ring or against another high-velocity stream of cells coming from the other direction, depending on the type of high-pressure homogenizer. The impingement design is more effective in machines than it is in those without it [30].

A combination of the enormous pressure drop, sharply focused turbulent eddies, and powerful shearing forces causes disruption of the cell wall. The rate of cell disruption is inversely related to the applied pressure and is roughly proportional to the third power of the turbulent velocity of the product flowing through the homogenizer channel. Therefore, the efficiency of disruption each pass through the machine increases as pressure increases. The following are the operational factors that affect high-pressure homogenizers' effectiveness:

- pressure,
- temperature,
- the number of passes,
- the design of the valve and impingement, and
- the flow rate [33].

The most effective method now available to mechanically disrupt nonfilamentous bacteria on a broad scale is high-pressure homogenizers. Animal tissue can also be processed, but first it needs to be homogenized using a paddle blender, rotor-stator homogenizer, or blade blender. Bead mill homogenizers are now challenging high-pressure homogenizers for their dominance in the disruption of microorganisms. The largest industrial high-pressure homogenizer models still beat bead mills in terms of throughput. The largest commercial equipment can handle 4,500 liters of high-pressure homogenizers versus 1200 liters of bead mills worth of microbial suspension each hour. High-pressure homogenizers with even greater capacities are available, however it is unknown how well they break up microbial cells [34].

The fact that most high-pressure homogenizers need multiple passes of the cell solution to produce significant levels of cell disruption, whereas bead mills usually need just one, lessens this throughput benefit considerably [35].

The French press, which generates pressures up to 40000 psi by using a motor-driven piston inside a steel cylinder, is a well-known commercial high-pressure homogenizer for the laboratory. At a rate of roughly 1 ml/min, pressurized sample suspensions up to 35 ml are bled through a needle valve. The sample, piston, and cylinder are often chilled beforehand because the procedure produces heat. For high efficiency of disruption, many runs through the press may be necessary at pressures between 8000 and 10000 psi [36].

In general, fewer passes are made when the pressure is stronger. There are 3.7 and 35 ml capacity pressure cells with a 20000 psi maximum rating, as well as a 35 ml capacity cell with a 40,000 psi rating [35].

The majority of high-pressure homogenizers in use today were modified from industrial machinery created to produce emulsions and homogenates in the food and pharmaceutical sectors. They combine an impingement valve with high pressure. Those with a 10,000 psi maximum pressure rating burst roughly 40% of the cells on the first pass, 60% during the second pass, and 85% during the fourth pass. At cell concentrations of 10-17% w/v, continuous homogenizers have capacities ranging from 55 to 4500 liters/hr. Multiple passes are required with the larger capacity machines to produce high quantities of disruption. A heat exchanger attached to the exit port is necessary since these homogenizers have the potential to generate a significant amount of heat while in use [36].

Mechanical Homogenizers

Mechanical homogenizers can be broken down into two (2) separate categories

a. rotor-stator homogenizers and
b. blade type homogenizers.

Homogenizers for Rotor-Stators

For plant and animal tissue, rotor-stator homogenizers—also known as colloid mills or Willems homogenizers—perform better than blenders with cutting blades. The rotor-stator homogenizer has been used successfully to destroy bacteria when combined with glass beads. However, the homogenized material is tainted with tiny glass and stainless-steel particles, and the rotor-stator homogenizer is experiencing unacceptable levels of abrasive wear. The rotor-stator homogenizer uses cavitation, mechanical and hydraulic shear to destroy cells. Some experts in the homogenizing sector also assert that high-energy sonic and ultrasonic pressure gradients play a little role in the process [37].

While some contest the idea that mechanical homogenizers are affected by high-energy sonic and ultrasonic pressure gradients. The main similarity between ultrasonic and mechanical (rotor-stator) homogenizing is the generation and usage of cavitation to varying degrees. Cavitation is the process by which low-pressure vapor cavities form and then collapse in a moving liquid. When you move a solid object quickly through a liquid, cavitation is produced. In ultrasonics, the probe is the object being moved, and cavitation is being produced as a result of the probe's rapid vibration. In mechanical homogenizing (rotor-stator), cavitation is produced when the blade (rotor) is pushed quickly through the liquid [38].

Most biological tissues are swiftly and fully homogenized with this device, which was originally created to manufacture dispersions and emulsions. A quickly revolving rotor (blade) set inside a stationary head or tube (stator) with slots or holes draws up the proper size of cellular material into the instrument. To leave through the slots or holes, the material is centrifugally propelled forth in a pump-like method. Because of the rotor's (blade's) high rotational speed and the combination of intense turbulence, cavitation, and mechanical shearing that takes place in the tiny space between the rotor and the stator, the tissue is rapidly shrunk in size. The product is repeatedly circulated because most rotor-stator homogenizers have an open layout. The procedure is quick, and depending on how hard the tissue sample is, the desired results are typically attained in 15 to 120 seconds. Shorter periods and a lower rotor speed are used for the recovery of intracellular organelles or receptor site complexes. The following factors need to be optimized for best effectiveness:

- rotor-stator (generator) size and design
- rotor tip speed
- initial sample size,
- medium viscosity,
- processing time or flow rate,
- volume of medium,
- sample concentration,
- vessel shape, and
- rotor-stator placement [39].

The diameter of the rotor-stator probe, also known as the generator, can range from a pencil's width for 0.01 to 10 ml sample volumes to much larger machines with 19000 liters of batch capacity or 68000 liters per hour for online systems. Large units have rotor speeds of 3000 rpm, whereas smaller units have rotor speeds of 8000–60000 rpm. The homogenizer's rotor speed should, in theory, double for every doubling of the rotor diameter. The crucial operational parameter is not the motor's rpms but rather the rotor's tip velocity. When choosing a rotor-stator homogenizer, it's also crucial to consider additional elements including the various rotor-stator designs, construction materials, and ease of leaning [38].

Rotor-stator homogenizers in the 0.01 ml to 20 liter range are used to process liquid samples in laboratories. The volume, viscosity, and kind of medium, as well as the kind and quantity of plant and animal tissue to be processed, should all be taken into account when determining the capacity of the rotor-stator. The volume range within which a given homogenizer rotor-stator size may function well is only approximately 10 fold, and utilizing a homogenizer that is too small will significantly reduce the speed and efficiency of homogenization. Additionally, the majority of laboratory-sized homogenizers only effectively process liquid samples with low to medium viscosities (100000 cps). The practical finding that concentrated samples are broken up more quickly because they collide more frequently must be weighed against this. Higher viscosity samples can be treated, but they need specialized rotor-stator configurations or homogenization vessels that are particularly designed. The sample must be of a size that can be pulled into the stator before being processed with the homogenizer. As a result, samples frequently need to be pre-cut, sliced, or fragmented [39].

Rotor-stator homogenizers can have issues with foaming and aerosols. The first issue can be solved by keeping the homogenizer's tip well submerged in the media and by using vessels of the appropriate size. Round vessels do not perform as well as square or fluted jars do, and it is advantageous to keep the immersed tip off-center. Utilizing covered vessels will help to reduce aerosols [37].

Aerosols are not produced by in-line homogenizers. No laboratory rotor-stator homogenizer is fully explosion-proof, despite some using sealed motors. When employing flammable organic solvents, proper caution should be used by conducting the homogenization in a well-ventilated hood. On the plus side, rotor-stator homogenizers produce little heat while processing, and this heat can be quickly dissipated by immersing the homogenization vessel in icy water [37].

The larger rotor-stator homogenizers are either in-line homogenizers or scaled-up versions of the laboratory units. The latter has teeth that are closely spaced apart and aligned with corresponding teeth in a static liner on the edge of a horizontally oriented, multi-bladed, high-speed impeller [38].

Homogenızers for Blade Types

Blade homogenizers, often known as blenders, have been used for many years to create fine powders and extracts from plant and animal tissue, despite being less effective than rotor-stator homogenizers. The cutting blades of this style of homogenizer rotate at rates ranging from 6000 to 50000 rpm and are either bottom- or top-driven. Unless glass beads or other abrasives are added to the medium, blenders are not suited for disrupting microorganisms. At that point, the same issues that were discussed above for rotor-stator homogenizers arise. Enzymatic browning, a biological oxidation process that frequently occurs in plant tissue homogenizers, might make it more difficult to perform following separation steps. Enzymatic browning is minimized by carrying out the extraction in the absence of oxygen or in the presence of thiol compounds such as mercaptoethanol. Sometimes the addition of polyethylene imine, metal chelators, or detergents such as Triton X-100 or Tween-80 also helps. Blade homogenizers are available for a range of liquid sample sizes from 0.01 ml to multi gallons. Some of the higher rpm homogenizers can reduce tissue samples to a consistent particulate size with distributions as small as 4um as determined by flow cytometric analysis [2].

Ultrasonic Homogenization

Ultrasonic disruption is one technique that is frequently used to damage cells. By creating powerful acoustic pressure waves in a liquid medium, these devices function. The pressure waves drive the liquid to stream and, under the correct circumstances, to rapidly produce micro-bubbles. These bubbles then develop and agglomerate to their resonant size, vibrate ferociously, and finally burst. This occurrence is known as *cavitation.* The vapor phase bubbles implode, creating a shock wave with enough energy to sever covalent bonds. Cells are damaged by shear from collapsing cavitation bubbles and eddying brought on by the vibrating sonic transducer. To effectively disrupt cells, a number of outside factors need to be optimized. The following are these variables:

- amplitude and intensity of the tip
- temperature,
- cell density,
- pressure,
- vessel size and form [40].

Modern ultrasonic processors use crystals composed of lead zirconate titanate as piezoelectric generators. The processor unit resonates between 15 and 25 kHz thanks to a titanium metal horn or probe that transmits the vibrations. Ultrasonic processors range in rated power from 10 to 375 Watts. Because lower power transducers are typically paired with probes having smaller tips, low power output does not always indicate that the cell disintegrator is less effective. What matters is the power density at the tip. To maintain the desired amplitude and intensity under circumstances of increased load, such as high viscosity or pressure, a higher output power is necessary. The volume of samples that can be processed increases with horn size since more power is needed to drive it. However, larger ultrasonic disintegrators produce a lot of heat while running, thus the sample will need to be aggressively cooled from the outside. The typical maximal tip amplitudes vary from 30 to 250 urn, with output intensities of 200 to 2000 W/square cm as a result [41].

The sample suspension's temperature needs to be as low as possible. Low media temperatures encourage high-intensity shock front propagation in addition to resolving the typical worries about the temperature sensitivity of proteins. Therefore, it is desirable to maintain the ultrasonicated fluid's temperature just above its freezing point. The ultrasonic disintegrator produces a lot of heat during processing, which makes things more difficult [42].

Higher hydrostatic pressure (usually 15–60 psi) and viscosity can potentially cause more disruption if the ultrasonic processor has enough power to handle the higher load demand and the sample heating issues are resolved. Glass beads in the 0.05 to 0.5 mm size range are added to microorganisms to increase cell disruption by concentrating energy from bubble implosions and through physical crushing. For yeast and spore disruption, beads are practically necessary. One volume of beads to two volumes of liquid is a decent ratio. Prior to ultrasonic treatment, tough tissues like skin and muscle should be macerated in a blender or similar device and contained within a tiny vessel. The tip shouldn't be inserted into the vessel so shallowly that foaming can occur. Avoid using anti-foaming chemicals or other substances that reduce surface tension. Finally, it is important to remember that free radicals might interact with biological components like proteins, polysaccharides, or nucleic acids because they are produced during ultrasonic activities. By adding scavengers like cysteine, dithiothreitol, or other SH compounds to the medium or saturating the sample with a shielding environment of helium or hydrogen gas, damage from oxidative free radicals can be reduced [43].

The tip diameter of an ultrasonic horn cannot be more than roughly 3 inches for practical reasons. This places a cap on how large these gadgets can be scaled up. By putting the probe tip in a chamber through which a stream of cells flows, standard-sized ultrasonic disrupters have been modified to operate continuously; however, cooling and free radical release are still issues [44].

Homogenization Applications

Applications of Homogenization in Food and Dairy

To make fluid products more stable and have a better texture, homogenization is the process of lowering the particle size of fluid products such as milk, fruit juice, and sauces under circumstances of severe pressure, shear, turbulence, acceleration, and impact. By applying intense pressure to a specific homogenizing valve, the desired effect is produced. Sizes of the particles entering the homogenizer typically range from 0.2 to 20 microns. Depending on the application, large particles are dispersed to create a product with particles that typically range from 0.4 to 1 micron [45].

Applications for Cosmetic Homogenizers

Since it is a very effective way to achieve the precise required product texture (via particle size reduction) and at the same time creating a very stable product compared to conventional devices like agitators, stirrers, rotor-stator devices, or colloid mills, high pressure homogenization is essential to the quality and stabilization of beauty creams, lotions, nail varnishes, shampoos, toothpastes, and emulsions containing different oils. By reducing particle size and homogeneity while under intense pressure, shear, and stress, this is accomplished [46].

The end result is a uniform, highly effective product with exceptional stability and shelf life. It is a well-known truth that the ability of various cosmetic products, such beauty creams, to penetrate the skin, limits how well they operate. The skin can absorb the active substances in these creams thanks to a significant reduction in particle size since following homogenization, the surface area increases (vs. volume ratio). The most effective utilization of the active ingredients is also made possible by high-pressure homogenization, also known as micronization, which lowers the price of the finished product [47].

Applications for Biotechnology Homogenizers

In order to extract intra-cell contents without employing solvents or other chemicals for the cell wall breakage, high-pressure homogenization is frequently employed for cell rupture (also known as cell rupturing, cell disruption) applications. The major goal is to release intercellular components, such as proteins, enzymes, and vitamins, which are concentrated in a variety of organisms, including bacteria, yeasts, algae, and other biological products. For cell rupture operations, high-pressure homogenizers are particularly effective and efficient and may frequently maximize the output from precious source material while maintaining a very high level of product quality [48].

Applications for Pharmaceutical Homogenizers

To make pharmaceutical goods more stable and clinically effective, homogenization—also referred to as micronization in the pharmaceutical industry—is the process of lowering their particle sizes under extremely high pressures, shear, turbulence, acceleration, and impact. The product's bioavailability rises, and some medications that would otherwise have borderline tolerance may become more tolerable [49].

The emulsion, suspension, or solution is pushed into the high-pressure homogenizer where it is driven at extremely high pressures (up to 1500 bar / 21750 PSI) through a specific homogenization valve. After treatment, the big particles are dispersed and reduced to sizes that typically range from 0.4 to 1 micron, depending on the application. The particles enter the homogenizer with maximum diameters of 500 m [2].

In comparison to conventional stirrers, rotor-stator devices, or colloid mills, high-pressure homogenization has been shown to produce more stable products with improved active ingredient dispersion. It accomplishes this by drastically lowering particle size and homogeneity under extremely high pressure [50].

Homogenization in Chemical Processing

Despite having numerous and varied uses, high-pressure pumping and homogenization in chemical processing attract much less attention than their usage in the food and dairy industries. It is best to develop stable emulsions, dispersions, and mixtures during the homogenization stage. The division of particles and droplets improves their surface area, which can speed up chemical reactions by lowering temperatures and reaction times while reducing or even doing away with the requirement for catalysts. High pressure is frequently used to improve extraction procedures and color intensity in various pigments and paintings. The petrochemical sector uses homogenization for optimal additive blending and viscosity control [51].

Milk Homogenization

Milk homogenization is the most common application of homogenizers in the food business.

Objectives

- To combat segregation, which is primarily creaming, the size of the fat globules should be drastically reduced. The user may find a cream layer or sediment in the product, especially if the package is opaque.
- Increasing stability toward partial coalescence. The homogenized fat globules' improved stability is due to their decreased diameter and newly acquired surface layer. Additionally, cream layers are particularly susceptible to partial coalescence, and cream layers form significantly more slowly in homogenized goods. Overall, the primary goal of homogenization is often to prevent partial coalescence [52].

Homogenization Effects

Milk is an emulsion of oil and water, with the fat globules spread throughout the continuous skim milk phase. However, if raw milk were allowed to stand, the fat would rise and form a layer of cream. Homogenization is a mechanical process that reduces the average diameter of milk's fat globules while increasing their number and surface area. It is accomplished by pushing milk under high pressure through a small opening. Practically speaking, the final consequence is a significantly decreased tendency for the creaming of fat globules. Three factors contribute to this enhanced stability of homogenized milk: a decrease in the mean diameter of the fat globules (a factor in Stokes Law), a decrease in the size distribution of the fat globules (causing the speed of rise to be similar for the majority of globules such that they don't tend to cluster during creaming), and an increase in density of the globules (bringing them closer to the continuous phase) owing to the adsorption of a protein membrane. The cryo-globulin complex, which tends to cluster fat globules and cause them to rise, is also broken down by heat pasteurization [53].

Positive and Negative Aspects

- Increased total fat globule surface reduces or postpones creaming of the fat and boosts light reflection
- Increased surface area for microbial lipase to assault, causing lipolytic alterations
- Improvement of flavor and texture
- Increased susceptibility to the effects of light, resulting in taste abnormalities such soupy, rancid, or oxidized
- An improvement in digestibility
- Expanded potential for microbial contamination
- The hue becomes whiter.
- Reduced protein thermal stability (homogenization is required after the UHT process for UHT milk)
- There is a slight rise in foam tendency.
- The texture becomes more even [54].

Mechanism for Homogenization

The theories of homogenization that have been put forth over the years have made an effort to connect the findings in the product to this high-energy transition in the valve. Many of these theories were formerly difficult to support or refute, but it now seems that turbulence is the main cause.

Since many people believe that shear is the primary mechanism of homogenization, it is important to start by discussing shear in the valve. The elongation and eventual breakdown of a scattered phase droplet as a result of various fluid velocities in its immediate vicinity is referred to as "shear" in this context. It is challenging to support the traditional notions of shear effects as they relate to the flow profile in the homogenizing valve. The viscosity of many fats and oils is higher than the maximum value permitted for shearing of the droplets, and the flow profile in the valve is such that the bulk of the liquid does not suffer significant velocity gradients because of the effects of boundary layers. These are two grounds against shear [55].

According to the cavitation theory, the large pressure drop via the valve causes the liquid to experience strong cavitation. When there is a significant enough pressure decrease, the liquid's vapor pressure rises above the atmospheric pressure, creating vapor bubbles (liquid cavities). Shock waves are created in the liquid when cavitation bubbles rupture (collapse of the cavities). The scattered droplets are split apart by these shock waves [30].

According to the second homogenization theory, which deals with turbulence, the liquid's energy loss causes strong eddies of turbulent motion. These eddies would match the typical droplet diameter in size. The droplets would be torn apart by the localized pressure differential and strong energy of the turbulence, which would shrink their average size. The droplet size decrease for emulsions is achieved by turbulent flow, according to flow visualization studies this author conducted using a modeled quartz homogenizing valve. Cavitation is prevented but turbulent flow is kept maintained by backpressure or second-stage pressure. Intense turbulent flow immediately disrupts any drops that are traveling through the turbulent jet at the discharge from the gap [29].

Homogenization in One Stage against Two Stages

Homogenization in one stage:

- Pressure: 2500psi
- Mean size: 0.5 μm
- Range: 0.2-2 μm
- Clustering takes place

Homogenization in one stage:

- Pressure used in:
  - 1st stage: 2000 psi
  - in 2nd stage: 500 psi
- No clustering takes place [2]

Homogenization in Two Stages

Gaulin invented the two-stage homogenizing valve, which has been around for a while. According to experimental results, the second-stage valve's main job is to control the first stage's homogenization effect's intensity [31].

Consequently, the second-stage valve "fine-tunes" the homogenization process. The second-stage backpressure for emulsions reduces cavitation, ends two-phase flow, and amplifies turbulence. It has been discovered that the appropriate second-stage pressure for oil-in-water emulsions should range between 10 and 15% of the overall homogenizing pressure. With milk, it has been shown that passing through the second-stage valve itself does not materially change the emulsion's state as long as the backpressure is kept low, as described above [32].

A sample of the milk taken both before and after the second step did not show a significant difference in the quality of the emulsion. It is simple to supply the proper backpressure to the first-stage homogenizing valve using the second-stage valve. The milk emulsion quality produced by a homogenizer using a two-stage valve and processing milk at 2000 psi (137.9 bar) may be comparable to that of a single-stage valve running at 2200 to 2300 pressure (151.7 to 158.6 bar) [2].

The two-stage valve configuration increases the achievable homogenization quality beyond that achieved by a single-stage valve set at the homogenizer's maximum operating pressure, which is 2000 psi. Other than basic emulsions, some items with intricate compositions could be impacted by the second step in terms of appearance or viscosity [33].

Homogenizing Efficiency Measurements

Setting up an evaluation procedure is the first step in figuring out the homogenizer's appropriate operating pressure for a specific product. A few examples of the procedure include viscosity rise or reduction, average particle size (as evaluated by instrument or under the microscope), rate of separation, grind gauge, and change in physical appearance. In the following phase, samples from the homogenizer are taken at various pressure settings, such as every 1000 psi (or fewer, if a limited pressure range is being tested). The evaluated parameter is then compared to or plotted against the homogenizing pressure using the appropriate analytical technique, and the lowest pressure that produces the desired product can then be chosen. It may be important to research the results of multipassing at the proper pressure for specific items [34].

Conclusion

There's little doubt most of us have heard the word: "homogenization" or some variation of it. We purchase homogenized milk at the store. We study homogenous mixtures like air and rain in school. We talk about globalization and cultural homogeneity as we discuss world events. But while we appreciate the word's roots in "sameness," "uniformity," "mixing" and "blending," when it comes to the actual scientific process of homogenization, there remains confusion even among experts. The objective of all mixing and processing equipment is to add ingredients together and create a more homogenous end product. Perhaps, for this reason, all sorts of equipment describe themselves as homogenizers. However, homogenization is a very specific and high-performance mixing technology.

References

[1] Wolfgang, M., & Karsten, M. (2012). Solid lipid nanoparticles: Production, characterization and applications. *Advanced Drug Delivery Reviews, 64 (Supplement),* 83-101.

[2] Poonam, D. (2014). Homogenization Fundamentals. *IOSR Journal of Engineering, 4(5),* 2278-8719.

[3] Xiang, L., Zhanli, L., Shaoqing, C., Chengcheng, L., Chenfeng, L., & Zhuo, Z. (2019). Predicting the effective mechanical property of heterogeneous materials by image based modeling and deep learning. *Computer Methods in Applied Mechanics and Engineering, 347,* 735-753.

[4] Gökmen, V., & Acar, J. (1995). Applications of High-pressure Technology in the Food Industry. *GIDA, 20 (3),* 167-172.

[5] Trujillo, A. J., Capellas, M., Saldo, J., Gervilla, R., & Guamis, B. (2002). Applications of high hydrostatic pressure on milk and dairy products: A review. *Inovative Food Science and Emerging Technologies, 3(4),* 295-307.

[6] Kınık, Ö., Kavas, G., Uysal, H., & Kesenkaş, H. (2004). The High Pressure Techniques in Dairy Industry. *GIDA. 29 (1),* 95-102.

[7] Deliza, R., Rosenthal, A., Abadio, F. B. D., Silva, C. H. O., & Castillo, C. (2005). Application of High-Pressure Technology in the Fruit Juice Processing: Benefits Perceived by Consumers. *Journal of Food Engineering, 67,* 241-246.

[8] Minerich, P. L., & Labuza, T. P. (2003). Development of a Pressure Indicator for High Hydrostatic Pressure Processing of Foods. *Innovative Food Science and Emerging Technologies, 4,* 235-243.

[9] Hoover, D. G., Metrick, C., Papineau, A. M., Farkas, D. F., & Knorr, D. (1989). Biological Efffects of High-Hydrostatic Pressure on Food Microorganisms. *Food Technology, 43,* 99-107.

[10] Şanal, İ. S., & Çalımlı, A. (2000). Application of High-hydrostatic Pressure Technology in the Food Industry. *GIDA, 25 (3),* 193-201.

[11] İbanoğlu, E., & İbanoğlu, Ş. (2003). Effect of High-hydrostatic Pressure on Gelatinization of Corn Starch. *GIDA, 28 (3),* 273-276.

[12] Mertens, B., & Deplace, G. (1993). Engineering Aspects of High-Pressure Technology in the Food Industry. *Food Technology,* 164-169.

[13] İbanoğlu, E. (2002). High-hydrostatic Pressure Applications in Foods. *GIDA, 27 (6),* 505-510.

[14] Brul, S., Rommens, A. J. M., & Verrips, C. T. (2000). Mechanistic Studies on the Inactivation of *Saccharomyces cerevisiae* by High-Pressure. *Innovative Food Science & Emerging Technologies, 1,* 99-108.

[15] O"Reilly, E. C., Kelly, A. L., Murphy, M. P., & Beresord, P. T. (2001). High pressure treatment: Applications in cheese manufacture and ripening. *Trend in Food Sci. Technology, 12(2),* 51-59.

[16] Wilkinson, N., Kurdziel, A. S., Langton, S., Needs, E., & Cook, N. (2001). Resistance of poliovirus to inactivation by high hydrostatic pressures. *Innovative Food Science & Emerging Technology, 2,* 95-98.

[17] Hugas, M., Garriga, M., & Monfort, J. M. (2002). New Mild Technologies in Meat Processing: High Pressure as a Model Technology. *Meat Science, 62,* 359-371.

[18] Reddy, N. R., Solomon, H. M., Tetzloff, R. C., & Rhodehamel, E. J. (2003). Inactivation of *Clostridium botulinum* Type A Spores by High Pressure Processing at Elevated Temperatures. *Journal of Food Protection, 66(8),* 1402-1407.

[19] Karakaya, M., Caner, C., & Sarıçoban, C. (2004). *The Use of High Hydrostatic Pressure in Meat Technology. Turkey 8th Food Congress.* Bursa, Turkey.

[20] Linton, M., Mc Clements, J. M. J., & Patterson, M. F. (2004). Changes in the Microbiological Quality of Vacuum-packaged, Minced Chicken Treated with High Hydrostatic Pressure. *Innovative Food Science and Emerging Technologies, 5,* 151-159.

[21] Yemencioğlu, A., & Özkan, M. (2004). *Methods of Preservation of Fruits and Vegetables and Products Obtained from These. Fruit and Vegetable Processing,* 226-227. Ankara. Turkey.

[22] Moerman, F. (2005). High Hydrostatic Pressure Inactivation of Vegetative Microorganisms, Aerobic and Anaerobic Spores in Pork Marengo, a Low Asidic Particulate Food Product. *Meat Science, 69,* 225-232.

[23] Tülek, Y., & Filizay, G. (2006). High Hydrostatic Pressure Systems Use in Food Industry. *Pamukkale University Journal of Engineering Sciences, 12(2),* 225-231.

[24] Özcan, T., & Kurtuldu, O. (2011). Alternative Methods on Production of ExtendedShelf Life Milk. *Journal of Agricultural Faculty of Bursa Uludag University, 25(1),* 119-129.

[25] Ohlsson, T., & Bengtsson, N. (2002). *Minimal Processing Technologies in the Food Industry.* Woodhead publishing limited. 41 s, England.

[26] Farkas, D. F., & Hoover, D. G. (2000). High Pressure Processing. In: Kinetics of Microbial Inactivation for Alternative Food Processing Technologies. *Journal of Food Science. Special Supplement,* 47-64.

[27] Russell, N. J. (2002). Bacterial membranes: the effects of chill storage and food processing: An overview. *International Journal of Food Microbiology, 79,* 27–34.

[28] Mackey, B. M., Forestiere, K., & Isaacs, N. (1995). Factors affecting the resistance of *Listeria monocytogenes* to high hydrostatic pressure. *Food Biotechnology, 9,* 1-11.

[29] Marquis, R. E., & Bender, G. R. (1987). *Barophysiology of prokaryots and proton-transiocating ATPases.* In: *Current Perspectives in High Pressure Biology* (Eds.: H. W. Jannisch, R. E. Marquis and A. M. Zimmermann). *Academic Press.* London. pp. 65-73.

[30] Patel, D. M., Patel, N. N., & Patel, J. K. (2021). *Nanomedicine Scale-Up Technologies: Feasibilities and Challenges.* In: Patel, J. K., Pathak, Y. V. (eds) *Emerging Technologies for Nanoparticle Manufacturing.* Springer, Cham.

[31] Muhammad B. R., Norela, J., Norasikin, O., Hilmi, A. R., Raja, N., Raja, S., Norul, F., & Mohamed, N. (2020). Stability of emulsion liquid membrane using bifunctional diluent and blended nonionic surfactant for phenol removal. *Chemical Engineering and Processing - Process Intensification, 148,* 107790.

[32] Mattanovich, D., Branduardi, P., Dato, L., Gasser, B., Sauer, M., & Porro, D. (2012). Recombinant Protein Production in Yeasts. In: Lorence, A. (eds) *Recombinant Gene Expression. Methods in Molecular Biology*, 824. Humana Press, Totowa, NJ.

[33] Diels, A. M., & Michiels, C. W. (2006). High-Pressure Homogenization as a Non-Thermal Technique for the Inactivation of Microorganisms. *Critical Reviews in Microbiology, 32(4),* 201-216.

[34] dos Santos Aguilar, J. G., Cristianini, M., & Sato, H. H. (2018). Modification of enzymes by use of high-pressure homogenization. *Food Research International, 109,* 120-125.

[35] Günerken, E., D'Hondt, M. H. M., Eppink, L., Garcia-Gonzalez, K., Elst, R. H. & Wijffels, (2015). Cell disruption for microalgae biorefineries. *Biotechnology Advances, 33(2),* 243-260.

[36] Chisti, Y., & Moo-Young, M. (1986). Disruption of microbial cells for intracellular products. *Enzyme and Microbial Technology, 8(4),* 194-204.

[37] El-Jaby, U., McKenna, T. F. L., & Cunningham, M. (2007), Miniemulsification: An Analysis of the Use of Rotor Stators as Emulsification Devices. *Macromolecular Symposia, 259,* 1-9.

[38] Kamaly, S. W., Tarleton, A. C., & Özcan-Taşkın, N. G. (2017). Dispersion of clusters of nanoscale silica particles using batch rotor-stators. *Advanced Powder Technology, 28(9),* 2357-2365.

[39] Trujillo-Cayado, L. A., Santos, J., Alfaro, M. C., Calero, N., & Muñoz, J. (2016). A Further Step in the Development of Oil-in-Water Emulsions Formulated with a Mixture of Green Solvents. *Industrial & Engineering Chemistry Research, 55 (27),* 7259-7266.

[40] Mehmood, T., Ahmad, A., Ahmed, A., & Ahmed, Z. (2017). Optimization of olive oil based O/W nanoemulsions prepared through ultrasonic homogenization: A response surface methodology approach. *Food Chemistry, 229,* 790-796.

[41] Aernouts, B., Beers, R. V., Watté, R., Huybrechts, T., Jordens, J., Vermeulen, D., Gerven, T. M., Lammertyn, J., & Saeys, W. (2015). Effect of ultrasonic homogenization on the Vis/NIR bulk optical properties of milk. *Colloids and Surfaces B: Biointerfaces, 126,* 510-519.

[42] Mehmood, T., Ahmed, A., Ahmad, A., Ahmad, M. S., & Sandhu, M. A. (2018). Optimization of mixed surfactants-based β-carotene nanoemulsions using response surface methodology: An ultrasonic homogenization approach. *Food Chemistry, 253,* 179-184.

[43] Shi-Ying, J., Jin, H., Shi-Xiao, J., Qing-Yuan, L. V., Jin-Xia, B., Hong-Ge, C., Rui-Sheng, L., Wei, W., & Hai-Long, Y. (2014). Characterization and evaluation in vivo of baicalin-nanocrystals prepared by an ultrasonic-homogenization-fluid bed drying method. *Chinese Journal of Natural Medicines, 12(1),* 71-80.

[44] Hassanshahi, N., Hu, G., Lee, K., & Li, J. (2023). Effect of ultrasonic homogenization on crude oil-water emulsion stability. *Journal of Environmental Science and Health, Part A, 58(3),* 211-221.

[45] Tobin, J., Heffernan, S. P., Mulvihill, D. M., Huppertz, T., & Kelly, A. L. (2015). Applications of High-Pressure Homogenization and Microfluidization for Milk and Dairy Products. In *Emerging Dairy Processing Technologies* (eds N. Datta and P. M. Tomasula).

[46] Yukuyama, M. N., Ghisleni, D. D. M., Pinto, T. J. A., & Bou-Chacra, N. A. (2016). Nanoemulsion: process selection and application in cosmetics – A review. *International Journal of Cosmetic Science, 38,* 13-24.

[47] Azmi, N. A. N., Elgharbawy, A. A. M., Motlagh, S. R., Samsudin, N., & Salleh, H. M. (2019). Nanoemulsions: Factory for Food, Pharmaceutical and Cosmetics. *Processes, 7,* 617.

[48] Almoselhy, R. I. M. (2022). High-Speed and High-Pressure Homogenization Techniques for Optimization of Food Processing, Quality, and Safety. *Open Access Journal of Microbiology & Biotechnology, 7(4),* 000243.

[49] Yadav, K. S., & Kale, K. (2020). High Pressure Homogenizer in Pharmaceuticals: Understanding Its Critical Processing Parameters and Applications. *Journal of Pharmaceutical Innovation, 15,* 690–701.

[50] Vinchhi, P., Patel, J. K., & Patel, M. M. (2021). High-Pressure Homogenization Techniques for Nanoparticles. In: Patel, J. K., Pathak, Y. V. (eds) *Emerging Technologies for Nanoparticle Manufacturing*. Springer, Cham.

[51] Salehi, F. (2020). Physico-chemical and rheological properties of fruit and vegetable juices as affected by high pressure homogenization: A review. *International Journal of Food Properties, 23(1),* 1136-1149.

[52] Pereda, J., Ferragut, V., Quevedo, J. M., Guamis, B., & Trujillo, A. J. (2007). Effects of Ultra-High Pressure Homogenization on Microbial and Physicochemical Shelf Life of Milk. *Journal of Dairy Science, 90(3),*1081-1093.

[53] Cano-Ruiz, M. E., & Richter, R. L. (1997). Effect of Homogenization Pressure on the Milk Fat Globule Membrane Proteins. *Journal of Dairy Science, 80(11),* 2732-2739.

[54] Michalski, M. C., & Januel, C. (2006). Does homogenization affect the human health properties of cow's milk? *Trends in Food Science & Technology, 17(8),* 423-437.

[55] Michalski, M. (2007). On the supposed influence of milk homogenization on the risk of CVD, diabetes and allergy. *British Journal of Nutrition, 97(4),* 598-610.

Chapter 2

Asymptotic Expansion Methods in Homogenization and Some Applications

Gennadiy V. Sandrakov*

Faculty of Computer Science and Cybernetics,
Taras Shevchenko National University of Kyiv, Ukraine
Pukhov Institute for Modelling in Energy Engineering, Kyiv, Ukraine

Abstract

Practically fifty years ago, homogenization methods were developed by I. Babushka, N. Bakhvalov, A. Bensoussan, J.-L. Lions, and G. Papanicolau to simplify computer modeling for boundary value problems in two-scale media with a periodic structure, which are similar to composite materials. The methods were based on asymptotic expansions in the microscale parameter for solutions to second-order equations with rapidly oscillating periodic coefficients.

The chapter is devoted to the general principles of constructing asymptotic expansions in small microscale for solutions of linear elliptic equations of arbitrary order with rapidly oscillating periodic coefficients. Such asymptotic expansions, after substitution into the equation, give a small discrepancy in such equations, which leads to an approximation of solutions, since, as a rule, relevant a priori estimates are satisfied for the equations. Examples will be given of applying the presented constructions to solutions of certain linear equations. In addition, such asymptotic constructions can be useful for understanding homogenized problems for nonlinear equations, since such problems can be linearized and formal asymptotics can be obtained in this case, which can be justified by passing to the limit. For example, the restriction

* Corresponding Author's Email Address: gsandrako@gmail.com.

In: A Closer Look at Homogenization
Editor: Gulcin Yildiz
ISBN: 979-8-89113-656-4

of the presented asymptotic expansions to ordinary differential equations of a particular form coincides with the Bogolyubov-Mitropolsky method. Also, such asymptotic constructions are applicable for pseudodifferential equations without changes. The simplest of these equations are equations with a shift. The application of such asymptotic constructions in this case coincides with the well-known Taylor expansion. The asymptotic constructions can be used to approximate solutions of parabolic and hyperbolic equations, since such equations are isomorphically equivalent to elliptic equations with a complex parameter due to the well-known approach that uses the Laplace transform. For example, applications of such asymptotic constructions to solutions of parabolic equations with contrasting (highly varying) periodic coefficients depending on additional parameters are given. The justification for some of the asymptotic constructions will also be formulated.

Keywords: multiscale media, homogenized problems, elliptic equations

AMS Subject Classification: 35B27, 35B40, 65N15

1. Introduction

Homogenization methods are a convenient tool for investigating problems for multiscale media. The most complete results here are obtained for two-scale composite media with a periodic structure. It should be emphasized that periodic structures are technologically convenient, since by choosing a cell of periodicity one can get a two-scale media through periodic continuation, just as city squares can be paved with tiles. Similarly, real ideal crystals are obtained by periodic continuation of a relevant periodicity cell. Such media usually have a finite volume, which determines the basic scale and variables, and a lot of periodicity cells, which determines the microscale and microvariables, just as a real crystal occupies a finite volume and consists of a very large number of atom cells. Hence, naturally there is at least one small parameter in problems for such media, which characterizes the microscale, but there may be other additional parameters. Thus, homogenization theory deals with composite multiscale media, which are analogous to crystals in a construction sense.

Initially, practically fifty years ago, homogenization methods were developed by I. Babushka, N.S. Bakhvalov, A. Bensoussan, J.-L. Lions, and G. Papanicolau to simplify computer simulations problem for composite two-scale media with a periodic structure according to [1]–[5]. The presence of a microscale in the media prevented the use of conventional numerical methods for solving problems for the composite media. On the other hand, at a very small microscale, homogenization methods make it possible to replace, for example, the modeling of thermal processes in the composite media with the modeling of thermal processes in homogeneous (homogenized) media with guaranteed accuracy, and the numerical simulation of such processes for the homogeneous media with guaranteed accuracy is standard. Thus, there is full control over the approximation accuracy of the simulation, which is required, for example, when modeling for construction in complex conditions. In a mathematical sense, this means that the solution of a heat conduction problem with rapidly oscillating periodic coefficients is guaranteed to be close to the solution of the heat conduction problem with constant coefficients for a small microscale parameter. Relevant accuracy estimates were proved in the works of the authors for regular data (suitable references can be found, for example, in [6] and [7]).

Also, it turned out that it is much easier to prove the convergence of solutions of problems with rapidly oscillating coefficients to solutions of problems with constant coefficients [6]. This approach, as a rule, does not require additional data regularity, but it does not guarantee the accuracy of the approximation, which may be very weak. Later, it also turned out that it is even easier to prove the multiscale convergence of solutions of the problems to solutions of certain multiscale problems [8]. The multiscale problems depend both on basic variables and microscale variables that are naturally present in the original problems. But, the type (in the classical sense) of such problems and the approximation accuracy are not at all clear without additional analysis and proofs, further details can be found, for example, in [8]–[14].

Initially, homogenization methods were based on asymptotic expansions in the microscale parameter. It turned out that several expansion terms are sufficient for an approximation of solutions to the multiscale problem. The construction of infinite asymptotic expansions, as a rule, is useless for improving approximation through complex boundary effects in real problems. Of course, asymptotic expansions for essentially non-linear problems are not always possible, and only convergence and multiscale convergence methods can be used to justify homogenized problems as, for example, in [15]–[17]. However, there are numerous papers for linear problems in which multiscale homogenized problems are presented as achievements, while such problems were obtained and justified earlier by asymptotic expansions methods.

The chapter is devoted to the general principles of constructing asymptotic expansions in small microscale for solutions of linear elliptic equations of arbitrary order with rapidly oscillating periodic coefficients [18]. Such asymptotic expansions, after substitution into the equation, give a small discrepancy in such equations, which leads to an approximation of solutions, since, as a rule, relevant a priori estimates are satisfied for the equations. The proposed principles were presented some time ago and are applicable to various problems, in particular to spectral and nonstationary problems. Such applications were carried out in dissertation [19], some of the results of which were not published. All hitherto known examples of constructing asymptotic expansions for solutions of such elliptic equations explicitly or implicitly use these principles. In addition, these asymptotic constructions can be used to approximate solutions of parabolic and hyperbolic equations, since such equations are isomorphically equivalent to elliptic equations with a complex parameter due to the well-known approach that uses the Laplace transform.

Applications of such constructions to solutions of parabolic equations with contrasting (highly varying) periodic coefficients depending on additional parameters are given. The justification for some of these constructions will also be formulated. The appearance of such equations with parameters is natural. Indeed, it is not very reasonable to design artificial composite media from materials with similar properties. But, when using materials with very different properties, the parameters that characterize these properties can be small in comparison with the microscale parameter, which is already present in the problem under consideration.

In addition, such asymptotic constructions can be useful for understanding homogenized problems for nonlinear equations, since such problems can be linearized and formal asymptotics can be obtained in this case, which can be justified by passing to the limit. For example, the restriction of the presented asymptotic expansions to ordinary differential equations of a particular form coincides with the Bogolyubov-Mitropolsky method, the justification of which is well known [20].

Also, such asymptotic constructions are applicable for pseudodifferential equations without changes. The simplest of the equations are equations with a shift; the application of such asymptotic constructions in the case coincides with the well-known Taylor expansions. The last example clarifies the problem of regularity in justifying homogenization, since the Taylor expansion provides a good approximation for smooth functions, while the convergence statement is the same as definition of continuous functions. This expansion may not give anything for discontinuous functions in the uniform norm, but in integral norms it is an approximation in the leading term of the expansion. These facts are well known and are regularly used in the construction of finite-dimensional approximations for partial differential equations when modeling various processes through computer simulations.

2. Elliptic Equations with Rapidly Oscillating Coefficients

In this section, general principles for constructing formal asymptotic expansions of solutions to elliptic equations with periodic rapidly oscillating coefficients will be described. It should be noted that all hitherto known examples of constructing asymptotic expansions for solutions of linear elliptic equations with periodic rapidly oscillating coefficients are covered by the algorithm presented here.

For an integer $n \geqslant 2$, assume that there are given an even integer $m \geq 0$, a bounded domain $\Omega \subset \mathbb{R}^n$ with a smooth (C^∞-class) boundary $\partial\Omega$ and a smooth complex-valued function $f \in C_0^\infty(\Omega; \mathbb{C})$. Here and below, function spaces are used, which are defined, for example, in [21], [22]. We will denote by $\alpha = (\alpha_1, \ldots, \alpha_n)$ multi-indices with non-negative integer components for which $|\alpha| = \sum_k \alpha_k$ and $D_x^\alpha = D_1^{\alpha_1} \dots D_n^{\alpha_n}$, where $D_k = i^{-1}\partial_k$ for $k = 1, \dots, n$. Here ∂_k denotes the partial derivative with respect to the variable x_k for $(x_1, \dots, x_k, , \dots, x_n) \in \mathbb{R}^n$ and $i = \sqrt{-1}$ is imaginary unit. We define the function $u = u(x)$ as a solution to the following Dirichlet problem for the m-order partial differential equation

$$\begin{aligned} P_\varepsilon(x/\varepsilon, \varepsilon D_x)\, u(x) &= a(x/\varepsilon)\, f(x) \quad \text{in} \quad \Omega, \\ \partial_\nu^s\, u(x) &= 0 \quad \text{on} \quad \partial\Omega, \end{aligned} \tag{1}$$

where ε is a small positive parameter, ∂_ν denotes the derivative with respect to the outer normal ν to the boundary $\partial\Omega$, $s = 0, \dots, m/2 - 1$, $\varepsilon^\alpha = \varepsilon^{|\alpha|}$ and

$$P_\varepsilon(x/\varepsilon, \varepsilon D_x)\, u(x) = \varepsilon^{-m} \sum_{|\alpha| \leqslant m} a_\alpha(x/\varepsilon)\, \varepsilon^\alpha D_x^\alpha u(x). \tag{2}$$

For each α with $|\alpha| \leqslant m$ coefficients $a_\alpha(y)$ and $a(y)$ of problem (1) and operator (2) are 1-periodic (periodic with period 1 for each $x_1, \dots, x_n$) complex-valued functions of class $C^\infty(\mathbb{R}^n; \mathbb{C})$. It is assumed that for the main part of the differential operator P_ε of problem (1) for $\varepsilon = 1$ is elliptic in the sense that

$$\sum_{|\alpha| = m} a_\alpha(y)\, \xi^\alpha \neq 0 \qquad \text{for} \quad y \in \mathbb{R}^n \quad \text{and} \quad \xi \in \mathbb{R}^n \setminus \{0\}, \tag{3}$$

where $\xi^\alpha = \xi_1^{\alpha_1} \dots \xi_n^{\alpha_n}$ for $\xi = (\xi_1, \dots, \xi_n) \in \mathbb{R}^n$. It is also assumed that for a sufficiently small fixed ε the solution to problem (1) is defined as an element of the Sobolev space $H_0^{m/2}(\Omega)$. Thus, one can try to construct asymptotic expansions for approximating the solution for the sufficiently small ε. Further, we will formulate general principles that allow us to construct such asymptotic expansions.

Separation of variables. A function of the form $F(x/\varepsilon, x)$ on the domain Ω will be called a *function with separated scale variables* if

$$F(x/\varepsilon, x) = N(x/\varepsilon)\, u(x),$$

where the function $u(x)$ does not depend on ε and is defined on Ω and $N(y)$ is 1-periodic function on $\mathbb{R}^n$. Thus $N(x/\varepsilon)$ is ε-periodic. Here we assume that $u(x)$ is extended by zero outside Ω. By definition, the right-hand side of equation (1) is the function with separated (scale) variables, and asymptotic expansions for the solution to problem (1) will be defined as an asymptotic sum whose terms are finite sums of functions with separated variables in the following form

$$u_p^l = \varepsilon^p \sum_{s=0}^{l} \varepsilon^s \Big(\sum_{\{\beta_s\}} N_{\beta_s}^s(x/\varepsilon)\, u_s^{\beta_s}(x) \Big), \tag{4}$$

where p and l denote integers with $l \geqslant 0$ and $\{\beta_s\}$ is some finite set of multi-indices, which is determined in further constructions.

Operator decomposition. For $\xi \in \mathbb{R}^n$ we define the symbol of the differential operator $P_\varepsilon = P_\varepsilon(x/\varepsilon, \varepsilon D_x)$ of problem (1) by the following equality

$$p_\varepsilon(x/\varepsilon, \varepsilon\,\xi) = \varepsilon^{-m} \sum_{|\alpha|\leqslant m} a_\alpha(x/\varepsilon)\,\varepsilon^\alpha \xi^\alpha = \varepsilon^{-m} \sum_{|\alpha|\leqslant m} a_\alpha(x/\varepsilon)\,(\varepsilon\,\xi)^\alpha,$$

where $\varepsilon^\alpha = \varepsilon^{|\alpha|}$. The symbol is obtained by formally replacing the derivatives D_x^α in $P_\varepsilon(x/\varepsilon, \varepsilon D_x)$ with ξ^α. We also define for a multi-index β and $\xi \in \mathbb{R}^n$ the symbol

$$p_\varepsilon^\beta(x/\varepsilon, \varepsilon\,\xi) = \partial_\xi^\beta\, p_\varepsilon(x/\varepsilon, \varepsilon\,\xi) \qquad \left(\partial_{\xi_k} = \frac{\partial}{\partial \xi_k}\right) \tag{5}$$

and the differential operator corresponding to this symbol will be denoted by $P_\varepsilon^\beta = P_\varepsilon^\beta(x/\varepsilon, \varepsilon D_x)$. The operator is obtained by replacing ξ^α in $p_\varepsilon^\beta(x/\varepsilon, \varepsilon\,\xi)$ with the derivatives D_x^α. It is useful to note that each differentiation in definition (5) removes the derivative with respect to x in P_ε, but leaves ε, since $\partial_\xi^\beta(\varepsilon\,\xi^\beta) = \varepsilon$ for $|\beta| = 1$. Here $p_\varepsilon^\beta = p_\varepsilon(x/\varepsilon, \varepsilon\,\xi)$ for $\beta = 0$ and we use terminology that is accepted in the pseudodifferential operator theory, for example, according to [23].

The following generalized Leibniz formula is well known in this theory

$$P_\varepsilon\big(N(x)\,u(x)\big) = \sum_{|\beta|\leqslant m} \big[P_\varepsilon^\beta(x/\varepsilon, \varepsilon D_x)\,N(x)\big] D_x^\beta\, u(x)/\beta!, \tag{6}$$

where $\beta! = \beta_1! \ldots \beta_n!$ for multi-index $\beta = (\beta_1, \ldots, \beta_n)$ and $0\,! = 1$ (see, for example, [23], (3.46)). This formula is a consequence of the usual Leibniz product differentiation formula, which holds, for example, when $N(x)$ is a smooth function and $u(x)$ is a distribution. Hence, the equality (6) holds for such $N(x)$ and $u(x)$.

The function $N(x)$ in the equality (6) can naturally depend on the parameter that is $N = N(x/\varepsilon)$. Let us now note that

$$\varepsilon^\beta\, D_x^\alpha N(x/\varepsilon) = \varepsilon^{\beta-\alpha}\, D_y^\alpha\, N(y) = \varepsilon^\beta\big(\varepsilon^{-\alpha} D_y^\alpha\big) N(y) \quad \text{for} \quad y = x/\varepsilon \tag{7}$$

and the equality (6) in the case under consideration takes the following form

$$P_\varepsilon\big(N(x/\varepsilon)\,u(x)\big) = \varepsilon^{-m} \sum_{|\beta|\leqslant m} \varepsilon^\beta \left[P^\beta(y, D_y)\,N(y)\right] D_x^\beta\, u(x)/\beta! =$$

$$= \varepsilon^{-m} \sum_{|\alpha|\leqslant m} \big[a_\alpha(y) D_y^\alpha N(y)\big] u(x) + \tag{8}$$

$$+\, \varepsilon^{-m+1} \sum_{|\beta|=1} \sum_{|\alpha|\leqslant m} \big[a_\alpha^\beta(y) D_y^\alpha N(y)\big] D_x^\beta\, u(x)/\beta! \, +$$

$$\ldots +\, \varepsilon^{-m+m} \sum_{|\beta|=m} \sum_{|\alpha|\leqslant m} \big[a_\alpha^\beta(y) D_y^\alpha N(y)\big] D_x^\beta\, u(x)/\beta!$$

for $y = x/\varepsilon$, where $P^\beta = P_1^\beta(y, D_y)$ denotes the differential operator of order $m - |\beta|$ with coefficients $a_\alpha^\beta(y)$, corresponding to the symbol (5) for $\varepsilon = 1$.

Using these equalities, we obtain the expansion of the operator P_ε acting on the function u_p^l, which is defined by the equality (4), in the following form

$$P_\varepsilon\, u_p^l \;=\; \varepsilon^p \sum_{s=0}^{l} \varepsilon^s \sum_{\{\beta_s\}} P_\varepsilon \left(N_{\beta_s}^s(x/\varepsilon)\, u_s^{\beta_s}(x)\right) \;=$$
$$=\; \varepsilon^{p-m} \sum_{s=0}^{l} \varepsilon^s \sum_{\{\beta_s\}} \Big(\sum_{|\beta|\leqslant m} \varepsilon^\beta \big[P^\beta(y, D_y) N_{\beta_s}^s(y)\big] D_x^\beta\, u_s^{\beta_s}(x)/\beta! \Big) \tag{9}$$

for $y = x/\varepsilon$. The main idea of constructing asymptotic expansions is to observe that the variables y and x are independent inside the brackets of the formula. In the sense that if $\sum_{\{\beta_0\}} \big[P(y, D_y) N_{\beta_0}^0(y)\big] u_0^{\beta_0}(x) = 0$, for example, in the formula for every y and x, then this is also true for $y = x/\varepsilon$. Here $P^\beta = P(y, D_y)$ for $\beta = 0$.

Zeroing expansion coefficients. By choosing $N_{\beta_s}^s$ and $u_s^{\beta_s}$ in the asymptotic representation (4), we will try to satisfy the asymptotic equation

$$P_\varepsilon(x/\varepsilon, \varepsilon D_x)\, u_p^l(x) \;=\; a(x/\varepsilon)\, f(x) \quad \text{in} \quad \Omega \tag{10}$$

with the greatest accuracy in the small parameter ε. To achieve this, let us equate the coefficients at the same powers ε to zero. Taking into account the expansion (9), we obtain the following relations

$$\sum_{\{\beta_0\}} \big[P(y, D_y)\, N_{\beta_0}^0(y)\big]\, u_0^{\beta_0}(x) \;=\; \delta_{mp}\, a(y)\, f(x),$$

$$\sum_{\{\beta_1\}} \big[P N_{\beta_1}^1(y)\big] u_1^{\beta_1}(x) + \sum_{\{\beta_0\}} \sum_{|\beta|=1} \big[P^\beta N_{\beta_0}^0(y)\big] D_x^\beta u_0^{\beta_0}(x)/\beta! = \delta_{m-1,p}\, a(y) f(x), \tag{11}$$

$$\ldots \qquad \ldots \qquad \ldots$$

$$\sum_{\{\beta_l\}} [PN^l_{\beta_l}(y)]u^{\beta_l}_l(x) + \sum_{\{\beta_{l-1}\}} \sum_{|\beta|=1} [P^\beta N^{l-1}_{\beta_{l-1}}(y)] D^\beta_x u^{\beta_{l-1}}_{l-1}(x)/\beta! +$$

$$+ \ldots + \sum_{\{\beta_0\}} \sum_{|\beta|=l} [P^\beta N^0_{\beta_0}(y)] D^\beta_x u^{\beta_0}_0(x)/\beta! = \delta_{m-l,p}\, a(y) f(x),$$

where δ_{mp} denotes the Kronecker symbol and it is convenient to assume that $P^\beta = 0$ for $|\beta| > m$ and $P^\beta = P(y, D_y)$ for $|\beta| = 0$. Note that for a fixed p in (4) only one of the equations in (11) has a non-zero right-hand side.

Investigation of the operator decomposition main part. Further constructions essentially depend on the properties of the operator $P(y) = P(y, D_y)$. This operator has 1-periodic coefficients and acts on 1-periodic functions in relations (11). Consequently, $P(y)$ can be considered as a differential operator on the n-dimensional torus T^n, which is a smooth closed manifold. Consider $P(y)$ as a bounded operator from $H^r(T^n)$ to $H^{r-m}(T^n)$ for some r and therefore for any real r (see, for example, [23]). It follows from (3) that this operator has a closed range and a finite-dimensional kernel and cokernel. It is also known that the operator $P(y)$ has zero index and is a Fredholm operator (see, for example, [23], [24]). In addition, we have the following inclusions

$$\ker P(y),\ \operatorname{coker} P(y) \subset C^\infty(T^n)$$

and therefore the subspaces do not depend on $r \in \mathbb{R}$ and are of the same dimensions.

Let $P(y)^*$ denote the operator conjugate to $P(y)$ with respect to the usual duality for complex-valued functions on T^n. Recall that the operator $P(y)^*$ acts on regular complex-valued functions according to the formula

$$P(y)^* u(y) = \sum_{|\alpha| \leqslant m} D^\alpha_y \left[\, \overline{a}_\alpha(y)\, u(y) \,\right]$$

and has a symbol

$$p^*(y, \xi) = \sum_{|\alpha| \leqslant m} \partial^\alpha_\xi D^\alpha_y \overline{p(y, \xi)}/\alpha!, \tag{12}$$

where the bar denotes complex conjugation. More precisely, the symbol $p^*(y, \xi)$ defines the formally conjugate operator $P^*(y, D_y)$ to $P(y)$, and to obtain from of this operator $P^*(y)$, we need to choose a suitable extension $P^*(y, D_y)$ in continuity with $C^\infty(T^n)$. In what follows, however, only the coincidence of $P^*(y, D_y)$ and $P^*(y)$ is used as operators on the space $C^\infty(T^n)$.

Thus, from assumption (3) we have $\ker P^*(y) = \operatorname{coker} P(y)$ and

$$d = \dim \ker P(y) = \dim \ker P^*(y) < \infty.$$

Suppose $d \geqslant 1$. Let us denote by $N_1(y), \ldots, N_d(y)$ and $M_1(y), \ldots, M_d(y)$ the generators of $\ker P(y)$ and $\ker P^*(y)$ respectively, normalized by the conditions

$$\langle N_k^2 \rangle = \langle M_k^2 \rangle = 1, \quad \langle N_k, N_h \rangle = \langle M_k, M_h \rangle = 0 \quad \text{for} \quad k \neq h, \quad (13)$$

where $k, h = 1, \ldots, d$. Here and further we will use the notation $\langle N \rangle = \langle N, 1 \rangle$ and $\langle N, M \rangle = \int_{T^n} N(y)\,\overline{M}(y)\,dy$. The Fredholm property of the operator $P(y)$ by definition means that for any $g \in C^\infty(T^n)$ the equation

$$P(y)\,w(y) = g(y) \quad \text{on} \quad T^n \quad (14)$$

is solvable if and only if

$$\{\, g \,\}_k \equiv \langle g, M_k \rangle = \int_{T^n} g(y)\,\overline{M}_k(y)\,dy = 0 \quad \text{for} \quad k = 1, \ldots, d. \quad (15)$$

If these conditions are satisfied, then the solution to the equation (14) from the space $C^\infty(T^n)$ has the form

$$w(y) = v(y) + C_1 N_1(y) + \cdots + C_d N_d(y),$$

where C_k are arbitrary constants and $v(y)$ is a particular solution of the equation (14), normalized by the conditions $\langle v, N_k \rangle = 0$ for $k = 1, \ldots, d$.

For $d = 0$ the equation (14) has a unique solution $w \in C^\infty(T^n)$ for any $g \in C^\infty(T^n)$. Further, in the construction of asymptotic expansions, options are possible, which we will begin to analyze.

3. Asymptotic Expansions of Taylor Type

In the case $d = 0$, further construction of asymptotic expansions will resemble expansions Taylor type, which will be confirmed below. We normalize u_p^l by choosing $p = m$ in (4), then the first relation in (11) takes the following form

$$\sum_{\{\beta_0\}} \left[P(y, D_y)\,N^0_{\beta_0}(y)\right] u_0^{\beta_0}(x) = a(y)\,f(x).$$

In order to satisfy this relation, it is sufficient to choose $\{\beta_0\} = \{0\}$, $u_0^0(x) = f(x)$ and define the function $N_0^0 \in C^\infty(T^n)$ as a solution to the equation

$$P(y, D_y)\,N_0^0(y) = a(y), \quad (16)$$

which can naturally be called *a cell problem.* In accordance with the assumption $d = 0$ the solution to the equation is uniquely determined.

To satisfy the second relation in (11), we take $\{\beta_1\}=\{\beta=(\beta_1,\ldots,\beta_n):|\beta|=1\}$ and $u_1^{\beta_1}(x)=D_x^{\beta_1}f(x)/\beta_1!$. Then this relation looks like

$$\sum_{\{\beta_1\}}[PN^1_{\beta_1}(y)]D_x^{\beta_1}f(x)/\beta_1! + \sum_{\{\beta_1\}}[P^{\beta_1}N^0_0(y)]D_x^{\beta_1}f(x)/\beta_1! = 0$$

and for the function $N^1_{\beta_1}(y)$ for each $\beta_1\in\{\beta_1\}$ we obtain the equations

$$P(y,D_y)\,N^1_{\beta_1}(y) = -P^{\beta_1}N^0_0(y), \tag{17}$$

from which the functions $N^1_{\beta_1}\in C^\infty(T^n)$ can be uniquely defined. Next, select

$$\{\beta_2\}=\{\beta=\beta^1+\beta_1:|\beta_1|=|\beta^1|=1\}\cup\{\beta:|\beta|=2\}$$

and $u_2^{\beta_2}=D_x^{\beta_2}f(x)/\beta_2!$. Then the third relation in (11) takes the form

$$\sum_{\{\beta_2\}}[PN^2_{\beta_2}(y)]D_x^{\beta_2}f(x)/\beta_2! + \sum_{\{\beta=\beta^1+\beta_1\}}[P^{\beta^1}N^1_{\beta_1}(y)]D_x^{\beta}f(x)/\beta! +$$

$$+\sum_{|\beta|=2}[P^{\beta}N^0_0(y)]D_x^{\beta}f(x)/\beta! = 0$$

and for the function $N^2_{\beta_2}(y)$ for each $\beta_2\in\{\beta_2\}$ we get the following equations

$$PN^2_{\beta_2}(y) = -P^{\beta^1}N^1_{\beta_1}(y), \qquad PN^2_{\beta_2}(y) = -P^{\beta_2}N^0_0(y), \tag{18}$$

from which the functions $N^2_{\beta_2}\in C^\infty(T^n)$ are uniquely determined.

Further construction of asymptotic expansions may be carried out by induction. To fulfill the l relation in (11) we select

$$\{\beta_l\}=\{\beta=\beta^1+\beta_{l-1}\}\cup\cdots\cup\{\beta=\beta^s+\beta_{l-s}\}\cup\cdots\cup\{\beta:|\beta|=l\}$$

with $|\beta^s|=|\beta_s|=s$ for $s=1,\ldots,l-1$ and $u_l^{\beta_l}(x)=D_x^{\beta_l}f(x)/\beta_l!$. Then for the function $N^l_{\beta_l}(y)$ for each $\beta_l\in\{\beta_l\}$ we obtain equations similar to (17) and (18), from which the functions $N^l_{\beta_l}\in C^\infty(T^n)$ can be uniquely defined.

Thus, the constructed asymptotic expansion

$$u^l_m = \varepsilon^m N^0_0(x/\varepsilon)\,f(x) + \varepsilon^{m+1}\sum_{|\beta|=1}N^1_\beta(x/\varepsilon)\,D_x^\beta f(x)/\beta! + \ldots \tag{19}$$

satisfies the boundary conditions from (1) and satisfies the equation (10) up to $O(\varepsilon^{l+1})$, where $O(\,\cdot\,)$ is calculated, for example, in the norm of the space

$C^0(\overline{\Omega})$. In the situation under consideration, for the solution to problem (1) the estimate

$$\| u \|_{H_0^{m/2}(\Omega)} + \varepsilon^{-m/2} \| u \|_{L^2(\Omega)} \leqslant C \varepsilon^{m/2} \| a f \|_{L^2(\Omega)} \qquad (20)$$

often holds with a constant C independent of ε. Taking into account the constructions given here and the linearity of problem (1), we have the following statement.

Theorem 1. *Let* $d = 0$ *and inequality (20) holds for problem (1). Then*

$$\| u - u_m^l \|_{H_0^{m/2}(\Omega)} + \varepsilon^{-m/2} \| u - u_m^l \|_{L^2(\Omega)} \leqslant C \varepsilon^{l+1+m/2},$$

where u *is the solution to problem (1),* u_m^l *is defined (19) and the constant* C *is independent of parameter* ε *for* $0 < \varepsilon \leqslant \varepsilon_0$ *with an appropriate positive* ε_0.

From estimate (20) it is not difficult to deduce that there is a unique solution to problem (1) for a fixed ε. Examples of equations for which the estimate is valid will be given below. Now remark that the asymptotic expansion (19) is very similar to the Taylor expansion. In fact, this is true for the original setting.

It is known that the generalized Leibniz formula (6) is also valid in the case when $P_\varepsilon = P(x/\varepsilon, \varepsilon D_x)$ is a pseudodifferential operator, with the only difference that the summation will be carried out over all possible multi-indices β according to [23]. The simplest elliptic pseudodifferential operator of zero order is the shift operator.

Let u be the solution to equation (1) for the operator in the following form

$$u(x - \varepsilon) = f(x) \quad \text{in} \quad \Omega,$$

where $a(y) = 1$ and $f \in C_0^\infty(\Omega)$. Hence, $u = f(x + \varepsilon)$ is a solution for small ε.

The symbols of the corresponding operators for this equation have the form

$$p(y, \xi) = e^{-i(\xi_1 + \ldots + \xi_n)}, \qquad p^\beta(y, \xi) = (-i)^{|\beta|} e^{-i(\xi_1 + \ldots + \xi_n)}.$$

in the notation from (5). Consequently, equation (16) becomes

$$P(y) N_0^0(y) \equiv N_0^0(y - 1) = 1.$$

There is a unique solution $N_0^0(y) = 1$ of this equation and therefore $d = 0$. For a similar reason, the equations (17) have solutions $N_\beta^1(y) = i$ for $|\beta| = 1$. Consequently, the asymptotic expansion of the solution $u = f(x + \varepsilon)$ to the equation under consideration is determined by the equality (19), which is of the form

$$u_0^l = f(x) + \varepsilon \sum_{|\beta|=1} \partial_x^\beta f(x)/\beta! + \ldots.$$

The interested reader may consider equation (18) and subsequent equations to obtain more accurate asymptotic expansions. Thus, general principles for constructing asymptotic expansions of solution $u = f(x+\varepsilon)$ to the equation with a small shift lead to the well-known and widely used Taylor formula for the smooth functions.

This example clearly shows *the problem of justifying asymptotics* (19) in the case under consideration. Indeed, the asymptotics u_0^l is a good approximation for $f(x+\varepsilon)$ up to $O(\varepsilon^{l+1})$, for example, in the uniform norm on $\overline{\Omega}$ for $f(x) \in C_0^\infty(\Omega)$. On the other hand, the approximation of $f(x+\varepsilon)$ by the asymptotics u_0^0 coincides with the definition of uniform continuity for the function $f(x) \in C_0^0(\Omega)$. While this approximation is meaningless for discontinuous functions, however, for such functions $f(x+\varepsilon)$ can converge to $f(x)$ for $\varepsilon \to 0$ in the relevant integral norms.

Let's fix $\lambda \in \mathbb{C}$ and note that the asymptotic construction for the solution to a seemingly similar equation

$$u\,(x-\varepsilon) + \lambda\, u\,(x) = a(x/\varepsilon) f(x) \qquad \text{in} \qquad \Omega,$$

can be much more complicated for $f \in C_0^\infty(\Omega)$. Indeed, the symbol of the corresponding operator for $\varepsilon = 1$ has the following form

$$p(y, \xi, \lambda) = \lambda + e^{-i\,(\xi_1 + \ldots + \xi_n)}$$

and may not be elliptic for $|\lambda| = 1$ and $\xi \in \mathbb{R}^n \setminus \{0\}$ in the sense that conditions (3) may not be met. Consequently, the equation (16) may not have a smooth solution and asymptotic constructions are impossible to carry out. On the other hand, if there is a smooth solution to this equation for every smooth $a(y)$, then the asymptotics (19) is defined and one can hope to justify the asymptotic expansion.

A description of the spectrum of this operator and a more general operator with the symbol $p(y,\xi,\lambda) = \lambda\, s(y) + e^{-i\,(\xi_1+\ldots+\xi_n)}$ can be found in [25] for relevant $s(y)$. Thus, outside the points of the spectrum one can hope to find the asymptotics in the form (19) for solutions to the corresponding equation with the operator. The considered case with a shift is also not so trivial, since everyone can consider, for example, shifts along the trajectories of relevant dynamical systems on manifolds. Definitions of some of these shifts can be found, for example, in [25].

Let's fix $\lambda \in \mathbb{C}$, a real smooth sign-definite function $s(y)$ and consider a more general equation than in (1) for differential operator with the following symbol

$$p(y, \xi, \lambda) = \sum_{|\alpha| \leqslant m} a_\alpha(y)\, \xi^\alpha + \lambda\, s(y). \tag{21}$$

Here it is convenient not to take into account the normalizing factor ε^{-m} (that is, choose $m = 0$), since this factor can be included in the asymptotic expansion (4) through renormalization of the exponent p. In addition, the factor was introduced following the tradition of writing second-order equations, which are discussed below.

It is known [26] that equation (16) with the operator defined by the symbol (21) is uniquely solvable for every λ except for at most a countable subset of $\mathbb{C}$, which is the spectrum of this operator. In addition, the conditions for the operator defined by the symbol $p(y, \xi, \lambda)$ are known [23], [26], under which an analogue of estimate (20) is valid (with $\varepsilon = 1$ and $f = 1$) for solutions of equation (16) and this estimate is satisfied for problem (1) for a fixed ε and relevant λ not from the spectrum.

Thus, the considered case is common (except countable set) for equations determined by elliptic operators with ε-periodic coefficients. Nevertheless, this case practically did not appear in publications, although it was discussed in sufficient detail in[18], and a special case of applying the algorithm to the shift operator was given in the dissertation [19] that is still available online. In fact, this case was used (starting from [27]–[29]) to construct asymptotic solution expansions for problems with contrast coefficients that depend on additional parameters and arise naturally in the theory of composites, metamaterials and photonic crystals. This asymptotic construction for second order partial differential equations will be explained further.

Asymptotic data. Some generalizations of the considered case can be obtained without significant changes in the above constructions.

Let the right-hand side for the equation of problem (1) have the form

$$a_0(x/\varepsilon)\, f_0(x) + \varepsilon\, a_1(x/\varepsilon)\, f_1(x) + \dots,$$

where the coefficients $a_0(y), a_1(y), \dots$ are 1-periodic smooth functions. Further in this case we will talk about *the asymptotic right-hand side*. Then, proceeding as above, we obtain that $u_0^0(x) = f_0(x)$ and the equation (16) retains its form that is $PN_0^0 = a_0$. Consequently, the second relation in (17) takes the form

$$\sum_{\{\beta_1\}} [PN^1_{\beta_1}(y)]\, u_1^{\beta_1}(x) + \sum_{|\beta|=1} [P^\beta N_0^0(y)] D_x^\beta f_0(x) \;=\; a_1(y)\, f_1(x).$$

In order to satisfy this equality, it is sufficient to choose $\{\beta_1\} = \{\beta : |\beta| \leqslant 1\}$, $u_1^\beta(x) = f_1(x)$ for $|\beta| = 0$ and $u_1^\beta(x) = D_x^\beta\, f_0(x)$ for $|\beta| = 1$. In this case, one more equation $PN^1_\beta = a_1$ will be added to the equations (17) for $|\beta| = 0$ and further construction is carried out obviously by induction.

Let the coefficients in the equation of problem (1) have the form

$$a_\alpha(x/\varepsilon) = a_\alpha^0(x/\varepsilon) + \varepsilon\, a_\alpha^1(x/\varepsilon) + \dots.$$

where the coefficients $a_\alpha^0(y), a_\alpha^1(y), \dots$ are 1-periodic smooth functions. In what follows, this variant for problem (1) will be called *the case of asymptotic coefficients*. This case corresponds to the expansion $P(y) = P_0(y) +$

$\varepsilon P_1(y) + \ldots$, which is easy to take into account in (9), as in the case of the asymptotic right-hand side. The given generalizations for $d = 0$ express some stability of the constructions with respect to the right-hand side and coefficients. In the case of $d > 0$ there may not be stability in terms of coefficients; corresponding examples will be given later.

Note that in the case of asymptotic coefficients and right-hand sides, it is possible to use expansions in non-integer powers, for example, in the sequences $\varepsilon, \varepsilon^{1+\sigma}, \varepsilon^{1+2\sigma}, \ldots$ (when $0 < \sigma < 1$) and $\varepsilon, \varepsilon^2 \ln \varepsilon, \ldots$, which are also not difficult to take into account in suitable expansions. The most significant at this point is the selection of the operator decomposition main part presented here according to (8).

The above constructions of asymptotic expansions are also simply generalized to the case of a vector small parameter $\vec{\varepsilon} = (\varepsilon_1, \ldots, \varepsilon_n)$ in the sense that the terms in operator definition (2) are in the following form

$$a_\alpha(x_1/\varepsilon_1, \ldots, x_n/\varepsilon_n)\,(\varepsilon_1 D_1 \ldots \varepsilon_n D_n)^\alpha \quad \text{for} \quad |\alpha| \leqslant m,$$

where all coefficients $a_\alpha(y_1, \ldots, y_n)$ are 1-periodic functions as before. Accordingly, in expansion (4) one should choose $N_0^0 = N_0^0(x_1/\varepsilon_1, \ldots, x_n/\varepsilon_n)$ and so on. Here it should be borne in mind, of course, that the main parts of the operator decomposition for different vector parameters $\vec{\varepsilon}_1$ and $\vec{\varepsilon}_2$ can differ significantly, for example, for $\vec{\varepsilon}_1 = (\varepsilon, \varepsilon, \ldots, \varepsilon)$ and $\vec{\varepsilon}_1 = (\varepsilon, \varepsilon^2, \ldots, \varepsilon^n)$. These observations make it possible to generalize the presented constructions to elliptic equations in thin domains.

An asymptotic expansion for solution to equation (1) with the right-hand side of the general form $F = F(x/\varepsilon, x)$ can be constructed as follows. Let us choose functions $N^s_{\beta_s}$ in expansion (4) in the same general form $N^s_{\beta_s} = N^s_{\beta_s}(x/\varepsilon, x)$. Then the differentiation formula (7) will take the following form

$$\varepsilon^\beta D_x^\alpha N(x/\varepsilon, x) = \varepsilon^\beta\big(\varepsilon^{-\alpha} D_y^\alpha + D_x^\alpha\big) N(y, x) \quad \text{for} \quad y = x/\varepsilon, \tag{22}$$

where the function $N(y, x)$ is 1-periodic in y for a fixed $x \in \Omega$ and smooth in (y, x). Taking into account this formula in (8), it can be noted that further construction continues as in the case of asymptotic coefficients. So, instead of equation (16), we obtain the equation $P(y, D_y)\, N_0^0(y, x) = F(y, x)$, in which x enters as a parameter. Solutions of elliptic equations retain a smooth dependence on the parameter. Hence, we can find the smooth solution $N_0^0 = N_0^0(y, x)$ of this equation that is 1-periodic in y for a fixed x and continue construction. Here it is assumed, of course, that the smooth function $F(y, x)$ is 1-periodic in y and $F \in C_0^\infty(\Omega; \mathbb{C})$ for a fixed y.

Similarly, one can try to construct the asymptotics of the solution to a more general equation of the following form

$$P_\varepsilon(x/\varepsilon, x, \varepsilon D_x)\, u(x) = F(x/\varepsilon, x) \quad \text{in} \quad \Omega \tag{23}$$

with boundary conditions from (1). In this case, we can also use $N^s_{\beta_s} = N^s_{\beta_s}(x/\varepsilon, x)$ in (4) and formula (22). However, the analogue of equation (16) will have the form $P(y, x, D_y)\, N^0_0(y, x) = F(y, x)$. Thus, it is necessary to assume that a smooth and 1-periodic solution to this equation exists and is uniquely defined for $F(y, x)$ as before. Then all constructions are saved. Here it is assumed that $d = 0$, otherwise the constructions can become significantly more complicated, since the dimension $\ker P(y, x)$ can depend on x. In the case, the dimension d can be undefined. However, in specific examples such constructions can also be performed.

Thus, starting from functions with separated variables, one can try to consider equations with common coefficients and right-hand sides, the variables for which are not separated. Note also that the 1-periodicity of the functions considered here means that *the periodicity cell* is the cube $Y = (0, 1)^n$. It would also be possible to choose a non-degenerate parallelepiped as the periodicity cell. However, by a non-degenerate variable change, the case can be reduced to the one under consideration, since elliptic operators remain as such after non-degenerate variable changes.

4. Elliptic Second Order Partial Differential Equations

Second order partial differential equations with ε-periodic coefficients are most often encountered in applications. As a rule, the coefficients of such equations are real and are determined by the characteristics of the relevant media and materials in which the corresponding processes are modeled. Thus, it is natural to consider such equations in more detail, since most works on homogenization are devoted to precisely such equations and some of their nonlinear analogues and generalizations.

Let $n = 2$ be given a bounded domain $\Omega \subset \mathbb{R}^2$ with a smooth boundary and a smooth real function $f \in C_0^\infty(\Omega)$. Let us define the function u as a solution to the Dirichlet problem for the following second-order equation

$$\begin{aligned} &- \partial_l\, (a_{lk}(x/\varepsilon)\partial_k\, u) + \varepsilon^{-1} a_k(x/\varepsilon)\, \partial_k\, u + \varepsilon^{-2} a_0(x/\varepsilon)\, u = \\ &\qquad = a(x/\varepsilon)\, f \quad \text{in} \quad \Omega, \qquad u = 0 \quad \text{on} \quad \partial\Omega, \end{aligned} \tag{24}$$

where ε is a small positive parameter. Here and below, the convention of summation over repeating indices k from 1 to 2 is accepted. So, for example, by definition the equality $a_k(x/\varepsilon)\, \partial_k = \sum_{k=1}^2 a_k(x/\varepsilon)\, \partial_k$ holds. The assumption $n = 2$ is made for simplicity of presentation, since as n increases, it is necessary to write out more symbols from (5). Suppose that for $l, k = 1, 2$ the coefficients $a_{lk}(y) = a_{kl}(y)$, $a_k(y)$ and $a_0(y)$ of the equation of problem (24) are 1-periodic (real-valued) functions of class $C^\infty(\mathbb{R}^2)$, which satisfy the following inequalities

$$\nu\,|\xi|^2 \leqslant a_{lk}(y)\,\xi_l\,\xi_k \leqslant \gamma\,|\xi|^2 \quad \text{for} \quad y \in \mathbb{R}^2 \quad \text{and} \quad \xi \in \mathbb{R}^2, \tag{25}$$

$$\partial_{y_k} a_k(y) \leqslant \sigma, \quad \sigma \leqslant a_0(y) \quad \text{for} \quad y \in \mathbb{R}^2, \tag{26}$$

where σ, ν and γ are given positive constants.

Let us prove that in the case under consideration, for solution to problem (24) the inequality (20) is valid. Indeed, multiplying scalarly in $L^2(\Omega)$ the equation of problem (24) by u and taking into account the boundary conditions, we get

$$\nu \,\|\, \nabla u \,\|^2_{L^2(\Omega)^2} - (2\varepsilon^2)^{-1}\int_\Omega (\partial_{y_k} a_k)(x/\varepsilon)\, u^2 dx \,+\, \sigma\,\varepsilon^{-2}\,\|\, u\,\|^2_{L^2(\Omega)} \leqslant$$

$$\leqslant \|\, a\, f\,\|_{L^2(\Omega)} \|\, u\,\|_{L^2(\Omega)} \leqslant \varepsilon^2\,\sigma^{-1}\,\|\, a\, f\,\|^2_{L^2(\Omega)} + \sigma(4\varepsilon^2)^{-1}\,\|\, u\,\|^2_{L^2(\Omega)},$$

where the integration by parts and the following equality are taken into account

$$\int_\Omega (a_k \partial_k u)u\, dx = -\int_\Omega u\,\partial_k\,(a_k u)\,dx = -\int_\Omega (a_k \partial_k u)u\, dx - \varepsilon^{-1}\int_\Omega (\partial_{y_k} a_k)u^2 dx.$$

In addition, we used (7) and the δ-inequality, written as $|ab| \leq (\delta/2)a^2 + b^2/(2\delta)$ for positive δ and arbitrary a, b. When the last inequalities were proven, $\delta = 2\varepsilon^2/\sigma$ was chosen. In the case of $\partial_{y_k} a_k(y) \leqslant 0$, the integral is not essential in the last inequalities, which implies the inequality (20). Therefore, separating the positive part of the function $\partial_{y_k} a_k(y)$, the last inequality can be rewritten as

$$\nu \,\|\, \nabla u \,\|^2_{L^2(\Omega)^2} + 3\,\sigma\,(4\varepsilon^2)^{-1}\,\|\, u\,\|^2_{L^2(\Omega)} \leqslant \varepsilon^2\,\sigma^{-1}\,\|\, a\, f\,\|^2_{L^2(\Omega)} +$$

$$+ (2\varepsilon^2)^{-1}\int_\Omega (\partial_{y_k} a_k)(x/\varepsilon)\, u^2\, dx \leqslant \varepsilon^2\,\sigma^{-1}\,\|\, a\, f\,\|^2_{L^2(\Omega)} + \sigma\,(2\varepsilon^2)^{-1}\,\|\, u\,\|^2_{L^2(\Omega)}.$$

Consequently, for solutions of problem (24) we have proven the inequality

$$\|\, \nabla u \,\|_{L^2(\Omega)^2} + \varepsilon^{-1}\,\|\, u\,\|_{L^2(\Omega)} \leqslant \varepsilon\, C\,\|\, a\, f\,\|_{L^2(\Omega)} \tag{27}$$

with the constant C independent of ε. From the inequality for a fixed ε it follows that a unique solution to problem (24) exists (for example, according to [22]).

Equation of problem (24) is written in traditional form. In order to use the notation from (2), this equation should be rewritten in the form

$$-\,a_{lk}(x/\varepsilon)\partial_l\,\partial_k\, u \,+\, \varepsilon^{-1} b_k(x/\varepsilon)\,\partial_k\, u \,+\, \varepsilon^{-2} a_0(x/\varepsilon)\, u \;=\; a(x/\varepsilon)\, f$$

where $b_k(y) = a_k(y) - \partial_{y_l} a_{lk}(y)$. Then, the relevant symbols from (5) are of the form

$$p(y,\xi) = a_{lk}(y)\,\xi_l\,\xi_k + b_k(y)\,(i\,\xi_k) + a_0(y),$$
$$p^{(1,0)}(y,\xi) = 2\,a_{11}(y)\,\xi_1 + 2\,a_{21}(y)\,\xi_2 + i\,b_1(y), \qquad (28)$$
$$p^{(0,1)}(y,\xi) = 2\,a_{22}(y)\,\xi_2 + 2\,a_{21}(y)\,\xi_1 + i\,b_2(y),$$
$$p^{(2,0)}(y,\xi) = 2\,a_{11}(y), \quad p^{(0,2)}(y,\xi) = 2\,a_{22}(y), \quad p^{(1,1)}(y,\xi) = 2\,a_{21}(y).$$

which should be multiplied by ε^{-2} in accordance with the notation from (2).

Repeating the proof of inequality (27) for 1-periodic functions with $\varepsilon = 1$, we conclude that $d = 0$. Therefore, the asymptotic expansion of the solution to problem (24) is determined by the equality (19) in the following form

$$u_2^l = \varepsilon^2 N_0^0(x/\varepsilon) f(x) + \varepsilon^3 N_{(1,0)}^1(x/\varepsilon)\,\partial_1 f(x) + \varepsilon^3 N_{(0,1)}^1(x/\varepsilon)\,\partial_2 f(x) + \ldots, \qquad (29)$$

where the functions $N_0^0(y)$, $N_{(1,0)}^1(y), \ldots$ are 1-periodic solutions of the equations

$$-\,\partial_{y_l}\left(a_{lk}(y)\,\partial_{y_k} N_0^0(y)\right) + a_k(y)\,\partial_{y_k} N_0^0(y) + a_0(y) N_0^0(y) = a(y),$$

$$-\,\partial_{y_l}\left(a_{lk}(y)\,\partial_{y_k} N_{(1,0)}^1(y)\right) + a_k(y)\,\partial_{y_k} N_{(1,0)}^1(y) + a_0(y) N_{(1,0)}^1(y) =$$
$$= -\,2\,a_{11}(y)\partial_{y_1} N_0^0(y) \;-\; 2\,a_{21}(y)\partial_{y_2} N_0^0(y) \;-\; b_1(y)\,N_0^0(y), \ldots,$$

since $\partial_k = iD_k$ and $\partial_{y_k} = iD_{y_k}$ by definition in asymptotic solution expansion (19).

Thus, for problem (24), Theorem 1 is completely proven, in which $m = 2$ and the asymptotic solution expansion has the form (29). It is easy to verify that in the theorem proof only inequalities (25), (26), the assumptions that $a_{lk}(y)$, $\partial_{y_k} a_k(y)$, $a_k(y)$, $a_0(y)$, $a(y)$ are elements of $L^\infty(Y)$ and $f \in L^2(\Omega)$ were used, where $l, k = 1, 2$ and $Y = (0,1)^2$ is usually called *a periodicity cell.* Note that in this case it is known [22] that the functions $N_0^0(y)$, $N_{(1,0)}^1(y), \ldots$ are Holder functions. But, to use asymptotics expansion (29), the f must be regular and compactly supported.

Problems for 1-periodic functions $N_0^0(y)$, $N_{(1,0)}^1(y), \ldots$, which are included in asymatotic expansion (29), are usually called *cell problems.* Such problems in practice are solved numerically with guaranteed accuracy through computer simulation, which is a standard problem in computational mathematics. This, together with Theorem 1, allows everyone to find a solution to problem (24) with guaranteed accuracy for small ε. Whereas a direct numerical solution of problem (24) practically requires enormous computer resources, since a very dense computing grid must be chosen to take into account the coefficient changes in every cell. This is precisely how the need arose to homogenize equations with oscillating coefficients.

Asymptotic data. Choosing $a_k(y) = \varepsilon\, a_k^1(y)$ for $k = 1,2$ in equation of problem (24) for comparison, we obtain an example of a problem with asymptotic coefficients. Assuming the fulfillment of the inequalities (25) and (26), we note that the inequality for $\partial_{y_k} a_k(y)$ now has the form $\varepsilon\, \partial_{y_k} a_k^1(y) \leqslant \sigma$ and is always satisfied for sufficiently small ε, for example, when $\partial_{y_k} a_k^1(y) \in L^\infty(Y)$. Let the remaining coefficients be the same as in the previous case. Then $P(y) = P_0(y) + \varepsilon\, P_1(y)$, where the operators $P_0(y)$ and $P_1(y)$ have the following symbols

$$p_0(y,\xi) = a_{lk}(y)\,\xi_l\,\xi_k - \partial_{y_l} a_{lk}(y)(i\,\xi_k) + a_0(y), \quad p_1(y,\xi) = a_k^1(y)(i\,\xi_k).$$

In this version of problem (24) the statements of Theorem 1 are also satisfied while preserving the proof. Therefore, the asymptotic expansion of the solution to the problem is defined in the following form

$$u_2^l = \varepsilon^2 N_0^0(x/\varepsilon) f(x) + \varepsilon^3 N_{(0,0)}^1(x/\varepsilon)\, f(x) + \varepsilon^3 N_{(1,0)}^1(x/\varepsilon)\, \partial_1 f(x) + \ldots,$$

where the functions $N_0^0(y)$, $N_{(0,0)}^1(y), \ldots$ are 1-periodic solutions of the equations

$$-\,\partial_{y_l}\left(a_{lk}(y)\,\partial_{y_k} N_0^0(y)\right) + a_0(y) N_0^0(y) = a(y),$$

$$-\,\partial_{y_l}\left(a_{lk}(y)\,\partial_{y_k} N_{(0,0)}^1(y)\right) + a_0(y) N_{(0,0)}^1(y) = -a_k^1(y)\,\partial_{y_k} N_0^0(y)$$

$$-\,\partial_{y_l}\left(a_{lk}(y)\,\partial_{y_k} N_{(1,0)}^1(y)\right) + a_0(y) N_{(1,0)}^1(y) =$$

$$= -\,2\,a_{11}(y)\partial_{y_1} N_0^0(y) - 2\,a_{21}(y)\partial_{y_2} N_0^0(y) + (\partial_{y_l} a_{l1}(y)) N_0^0(y), \ldots .$$

Here it should be emphasized that choosing $a_0 = \varepsilon^\nu a_0^\nu(y)$ for a positive ν in problem (24) will lead to the case $d > 0$, since the main symbol will take the form

$$p_0(y,\xi) = a_{lk}(y)\,\xi_l\,\xi_k + b_k(y)\,(i\,\xi_k),$$

for $b_k(y) = a_k(y) - \partial_{y_l} a_{lk}(y)$ and the function $N(y) = 1$ will be included in the nontrivial kernel of the relevant operator. This is the specificity of such operators, which are usually called *second-order divergent operators*. These operators are the most thoroughly studied in homogenization theory.

Weak diffusion by microneedles. From inequality (27) it follows that the solution to problem (24) is very small for small ε, which is obvious from the asymptotic expansion (29). In order to isolate the main part of the solution asymptotics, we renormalize the right-hand side. Let the right-hand side in (24) be the form

$$\begin{aligned} af &= \varepsilon^{-2}\hat{a}(x/\varepsilon) f(x) + \varepsilon^{-1}\partial_k\left(\hat{a}^k(x/\varepsilon)\hat{f}(x)\right) \\ &= \varepsilon^{-2}\hat{a}(y) f(x) + \varepsilon^{-2}\left(\partial_{y_k}\hat{a}^k(y)\right)\hat{f}(x) + \varepsilon^{-1}\hat{a}^k(y)\partial_k\hat{f}(x) \quad \text{for} \quad y = x/\varepsilon \end{aligned} \tag{30}$$

and the coefficients are of general types. Then this problem is the following form

$$-\varepsilon^2\,\partial_l\big(a_{lk}(x,x/\varepsilon)\partial_k\,u\big)\;+\;\varepsilon\,a_k(x,x/\varepsilon)\,\partial_k\,u\;+\;a_0(x,x/\varepsilon)\,u\;=$$

$$=\;\hat{a}(x/\varepsilon)f(x)+\varepsilon\,\partial_k\big(\hat{a}^k(x/\varepsilon)\hat{f}(x)\big)\quad\text{in}\quad\Omega,\qquad u=0\quad\text{on}\quad\partial\Omega,\tag{31}$$

where the coefficients $a_{lk}(x,y)=a_{kl}(x,y),\ldots,a_0(x,y)$ are smooth and 1-periodic in y for $l,k=1,2$. Thus, we renormalized problem (24) (multiplying by ε^2 in fact, so $p=0$ in (4), (29)) and selected general data. Assuming that (25) and (26) are valid for all $(x,y)\in\Omega\times Y$, we can repeat the proof of (27) and get the estimate

$$\varepsilon\,\big\|\,\nabla u\,\big\|_{L^2(\Omega)^2}+\big\|\,u\,\big\|_{L^2(\Omega)}\leqslant C\,\big\|\hat{a}\,f\big\|_{L^2(\Omega)}+C\,\big\|(\hat{a}^1+\hat{a}^2)\,\hat{f}\big\|_{L^2(\Omega)}$$

with the constant C independent of ε. From this estimate follows the estimate

$$\varepsilon\,\big\|\nabla u\big\|_{L^{2,\infty}(\Omega)^2}+\big\|u\big\|_{L^{2,\infty}(\Omega)}\leqslant C\big\|\hat{a}f\big\|_{L^{2,\infty}(\Omega)}+\;C\big\|(\hat{a}^1+\hat{a}^2)\hat{f}\big\|_{L^{2,\infty}(\Omega)}\tag{32}$$

in accordance with [30], where $L^{2,\infty}(\Omega)$ is a slightly wider space, in the sense that $L^2(\Omega)\subset L^{2,\infty}(\Omega)\subset L^{2-\nu}(\Omega)$ for positive $\nu\leqslant 1$. This space is called a weak Lebesgue space and is notable for the fact that it contains an "antiderivative" of the δ-function. It is known [30], that if $\hat{x}\in\Omega$ then $(x-\hat{x})|x-\hat{x}|^{-2}\in L^{2,\infty}(\Omega)^2$ and

$$\partial_k\big((x_k-\hat{x}_k)/(2\pi|x-\hat{x}|^2)\big)=\delta_2(x-\hat{x}).$$

From inequality (32) for a fixed ε it follows that a unique solution to problem (31) exists for the appropriate right-hand sides (for example, according to [31], [32]).

Now we will consider problem (31) with periodic δ-functions. This problem arises when microneedle arrays for transdermal (subcutaneous) drug injections are used in modern medicine. Such arrays are formed by a large number of very thin microneedles attached in a periodic manner to a flat base. The microneedles of such arrays are either conventional medical needles, manufactured as hollow cylinders for administering drugs under pressure, or are synthesized from biosoluble polymers that dissolve at a prescribed rate after injection. Further details can be found in reviews [33], [34]; images of such arrays are given, for example, in [35], [36], where the problem of elastic resistance of shells when interacting with microneedle arrays was considered as a problem for the variational inequality with obstacles.

In the first case, a schematic distribution graph of such injections is presented in the first image of Fig. 1. Here, at a large number of the boundary points $x_2^1, x_2^2, \ldots, x_2^\kappa$ located ε-periodically, microneedles of length l_1 are inserted from the ends of which a certain volume of drugs is uniformly introduced, which is represented by δ-functions. Thus, considering such a strip of δ-functions as the right side in the equation of problem (31), one can obtain an image of the volume distribution of drugs over the domain Ω. The image will correspond to very weak diffusion and transfer, since the corresponding coefficients are very small for sufficiently small ε.

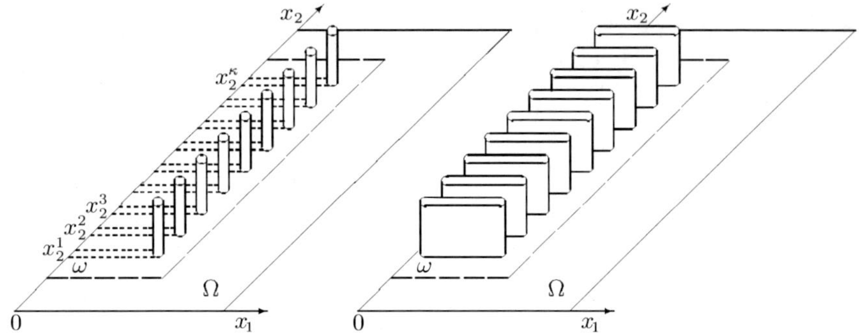

Figure 1. Graphs of drug volume distribution functions.

More precisely, let $\Omega = (0, L_1)\times(0, L_2)$ be a sufficiently large rectangle containing the rectangle $\omega = (0, l_1)\times(l_2^0, l_2 - l_2^0)$ corresponding to the region containing the microneedles. Here l_1, l_2^0, l_2, $l_2 - l_2^0$ are positive numbers. Let $\kappa - 1 = (l_2 - l_2^0)/\varepsilon$ be an integer and $x_2^1 = l_2^0$, $x_2^2 = x_2^1 + \varepsilon, \ldots x_2^\kappa = x_2^{\kappa-1} + \varepsilon = x_2^1 + (\kappa - 1)\,\varepsilon = l_2$. Thus, δ-functions are concentrated at points $\hat{x}^1 = (l_1, x_2^1)$, $\hat{x}^2 = (l_1, x_2^2), \ldots, \hat{x}^\kappa = (l_1, x_2^\kappa)$.

The δ-functions are ε-periodic with respect to x_2. Let us first define these functions exactly on the cell $Y = (0, 1)^2$. To do this, we denote by $\hat{y}$ the center of this cell and define a smooth function $\varrho(y)$ equal to 1 on the circle of radius 0.1 and 0 outside the circle of radius 0.2 with centers at the point $\hat{y}$. Let the smooth function $\hat{f}_1 = \hat{f}_1(x_1)$ be equal to 1 on the segment $[\,0.3\,\varepsilon, l_1 + 0.1\,\varepsilon\,]$ and 0 outside the segment $[\,0.1\,\varepsilon, l_1 + 0.3\,\varepsilon\,]$, and the smooth function $\hat{f}_2 = \hat{f}_2(x_2)$ be equal to 1 on the segment $[\,l_2^0 - 0.1\,\varepsilon, l_2 + 0.1\,\varepsilon\,]$ and 0 outside the segment $[\,l_2^0 - 0.3\,\varepsilon, l_2 + 0.3\,\varepsilon\,]$. The functions exist and $|\hat{f}_k| \leqslant 1$, $k = 1, 2$, for example, according to [37].

Next, we define the functions $\hat{f}(x) = \hat{f}_1(x_1)\hat{f}_2(x_2)$ and

$$\hat{a}^k(y) = \varrho(y - \hat{y})(y_k - \hat{y}_k)/(2\pi|y - \hat{y}|^2)$$

for $k = 1, 2$. The functions $\hat{a}^k(y)$ will be continued as 1-periodic functions with respect to y_2 and by zero relative to y_1. Hence, taking into account (30) we get

$$\varepsilon\,\partial_k\big(\hat a^k(x/\varepsilon)\hat f(x)\big) = \sum_{i=1}^{\kappa} \delta_2((x-\hat x^i)/\varepsilon)$$

$$+\sum_{i=1}^{\kappa} \big(\partial_{y_k}\varrho(y^i)\big)y^i_k/(2\pi|y^i|^2)\hat f(x) + \sum_{i=1}^{\kappa} \big(\varepsilon\,\partial_k \hat f(x)\big)\varrho(y^i)\,y^i_k/(2\pi|y^i|^2),$$

for $y^i = (x-\hat x^i)/\varepsilon$, where only one term in these sums is non-zero and this term has the number i if and only if $x\in\varepsilon Y_i$ for $i=1,2,\dots,\kappa$. Here εY_i denotes a cell of the form $(0,\varepsilon)^2$ whose center coincides with $\hat x^i$.

Denote the last two terms in this equality with a minus sign by $\hat a(x/\varepsilon)f(x)$ in the equation of problem (31). Then the right-hand side of the equation coincides precisely with the δ-functions concentrated at the points $\hat x^1$, $\hat x^2$, ..., $\hat x^\kappa$, since $f=\hat f=1$ and $\varrho((x-\hat x^i)/\varepsilon)=1$ on the supports of these functions.

Let us check the fulfillment of inequality (32). For example, for the function $\hat a^k(x/\varepsilon)\hat f(x)$, $k=1,2$, we notice that $|\hat f(x)|\leqslant 1$ and only one term in the relevant sum is non-zero for $x\in\varepsilon Y_i$ for $i=1,2,\dots,\kappa$. In addition, we receive

$$\big\|(\hat a^1+\hat a^2)\hat f\big\|_{L^{2,\infty}(\varepsilon Y_i)} \leqslant \big\|\hat a^1+\hat a^2\big\|_{L^{2,\infty}(\varepsilon Y_i)} = \varepsilon\,\big\|\hat a^1+\hat a^2\big\|_{L^{2,\infty}(Y)} \leqslant \varepsilon\,C \tag{33}$$

according to [30], (1), since, as already noted, $\hat a^k(y) = \varrho(y-\hat y)(y_k-\hat y_k)/(2\pi|y-\hat y|^2)$ is an element of $L^{2,\infty}(Y)$ for $k=1,2$. By definition, $\kappa-1=(l_2-l_2^0)/\varepsilon$. Hence, summing up (33), we get $\big\|(\hat a^1+\hat a^2)\hat f\big\|_{L^{2,\infty}(\Omega)}\leqslant C$, where the constant does not depend on ε. Here it is assumed that a^1, a^2 is extended by zero with respect to x_1.

Similarly, one can estimate $\|\hat a f\|_{L^{2,\infty}(\Omega)}$. It should be kept in mind here that the function $\hat f$ depends on ε. Hence, for example, the derivative $\partial_2\hat f(x)$ may be unbounded, but $\varepsilon\,\partial_2\hat f(x)$ is a bounded function (which follows directly from the definition) and is different from zero only for $x\in\varepsilon Y_1\cup\varepsilon Y_\kappa$. Hence, the relevant term is of order ε. Thus, for a fixed ε, the unique solution to problem (31) is defined to be bounded in the sense of energy inequality (32), where instead of $\|u\|_{L^{2,\infty}(\Omega)}$ one can take $\|u\|_{L^p(\Omega)}$ for a fixed $p\geqslant 2$ due to Soblev's inequality according to [30].

Let us choose as an approximation for solution to problem (31) only the one term $u_0^0 = N(x,x/\varepsilon)\hat f(x)$ from expansion (19) for simplicity of presentation, where $N=\hat N(x,y-\hat y)$ is a 1-periodic with respect to y_2 solution of the following problem

$$-\,\partial_{y_l}\big(a_{lk}(x,y)\partial_{y_k}\hat N\big) + a_k(x,y)\,\partial_{y_k}\hat N + a_0(x,y)\hat N$$
$$= \delta_2(y-\hat y)\quad\text{in}\quad Y,\quad \hat N=0\quad\text{for}\quad \{y_1=0\}\ \text{and}\ \{y_1=1\} \tag{34}$$

for fixed $x \in \Omega$. Following [31] and [32], one can verify that a solution to this problem exists and satisfies an energy estimate similar to (32) for $\varepsilon = 1$ and $\Omega = Y$. Let us continue this solution in the 1-periodic manner for y_2 and by zero for y_1. One can also check that such a solution depends smoothly on $x \in \Omega$. To do this, one can formally differentiate the equation of problem (34) with respect to x and, using the uniqueness of the solution to the resulting problem, conclude that $\partial_k N$ and $\partial_l \partial_k N$ satisfy the estimates similar to (32), which will be sufficient for what follows.

Using, for example, expansion (8) with $u = \hat{f}(x)$, we estimate the residual in the corresponding equation for $r = u - u_0^0$. Consider, for example, the last term in (8). This term is zero if it contains the differentiation of $\hat{f}$ with respect to x_1. For a term containing $D_2^2 \hat{f}$, for example, it should be noted that $\varepsilon^2 D_2^2 \hat{f}(x)$ is a bounded function and is different from zero only for $x \in \varepsilon Y_1 \cup \varepsilon Y_\kappa$.

In addition, in the term $N((x - \hat{x}^1)/\varepsilon) + N((x - \hat{x}^\kappa)/\varepsilon)$ is multiplied by the bounded function that corresponds to one of the coefficients of equation (31). Consequently, using the inequality similar to (33), we obtain

$$\left\| \varepsilon^2 D_2^2 \hat{f} \left[N((x - \hat{x}^1)/\varepsilon) + N((x - \hat{x}^\kappa)/\varepsilon) \right] \right\|_{L^{2,\infty}(\Omega)} \leqslant \varepsilon\, C,$$

where the constant does not depend on ε. Similarly, terms without differentiations are $\varepsilon \sum_{i=1}^{\kappa} N((x - \hat{x}^i)/\varepsilon)$ multiplied by the bounded functions. Therefore, we can again use the inequality similar to (33). In the same way, we can estimate the remaining terms in (8) and conclude that for solution to problem (31) (with the right-hand side $\sum_{i=1}^{\kappa} \delta_2((x - \hat{x}^i)/\varepsilon)$) the following accuracy estimate

$$\varepsilon \left\| \nabla u - \nabla u_0^0 \right\|_{L^{2,\infty}(\Omega)^2} + \left\| u - u_0^0 \right\|_{L^{2,\infty}(\Omega)} \leqslant \varepsilon\, C \tag{35}$$

is proven, where the constant does not depend on ε. By choosing several terms in the asymptotic expansion (19) and adding auxiliary terms in the extreme cells εY_1 and εY_κ, one can improve estimate (35) and thus prove Theorem 1. It is possible to similarly consider problem (31) in the case $n > 2$, if the space $L^{2,\infty}(\Omega)$ in energy inequality (32) is replaced by the space $L^{n/(n-1),\infty}(\Omega)$ according to [31] and [32].

In the case when microneedles of such arrays are synthesized from biosoluble polymers, drugs are injected evenly along the entire length of the needles, which can be modeled by the sum of one-dimensional δ-functions $\sum_{i=1}^{\kappa} \delta_1(x_2 - x_2^i)$ that are limited to the domain ω. In the case, a schematic distribution graph is presented in the second image of Fig. 1. Here, at a large number of the boundary points $x_2^1, x_2^2, \ldots, x_2^\kappa$ located ε-periodically as in the previous one, microneedles of length l_1 are inserted and dissolve along the entire length, inducing drug injections.

In fact, the needles are usually sharp at the end and therefore the injection magnitudes are declining. Similarly, the injections near the surface are not realized due to the hardness of the boundary layer, after which the magnitudes begin to increase. Thus, in the notation of the previous case, we can multiply the δ-functions by a smooth function $\hat{f} = \hat{f}(x)$ that realizes such a decline and increase.

Let $\hat{y} = (\hat{y}_1, \hat{y}_2)$ be the center of the periodicity cell Y as in the previous case. We define the functions $\hat{a}^1(y) = 0$ and

$$\hat{a}^2(y) = -y_2 \quad \text{for} \quad y_2 \in (0, \hat{y}_2\,] \quad \hat{a}^2(y) = 1 - y_2 \quad \text{for} \quad y_2 \in (\,\hat{y}_2, 1].$$

Then, we have $\partial_{y_k} \hat{a}^k = \delta_1(y_2 - \hat{y}_2) - 1$, for example, according to [37]. The function $\hat{a}^2(y)$ will be continued as 1-periodic function with respect to y_2. Thus, we get

$$\varepsilon\, \partial_k \big(\hat{a}^k(x/\varepsilon)\hat{f}(x)\big) = \sum_{i=1}^{\kappa} \hat{f}(x)\, \delta_1(y^i) - \hat{f}(x) + \big(\varepsilon\, \partial_2 \hat{f}(x)\big)\hat{a}^2(x/\varepsilon)$$

for $y^i = (x_2 - x_2^i)/\varepsilon$. Denote the last two terms in this equality with a minus sign by $\hat{a}(x/\varepsilon) f(x)$ in the problem equation (31). Then the right-hand side of the equation is precisely the sum $\sum_{i=1}^{\kappa} \hat{f}(x)\, \delta_1((x_2 - x_2^i)/\varepsilon) = \sum_{i=1}^{\kappa} \hat{f}_1(x_1)\, \delta_1((x_2 - x_2^i)/\varepsilon)$, since $\hat{f}_2(x_2) = 1$ on the support the function $\sum_{i=1}^{\kappa} \delta_1((x_2 - x_2^i)/\varepsilon)$.

In the case under consideration, $\hat{a}^1, \hat{a}^2 \in L^\infty(Y)$, therefore inequality (32) is satisfied for $L^2(\Omega)$-norms while preserving the proof. Let us choose as an approximation for solution to problem (31) only the one term $u_0^0 = N(x, x/\varepsilon)\hat{f}(x)$ from (19), where $N = \hat{N}(x, y - \hat{y})$ is a 1-periodic with respect to y solution of the problem

$$-\,\partial_{y_l}\big(a_{lk}(x,y)\partial_{y_k}\hat{N}\big) + a_k(x,y)\,\partial_{y_k}\hat{N} + a_0(x,y)\hat{N} = \delta_1(y_2 - \hat{y}_2) \ \text{ in } Y, \tag{36}$$

for fixed $x \in \Omega$. It can be verified that a suitable solution to this problem exists and satisfies inequality (32) for $L^2(Y)$-norms, since $\delta_1(y_2 - \hat{y}_2) = \partial_{y_2}\hat{a}^2 + 1$.

Thus, repeating the above proofs, one can conclude that for solution to problem (31) (with the right-hand side $\sum_{i=1}^{\kappa} \hat{f}(x)\, \delta_1((x_2 - x_2^i)/\varepsilon)$) the accuracy estimate (35) is valid with a constant independent of ε. By choosing several terms in expansion (19) and adding auxiliary terms in the extreme stripes $\varepsilon Y_1 \times (0, l_1)$ and $\varepsilon Y_\kappa \times (0, l_1)$, one can improve estimate (35) and thus prove Theorem 1.

It is possible to similarly consider problem (31) in the case $n = 3$, when the right-hand side has the form either $\sum_{i=1}^{\kappa} \hat{f}(x_1, x_2, x_3)\, \delta_1((x_2 - x_2^i)/\varepsilon)$ or

$$\sum_{i,j=1}^{\kappa_2, \kappa_3} \hat{f}(x_1, x_2, x_3)\, \delta_2((x_2 - x_2^i)/\varepsilon, (x_3 - x_3^j)/\varepsilon)$$

in natural notations. For $n > 3$, it may be helpful to use the spaces $L^{s/(s-1),\infty}(\Omega)$ when considering δ-functions $\delta_s(x)$ of the dimension s according to [31] and [32].

In a sense, the case of steady-state weak diffusion is considered here. The cases of non-weak and non-stationary diffusions are also interesting and will be considered in the future. Non-stationary diffusion is more realistic, since in this case $f = f(t,x)$ and $\hat{f} = \hat{f}(t,x)$ are decreasing in time functions in an analogue of problem (31), which take into account the gradualness of drug injections. The considered problems with delta functions also model the problems of irrigation, which significantly affects environmental safety in the event of irrational use. Examples of direct numerical modeling and optimization of such problems with several delta functions are given in [38] and [39]. This approach allows the use of parallel computing algorithms [40], which is very essential when modeling environmental problems with very large data, as noted in [41]. The approach presented here also allows the use of parallel algorithms from [40], since cell problems can be solved using such algorithms.

5. Asymptotic Expansions of First Order Type

In this section the dimension case $d = 1$ in solvability conditions (15) is considered. Thus, the kernel and cokernel are non-trivial. Let $N = N_1(y)$ and $M = M_1(y)$ denote the generators of $\ker P(y)$ and $\ker P^*(y)$, respectively. Let us choose $p < m - 1$ in the expansion (4). Then the first equality in (11) is the following form

$$\sum_{\{\beta_0\}} \left[P(y, D_y)\, N^0_{\beta_0}(y)\right] u_0^{\beta_0}(x) \;=\; 0.$$

To satisfy this equality, it is enough to choose $\{\beta_0\} = \{0\}$ and $N^0_0 = C^0_0\, N(y)$ ($u^0_0(x)$ not yet defined). Then the second equality in (11) takes the form

$$\sum_{\{\beta_1\}} [PN^1_{\beta_1}(y)]\, u_1^{\beta_1}(x) + C^0_0 \sum_{|\beta|=1} \left[P^\beta N(y)\right] D_x^\beta\, u^0_0(x) = 0\,. \tag{37}$$

Applying to this equality the operator for calculating the weighted average of the form $\{\,\cdot\,\} = \{\,\cdot\,\}_1$, defined by the equality (15), we obtain

$$C^0_0 \sum_{|\beta|=1} \{P^\beta N\} D_x^\beta\, u^0_0(x) \;= 0\,,$$

where it is taken into account that $\{PN^1_{\beta_1}\} = \langle PN^1_{\beta_1}, M\rangle = \langle N^1_{\beta_1}, P^*M\rangle = 0$. Let us denote the constants $A^1_\beta = \{P^\beta N\}$ for multi-indices β with $|\beta| = 1$.

Homogenized first order equation. Suppose that $A^1_\beta \neq 0$ for some β with $|\beta| = 1$. We normalize u^l_p by choosing $p = m-1$ in (4), Then the second equality in (11) takes the following form

$$\sum_{\{\beta_1\}} [PN^1_{\beta_1}(y)]u_1^{\beta_1}(x) + C^0_0 \sum_{|\beta|=1} [P^\beta N(y)] D^\beta_x u^0_0(x) = a(y)f(x). \quad (38)$$

Applying to this equality the weighted average operator $\{\,\cdot\,\}$, we obtain

$$C^0_0 \sum_{|\beta|=1} \{P^\beta N\} D^\beta_x u^0_0(x) = \{a\}\, f(x). \quad (39)$$

Let us denote $\hat{u}^0_0 = C^0_0 u^0_0$. By definition $u^l_{m-1} = \varepsilon^{m-1} C^0_0 N u^0_0 + \ldots = \varepsilon^{m-1} N \hat{u}^0_0 + \ldots$ and therefore we can assume that $C^0_0 = 1$ in (39).

Thus, for u^0_0 we obtain the first-order equation with constant coefficients

$$\sum_{|\beta|=1} A^1_\beta\, D^\beta_x\, u^0_0(x) = \{a\}\, f(x). \quad (40)$$

Multiplying (40) by $M(y)$ and subtracting the resulting relation from (38), we have

$$\sum_{\{\beta_1\}} [PN^1_{\beta_1}] u_1^{\beta_1} + \sum_{|\beta|=1} [P^\beta N - A^1_\beta M] D^\beta_x u^0_0 = (a - \{a\}M\,) f. \quad (41)$$

To satisfy the equality (41) it is enough to choose $\{\beta_1\} = \{\beta : |\beta| \leqslant 1\} \cup \{0\}$, $u_1^{\beta_1} = f(x)$ for $|\beta_1| = 0$, $u_1^{\beta_1} = D^{\beta_1}_x u^0_0(x)$ for $|\beta_1| = 1$ and $N^1_0 = C^1_0 N(y)$ ($u^0_1(x)$ not yet defined). Then for $N^1_{\beta_1}$ with $|\beta_1| \leqslant 1$ from (41) we get the equations

$$\begin{aligned} P(y, D_y)\, N^1_{\beta_1}(y) &= a(y) - \{a\}M(y) \quad \text{for} \quad |\beta_1| = 0, \\ P(y, D_y)\, N^1_{\beta_1}(y) &= -\big(P^{\beta_1} N(y) - A^1_\beta\, M(y)\big) \quad \text{for} \quad |\beta_1| = 1. \end{aligned} \quad (42)$$

The last equations are solvable because, for example, one has

$$\{a - \{a\}M\} = \{a\} - \{a\}\{M\} = \{a\} - \{a\}\langle M^2 \rangle = 0$$

due to (13). The solutions to the equations belong to $C^\infty(T^n)$ and have the form

$$N^1_{\beta_1}(y) = N_{\beta_1}(y) + C^1_{\beta_1} N(y), \qquad \langle N_{\beta_1}, N \rangle = 0. \quad (43)$$

Remark that if $\langle M \rangle \neq 0$, then the equation (40) can be subtracted directly from (38). In this case, we should assume that $M(y)$ is normalized by the condition $\langle M \rangle = 1$ and for $N^1_{\beta_1}$ we obtain solvable equations similar to (42).

Further, the third equality in (11) can be rewritten as

$$\sum_{\{\beta_2\}}[PN^2_{\beta_2}]u_2^{\beta_2} + \sum_{|\beta_1|=0}\sum_{|\beta|=1}[P^\beta(N_{\beta_1}+C^1_{\beta_1}N)]D^\beta_x f + \\ + \sum_{|\beta_1|=1}\sum_{|\beta|=1}[P^\beta(N_{\beta_1}+C^1_{\beta_1}N)]D^{\beta_1+\beta}_x u^0_0 + \\ + C^1_0\sum_{|\beta|=1}[P^\beta N]D^\beta_x u^0_1 + \sum_{|\beta|=2}[P^\beta N]D^\beta_x u^0_0/\beta! = 0. \tag{44}$$

Applying the weighted average operator $\{\cdot\}$ to this equality and choosing $C^1_0 = 1$, we obtain for u^0_1 the following first-order equation with constant coefficients

$$\sum_{|\beta|=1} A^1_\beta D^\beta_x u^0_1 = -\sum_{|\beta_1|=0}\sum_{|\beta|=1}[A^{10}_{\beta\beta_1} + C^1_{\beta_1}A^1_\beta]D^\beta_x f - \\ -\sum_{|\beta_1|=1}\sum_{|\beta|=1}[A^{11}_{\beta\beta_1} + C^1_{\beta_1}A^1_\beta]D^{\beta_1+\beta}_x u^0_0 - \sum_{|\beta|=2}A^2_\beta D^\beta_x u^0_0/\beta!\,. \tag{45}$$

Here we have fixed some constants $C^1_{\beta_1}$ for $|\beta_1| \leqslant 1$ and use the notations

$$A^{11}_{\beta\beta_1} = \{P^\beta N_{\beta_1}\} \quad \text{for} \quad |\beta| = |\beta_1| = 1,$$

$$A^{10}_{\beta\beta_1} = \{P^\beta N_{\beta_1}\} \quad \text{for} \quad |\beta| = 1, |\beta_1| = 0, A^2_\beta = \{P^\beta N\} \quad \text{for} \quad |\beta| = 2.$$

Multiplying (45) by $M(y)$ and subtracting the resulting relation from (44), we get an equality similar to (41). From the equality, with an appropriate choice of $\{\beta_2\}$ and $u_2^{\beta_2}$, we have equations for $N^2_{\beta_2}$. These equations are solvable due to the choice of the constants $A^{10}_{\beta\beta_1}$, $A^{11}_{\beta\beta_1}$, A^2_β, and further construction is carried out by induction.

Equations of the form (40) and (45) can be integrated if homogeneous boundary conditions are specified on the 'half' of the boundary $\partial\Omega$, from which the characteristics are leaving (direct lines defined by the direction of the vector A^1_β for $|\beta| = 1$), for example, according to [42]. Under the additional condition that the solutions to the problems are sufficiently smooth, the constructed expansion u^l_{m-1} satisfies the equation (10) up to $O(\varepsilon^l)$ and only partially the boundary conditions from (1).

Transfer and weak diffusion by microneedles. The accuracy estimate (35) indicates weak diffusion and transfer in problem (31), which is actually associated with a large multiplier for the coefficient $a_0(y)$ in (24). Here we will reduce this coefficient. This necessitates renormalization of the right-hand side in (24) to isolate the main part of the asymptotics. Let in (24) we have

$$af = \varepsilon^{-1}\,\hat{a}(x/\varepsilon)f(x) + \partial_k\big(\hat{a}^k(x/\varepsilon)\hat{f}(x)\big)$$

$$= \varepsilon^{-1}\,\hat{a}(y)f(x) + \varepsilon^{-1}\big(\partial_{y_k}\hat{a}^k(y)\big)\hat{f}(x) + \hat{a}^k(y)\partial_k\hat{f}(x) \quad \text{for} \quad y = x/\varepsilon, \tag{46}$$

coefficients are of general types and $a_0 = \varepsilon\, a_0^1(x,y)$. Then the problem is like

$$-\,\varepsilon\,\partial_l\big(a_{lk}(x,x/\varepsilon)\partial_k\, u\big) \;+\; a_k(x,x/\varepsilon)\,\partial_k\, u \;+\; a_0^1(x,x/\varepsilon)\, u \;=$$

$$= \hat{a}(x/\varepsilon)f(x) + \varepsilon\,\partial_k\big(\hat{a}^k(x/\varepsilon)\hat{f}(x)\big) \quad \text{in} \quad \Omega, \quad u = 0 \quad \text{on} \quad \partial\Omega, \tag{47}$$

where the coefficients $a_{lk}(x,y) = a_{kl}(x,y), \ldots, a_0^1(x,y)$ are smooth and 1-periodic in y for $l,k = 1,2$. Thus, we renormalized problem (24) (multiplying by ε in fact, so $p = 0$ in (4)) and selected general data with asymptotic coefficient $a_0 = \varepsilon\, a_0^1(x,y)$.

This problem is not so simple even for the case when the coefficients of the problem equation do not depend on $y \in Y$ at all. Starting with the works of Levinson [42] and Oleinik [43] for such coefficients, it was shown that the solution to the problem converges in some sense to a suitable solution of the equation

$$a_k(x)\,\partial_k\, u_0 \;+\; a_0^1(x)\, u_0 \;=\; \hat{a}\, f(x). \tag{48}$$

In order to determine this solution, the dynamical system

$$(\,x_1)'_t = a_1(x_1,x_2), \quad (\,x_2)'_t = a_2(x_1,x_2)$$

was considered in [42] and [44]. The arc length of the integral curve for the system was denoted by τ under the assumption that the length increases with increasing parameter t. Along integral curves as characteristics, equation (48) is of the form

$$\big(a_1^2 + a_2^2\big)^{1/2}\,\frac{d\,u_0}{d\,\tau} + a_0^1\, u_0 \;=\; \hat{a}\, f(x).$$

Thus, special research is necessary for singular points and closed integral curves of the vector field (a_1^2, a_2^2). In part, the research was carried out in [44] and [45], where problem (47) was studied (with coefficients depending only on $x \in \Omega$) under various additional assumptions on these integral curves.

Here we will exclude the cases of singular points and closed curves, since our main goal is to consider equation (40), where the coefficients are constant and nonsingular by definition. In the case, one can assume that equation (48) has the form

$$\partial_1\, u_0 \;+\; a_0^1(x)\, u_0 \;=\; \hat{a}\, f(x).$$

In addition, by replacing the unknown function $u_0 = u_0^0 \exp(-\int^{x_1} a_0^1(\tau, x_2)\, d\tau)$, this equation is reduced to the equation $\partial_1 u_0^0 = \hat{a}\, \hat{f}(x)$. Hence

$$u_0^0 = \int^{x_1} \hat{a}\, \hat{f}(\tau, x_2)\, d\tau$$

and it is only necessary to indicate the lower limit of integration. The limit is the $\tau = 0$ for the domain Ω presented in Fig. 1. In the general case, as the limit it is necessary to take those points in $\partial\Omega$, where $(e_1, \nu) > 0$ for the scalar product between $e_1 = (1, 0)$ and the internal normal ν for the boundary $\partial\Omega$.

In the situation, problem (47) with constant coefficients for $(a_1, a_2) = (1, 0)$, $a_0^1 = 0$, $(\hat{a}^1, \hat{a}^2) = (0, 0)$ was considered in papers [46] and [47], where asymptotic solution expansions were constructed and justified. The asymptotic expansions are quite complex and contain boundary layer terms of various types, since the boundary points at which $(e_1, \nu) = 0$ can be singular in a certain sense. Such asymptotics for the problem in domains with corner points were studied, for example, in [46], where additional references can be found.

Returning to problem (47) with general data and the asymptotic coefficient $a_0 = \varepsilon\, a_0^1(x, y)$, we note that the unique solution to this problem exists for a fixed ε by virtue of the well-known Schauder theorem [21]. In order to obtain an energy estimate that is exactly dependent on ε, we assume that $\partial_{y_k} a_k(x, y) = 0$ and inequality (25) is valid for all $(x, y) \in \Omega \times Y$. In addition, analogues of inequalities (26) are satisfied in the form $\partial_k a_k(x, y) \leqslant \sigma$ and $\sigma \leqslant a_0^1(x, y)$ for a positive σ.

Then, we can repeat the proof of (27) and get for solution to (47) the estimate

$$\varepsilon^{1/2} \,\|\, \nabla u \,\|_{L^2(\Omega)^2} + \|\, u \,\|_{L^2(\Omega)} \leqslant C \,\|\, \hat{a}\, f \,\|_{L^2(\Omega)} + \varepsilon^{1/2}\, C \,\|\, (\hat{a}^1 + \hat{a}^2)\, \hat{f} \,\|_{L^2(\Omega)} \tag{49}$$

and relevant analogue of estimate (32) with the constants C independent of ε.

In the situation under consideration, the operator of problem (47) is of the form $P(y) = P_0(y) + \varepsilon\, P_1(y)$, where the operators $P_0(y)$ and $P_1(y)$ have the symbols

$$p_0(y, \xi) = \varepsilon^{-1} a_{lk}(x, y)\, \xi_l\, \xi_k + \varepsilon^{-1} b_k(x, y)(i\, \xi_k), \quad p_1(y, \xi) = b_0^1(x, y, \xi),$$

with $b_k(x, y) = a_k(x, y) - \partial_{y_l} a_{lk}(x, y)$ and $b_0^1(x, y, \xi) = -\,\partial_l\, a_{lk}(x, y)(i\, \xi_k) + a_0^1(x, y)$. Recall that in definition (2) the differentiation $D_k = i^{-1}\, \partial_k$ is multiplied by ε. The symbols $p_0^\beta(y, \xi)$ and $p_1^\beta(y, \xi)$ for the operators can be calculated similarly (28).

Let us denote by $Q(y)$ the operator conjugate to $P(y)$ on $C^\infty(T^n)$. Then $Q(y) = Q_0(y) + \varepsilon\, Q_1(y)$, where the operators $Q_0(y)$ and $Q_1(y)$ have the symbols

$$q_0(y,\xi) = \varepsilon^{-1} a_{lk}(x,y)\, \xi_l\, \xi_k - \varepsilon^{-1} q_k^0(x,y)(i\,\xi_k), \quad q_1(y,\xi) = q_0^1(x,y,\xi),$$

with $q_k^0(x,y) = a_k(x,y) + \partial_{y_l} a_{lk}(x,y)$, $q_0^1(x,y,\xi) = b_0^1(x,y,\xi) - \partial_k a_k(x,y)$. These equalities follow from (12) and can be verified directly.

Thus, we have $N = 1$ and $M = 1$. Let us denote

$$A_1(x) = \langle a_1(x,y)\rangle, \quad A_2(x) = \langle a_2(x,y)\rangle,$$

and assume that $A_1(x)^2 + A_2(x)^2 \neq 0$, then equation (40) has the form

$$A_k(x)\, \partial_k\, u_0^0 \,+\, A_0(x)\, u_0^0 \,=\, \hat{A}\, f(x). \tag{50}$$

where $\hat{A} = \langle\, \hat{a}\,\rangle$, $A_0(x) = \langle\, a_0^1(x,y)\,\rangle$ and $A_k(x) = \langle b_k(x,y)\rangle$ for $k = 1, 2$, since

$$\langle \partial_{y_l} a_{lk}(x,y)\rangle = -\langle a_{lk}(x,y)\partial_{y_l} 1\rangle = 0 \quad \text{and} \quad \langle \partial_{y_k} \hat{a}^k(y)\rangle = -\langle \hat{a}^k(y)\partial_{y_l} 1\rangle = 0$$

due to integration by parts. In addition, we have $\langle\, b_0^1(x,y,D_y)N\,\rangle = \langle\, a_0^1(x,y)\,\rangle$, because the first term of b_0^1 contains differentiation.

We emphasize that the case $A_1(x)^2 + A_2(x)^2 \neq 0$ is common even if the equation coefficients in (47) are constants with $a_1^2 + a_2^2 \neq 0$ and ellipticity (25) is valid.

Knowing the symbol of the conjugate operator $P_\varepsilon^* = P_\varepsilon^*(x/\varepsilon, x, \varepsilon D)$, we can consider (conjugate) problem (47) with the operator of the following form

$$P_\varepsilon^* v = \hat{a}(x/\varepsilon) f(x) + \varepsilon\, \partial_k\big(\hat{a}^k(x/\varepsilon)\hat{f}(x)\big) \quad \text{in} \quad \Omega, \quad u = 0 \quad \text{on} \quad \partial\Omega. \tag{51}$$

Consequently, repeating the conclusion (38)–(45), we get equation (40) in the form

$$-\, A_k(x)\, \partial_k\, v_0^0 \,+\, B_0(x)\, v_0^0 \,=\, \hat{A}\, f(x),$$

where $B_0(x) = A_0(x) - \partial_k A_k(x)$, because the equation coefficients of problem (47) are real and smooth in x. This equation can be rewritten like this

$$-\, \partial_k\, \big(A_k(x)\, v_0^0\big) \,+\, A_0(x)\, v_0^0 \,=\, \hat{A}\, f(x). \tag{52}$$

Following [18] and [19], we will now prove the weak convergence of solutions to problem (47) to the solution of equation (50). In order to do this, we take the terms for ε^0 and ε^1 in the asymptotic expansion (4) with the notation from (38)–(45) for the solution to conjugate problem (51). This expansion have the form

$$v_0^1 = v_0^0 + \varepsilon\, M_0^1 f + \varepsilon\, M_1^1 \hat{f} + \varepsilon\, M_{\beta_0}^1 v_0^0 + \varepsilon \sum_{|\beta|=1} M_\beta^1 D_x^\beta v_0^0, \qquad (53)$$

where

$$Q(y)M_0^1 = \hat{a} - \hat{A}, \quad Q(y)M_1^1 = \partial_{y_k} \hat{a}^k, \quad Q(y)M_{\beta_0}^1 = B_0 - a_0^1 + \partial_k a_k, \qquad (54)$$
$$Q(y)M_{(1,0)}^1 = ia_1 + i\,\partial_{y_l} a_{l1} - i A_1, \quad Q(y)M_{(0,1)}^1 = ia_2 + i\,\partial_{y_l} a_{l2} - i A_2,$$

in according to (42). Here it is used that $Q(y) = Q(y, D_y) = Q_0(y, D_y) + \varepsilon\, Q_1(y, D_y)$ for the asymptotic coefficients and the asymptotic right-hand side by (46).

For problem (47) with smooth data, all 1-periodic problems (54) have smooth solutions due to the choice of coefficients in the right-hand sides. Using, for example, expansion (8) and fixing arbitrary $v_0^0, f, \hat{f} \in C_0^\infty(\Omega)$ in representation (53), we get

$$P_\varepsilon^* v_0^1 = (\hat{a} - \hat{A}) f + \big(\partial_{y_k} \hat{a}^k\big) \hat{f} - A_k \partial_k v_0^0 + B_0\, v_0^0 + \varepsilon\, r(x, y) \qquad (55)$$

for $y = x/\varepsilon$, where $|\, r(x, x/\varepsilon)| \leqslant C$ with a constant independent of ε, since all data are smooth functions in accordance with expansion (9).

Multiplying the equation of problem (47) by such v_0^1 and integrating the resulting equality over Ω and by parts, we choose $f = \hat{f} = 0$ in (55) and obtain

$$\int_\Omega u\, P_\varepsilon^* v_0^1\, dx = \int_\Omega u\, (-A_k \partial_k + B_0)\, v_0^0\, dx = \int_\Omega \hat{a}\, f\, v_0^0\, dx + \varepsilon \int_\Omega u\, \hat{r}\, dx,$$

where $|\, \hat{r}(x, x/\varepsilon)| \leqslant C$ with a constant independent of ε.

The solution $u = u_\varepsilon$ to problem (47), of course, depends on the parameter ε and is uniformly bounded in the space $L^2(\Omega)$ with respect to this parameter due to inequality (49). Hence, there are a subsequence $u_{\varepsilon'}$ and $u_0 \in L^2(\Omega)$ such that

$$u_{\varepsilon'} \rightharpoonup u_0 \qquad \text{weekly in} \quad L^2(\Omega)$$

and

$$\int_\Omega u_0\, (-A_k \partial_k + B_0)\, v_0^0\, dx = \int_\Omega \hat{A}\, f\, v_0^0\, dx \qquad (56)$$

for every $v_0^0 \in C_0^\infty(\Omega)$, since $\hat{a}(x/\varepsilon) \rightharpoonup \hat{A}$ as is well known [6]. By virtue of (52), the last identity means that u_0 satisfies equation (50) in the sense of distributions.

By choosing in (56) functions $v_0^0 \in C^\infty(\overline{\Omega})$ that are equal to zero in the neighborhood of the part $S_2 = \{x \in \partial\Omega : A_k(x)\,\nu^k(x) < 0\} \subset \partial\Omega$, it was proposed in [45] to take integral identity (56) as the definition of a solution to equation (50). In addition, it was proven in [45] that the solution is unique under an additional condition on the trajectory of the vector field $(A_1(x), A_2(x))$ and for $B_0 \geqslant \sigma > 0$.

The condition is as following: for the part $S_1 = \{x \in \partial\Omega : A_k(x)\,\nu^k(x) > 0\} \subset \partial\Omega$ there is such an open neighborhood $O(S_1) \supset S_1$ that trajectories starting at the point $s \in O(S_1) \cap \Omega$ intersect the boundary $\partial\Omega$ at internal points of $S_1 \cup S_0$ for almost all points of $O(S_1)$. Here $S_0 = \{x \in \partial\Omega : A_k(x)\,\nu^k(x) = 0\} \subset \partial\Omega$ and ν is the internal normal for the boundary $\partial\Omega$.

Remark that these conditions are fulfilled for equation (40) for general domains, since the coefficients of the equation are constant and, as already noted, by replacing the unknown function, one can always add (or remove) a zero-order term of the desired sign in the equation. Note also that the statement about the weak solution convergence for problem (1) to the solution of equation (40) in the case when $d = 1$ and $A^1_\beta \neq 0$ for $|\beta| = 1$ was proved in [18], Theorem 3.1.

Similarly, for the conjugate problem (51) it can be proven that the solutions $v = v_\varepsilon$ weakly in $L^2(\Omega)$ converge to the solution v_0 of equation (52). Note also that, as verified in [18], all constants encountered in the construction of expansion (4) for equation (1) (for example, A^1_β, $\{a\}$ in (40) and $A^{11}_{\beta\beta_1}$, $A^{10}_{\beta\beta_1}$, A^2_β in (45)) are mutually complexly conjugate with the constants arising in the construction of expansion (4) for equation (1) with the conjugate operator $P^*_\varepsilon = P^*_\varepsilon(x/\varepsilon, \varepsilon D_x)$. For example, equation (40) for the conjugate problem has the form

$$\sum_{|\beta|=1} \bar{A}^1_\beta\, \bar{D}^\beta_x\, v^0_0 \;=\; \{\bar{a}\}\, f$$

and therefore $\bar{A}^1_\beta \neq 0$ if and only if $A^1_\beta \neq 0$ for $|\beta| = 1$.

Now we can consider problem (47) with periodic δ-functions as for problem (31). Indeed, if in the notation for (31) the right-hand side of the problem equation (47) is $\sum_{i=1}^{\kappa} \delta_2((x - \hat{x}^i)/\varepsilon)$, then estimate (49) for $L^{2,\infty}(\Omega)$-norms is satisfied. By estimating the remainder $r(x, y)$ in equality (55) as well as in inequality (33) and similar inequalities, we can pass to the limit in the integral identity for $P^*_\varepsilon v^1_0$. Hence by the Banach-Alaoglu theorem, there are a subsequence $u_{\varepsilon'}$ and $u_0 \in L^{2,\infty}(\Omega)$ such that

$$u_{\varepsilon'} \rightharpoonup u_0 \qquad \text{weekly-*} \quad \text{in} \quad L^{2,\infty}(\Omega),$$

where u_0 satisfies (56) and so equation (50) is satisfied in the sense of distributions. Here it is used that $L^{2,\infty}(\Omega)$ is conjugate to $L^{2,1}(\Omega) \subset L^{2,\infty}(\Omega)$ (see [31] and [32]).

Let now $v_0^0 = v_0^0(x)$ in representation (53) be a solution to equation (52). Then, comparing (51)and (55), we conclude that

$$P^*_\varepsilon(\, v_0^1 - v\,) \;=\; \varepsilon\, r(x, x/\varepsilon). \tag{57}$$

Similarly, using the asymptotic expansion (53) for solution to problem (47), where u_0^0 is a solution to equation (50) and $N_0^1,\ N_1^1, \ldots$ satisfy analogues of equations (54) according to (42), we obtain $P_\varepsilon(\, u_0^1 - u\,) \;=\; \varepsilon\, r(x, x/\varepsilon)$.

These relations do not yet imply accuracy estimates for solutions to the problems, even if the coefficients of the problem equations are constant, for example, in accordance with [46] and [47] through boundary layer effects.

Nevertheless, there is a natural case when such accuracy estimates are correct. In fact, suppose that $A_1 > 0, A_2 = 0$ for equation (50) and the Direchlet boundary condition in problem (47) is replaced on part of the boundary $\{x_1 = L_1\}$ by the Neumann boundary condition $\partial_1 u = 0$. Here $\Omega = (0, L_1) \times (0, L_2)$ in problem (47) by definition as for the images in Fig. 1. In this case, we can assume that

$$u_0^0 = \int_0^{x_1} \hat{a}\, \hat{f}(\tau, x_2)\, d\tau$$

as already noted for the solution to equation (50). Hence, we have $\partial_1 u_0^0 = 0$ on the boundary $\{x_1 = L_1\}$ and $u_0^0 = 0$ on the remaining boundaries, since $\hat{f}$ is a compactly supported function on the domain Ω.

Thus, if for problem (47) with such Direchlet-Neumann boundary condition (and with the right-hand side $\sum_{i=1}^{\kappa} \delta_2((x - \hat{x}^i)/\varepsilon))$ the energy estimate (49) is fulfilled for $L^{2,\infty}(\Omega)$-norms, then the following accuracy estimate

$$\varepsilon^{1/2}\, \|\nabla u - \nabla u_0^1\|_{L^{2,\infty}(\Omega)^2} + \|u - u_0^1\|_{L^{2,\infty}(\Omega)} \leqslant \varepsilon\, C \tag{58}$$

is valid due to (57), where the constant does not depend on ε and u_0^1 is determined by the asymptotic representation (53) with the corresponding replacements of equations (54) according to (42). For example, the problem for $N_0^1 = N_0^1(x, x/\varepsilon)$ is problem (34), where $a_0 = 0$ and the right hand side is $\delta_2(y - \hat{y}) - 1$.

Similarly, repeating the above arguments, one can conclude that for solution to problem (47) with the Direchlet-Neumann boundary condition and the right-hand side $\sum_{i=1}^{\kappa} \hat{f}(x)\, \delta_1((x_2 - x_2^i)/\varepsilon)$ the accuracy estimate (58) for $L^2(\Omega)$-norms is valid with a constant independent of ε. Here, the problem for $N_0^1 = N_0^1(x, x/\varepsilon)$ is problem (36), where $a_0 = 0$ and the right hand side is $\delta_1(y_2 - \hat{y}_2) - 1$. Note that graphs of such periodic functions $N_0^1 = N_0^1(x, x/\varepsilon)$ can be found in [48].

In the case under consideration, we note that the Dirichlet-Neumann boundary condition is more natural for obtaining a volume distribution graph of drugs over the domain Ω. Indeed, this graph corresponds to transport and weak diffusion through the order of the coefficients in the problem equation (47). Thus, according to (53) for u_0^1, transfer with homogenized speed $A_1(x)$ from the hard boundary inward is the main one and is described by the transfer equation (50). On the other hand, a homogeneous Dirichlet boundary condition at the boundary $\{x_1 = L_1\}$ may be an obstacle to such transfer, while the Neumann condition allows free transfer. A similar remark applies to the case presented in the second image in Fig. 1.

Homogenized zero order equation. The case of problem (47) considered in the previous section is quite general. However, the case when $A_1 = A_2 = 0$ is also possible and is quite significant for spectral problems. Let's look at this case in the simplest version of problem (24) for $a_1 = a_2 = 0$, $a_0 = \varepsilon\, a_0^1(y)$ and diffusion determined by the identity matrix. Thus, to simplify the presentation, we define the function u as the solution to the Dirichlet problem for the second-order equation

$$-\Delta\, u + \varepsilon^{-1} a_0^1(x/\varepsilon)\, u = a(x/\varepsilon)\, f \quad \text{in} \quad \Omega, \quad u = 0 \quad \text{on} \quad \partial\Omega, \qquad (59)$$

where, as before, $f \in C_0^\infty(\Omega)$, the domain $\Omega \subset \mathbb{R}^2$, $a_0^1(y)$ and $a(y)$ are 1-periodic smooth (real-valued) functions. For $a_0^1(y)$, condition (26) is assumed to be satisfied.

In the case, $N(y) = 1$ and $M(y) = 1$. In addition, $P(y) = P_0(y) + \varepsilon\, P_1(y)$, where $P_0(y) = -\Delta_y$, $P_1(y) = a_0^1(y)$ and therefore $d = 1$ for the operator $P_0(y)$. Further, for the symbols of the operators $P_0(y)$ and $P_1(y)$ we have

$$p_0(y,\xi) = |\xi|^2, \qquad p_0^{(1,0)}(y,\xi) = 2\,\xi_1, \qquad p_0^{(0,1)}(y,\xi) = 2\,\xi_2,$$

$$p_0^{(2,0)}(y,\xi) = 2, \qquad p_0^{(0,2)}(y,\xi) = 2, \qquad p_1(y,\xi) = a_0^1(y),$$

which should be multiplied by ε^{-2} due to definitions (5). Consequently we get

$$A_\beta^1 = \langle P_0^\beta N \rangle = 0 \quad \text{for} \quad |\beta| = 1, \qquad A_0^0 = \langle P_1 N \rangle = \langle\, a_0^1 \,\rangle,$$

according to (39), where $p = 1$, $P_0^{(1,0)} = 2D_1$ and $P_0^{(0,1)} = 2D_2$.

From condition (26) it follows that $\langle\, a_0^1 \rangle > 0$ and for u_0^0 we have the following zero-order equation with constant coefficients

$$\langle\, a_0^1 \,\rangle\, u_0^0(x) = \langle\, a \,\rangle\, f(x) \qquad \text{in} \qquad \Omega. \qquad (60)$$

To explain further constructions, we write out several equalities (11) in the form

$$\sum_{\{\beta_1\}} \big[P_0 N_{\beta_1}^1(y)\big] u_1^{\beta_1}(x) + \sum_{|\beta|=1} \big[P_0^\beta N(y)\big] D_x^\beta u_0^0(x) + \big[P_1 N(y)\big] u_0^0(x) = a(y) f(x),$$

$$\sum_{\{\beta_2\}} [P_0 N^2_{\beta_2}(y)] u_2^{\beta_2}(x) + \sum_{\{\beta_1\}} \sum_{|\beta|=1} [P_0^\beta N^1_{\beta_1}(y)] D_x^\beta u_1^{\beta_1}(x) \tag{61}$$

$$+ \sum_{|\beta|=2} [P_0^\beta N(y)] D_x^\beta u_0^0(x)/\beta! + \sum_{\{\beta_1\}} [P_1 N^1_{\beta_1}(y)] u_1^{\beta_1}(x) = 0,$$

where $\{\beta_0\} = 0$ was chosen. Due to (60) we have $u_0^0 = \langle a \rangle \langle a_0^1 \rangle^{-1} f(x)$. Therefore, the first equality (61) can be rewritten in the form

$$\sum_{\{\beta_1\}} [P_0 N^1_{\beta_1}(y)] u_1^{\beta_1}(x) + \big(a_0^1(y) \langle a \rangle \langle a_0^1 \rangle^{-1} - a(y)\big) f(x) = 0.$$

In order to satisfy the equality, it is enough to choose $\{\beta_1\} = \{0\} \cup \{1\}$, $u_1^1 = f(x)$ and $N_0^1 = 1$ ($u_1^0(x)$ not yet defined). Then for N_1^1 we obtain the equation

$$-\Delta_y N_1^1(y) = a(y) - a_0^1(y) \langle a_0^1 \rangle^{-1} \langle a \rangle, \tag{62}$$

which is solvable due to the choice of the appropriate constants.

Further, the next equality from (61) takes the form

$$-\sum_{\{\beta_2\}} [\Delta_y N^2_{\beta_2}(y)] u_2^{\beta_2}(x) + [2D_{y_1} N_1^1(y)] D_1 f(x) + [2D_{y_2} N_1^1(y)] D_2 f(x)$$

$$+ D_1^2 u_0^0(x) + D_2^2 u_0^0(x) + [a_0^1(y) N_1^1(y)] f(x) + [a_0^1(y) N_0^1(y)] u_1^0(x) = 0.$$

Applying the average operator to the last equality, we get for u_1^0 the equation

$$\langle a_0^1 \rangle u_1^0(x) = -D_1^2 u_0^0(x) - D_2^2 u_0^0(x) - \langle a_0^1 N_1^1 \rangle f(x) \quad \text{in} \quad \Omega.$$

Subtracting the last equality from the penultimate one, we obtain

$$-\sum_{\{\beta_2\}} [\Delta_y N^2_{\beta_2}(y)] u_2^{\beta_2}(x) + [2D_{y_1} N_1^1(y)] D_1 f(x) + [2D_{y_2} N_1^1(y)] D_2 f(x)$$

$$+ [a_0^1(y) N_1^1(y) - \langle a_0^1 N_1^1 \rangle] f(x) + [a_0^1(y) N_0^1(y) - \langle a_0^1 \rangle] u_1^0(x) = 0.$$

To satisfy this equality, it is enough to choose $\{\beta_2\} = \{\beta : |\beta| \leqslant 1\} \cup \{0\} \cup \{1\}$, $u_2^\beta = D_x^\beta f(x)$ for $|\beta| = 1$, $u_2^\beta = f(x)$ for $|\beta| = 0$ and $u_2^1 = u_1^0(x)$ ($u_2^0(x)$ not yet defined). Then for N_β^2 we get the equations

$$\Delta_y N_\beta^2(y) = 2D_y^\beta N_1^1(y) \quad \text{for} \quad |\beta| = 1,$$

$$\Delta_y N_\beta^2(y) = a_0^1(y) N_1^1(y) - \langle a_0^1 N_1^1 \rangle \quad \text{for} \quad |\beta| = 0,$$

$$\Delta_y N_1^2(y) = a_0^1(y) - \langle a_0^1 \rangle, \qquad \Delta_y N_0^2(y) = 0,$$

which is solvable due to the choice of the appropriate constants.

Continuing by induction, we obtain the asymptotic expansion in the form

$$u_1^l = \varepsilon \langle a \rangle \langle a_0^1 \rangle^{-1} f(x) + \varepsilon^2 N_1^1(x/\varepsilon) f(x) + \dots, \tag{63}$$

for the solution to problem (59). In addition, we conclude that

$$P_\varepsilon (u - u_1^l) = O(\varepsilon^l) \quad \text{in} \quad \Omega, \qquad u - u_1^l = 0 \quad \text{on} \quad \partial\Omega.$$

Thus, repeating proof (27), we have the following accuracy estimate

$$\| u - u_1^l \|_{H_0^1(\Omega)} + \varepsilon^{-1/2} \| u - u_1^l \|_{L^2(\Omega)} \leqslant C \varepsilon^{l+1/2},$$

where the constant C does not depend on ε and u is a solution to problem (59).

The interested reader may have the opportunity to consider problem (47) under the assumption that $A_1(x) = A_2(x) = 0$ and $A_0(x) > 0$ for $x \in \bar{\Omega}$ in (50). For the solution to problem (47) with the right-hand side $\sum_{i=1}^{\kappa} \delta_2((x - \hat{x}^i)/\varepsilon)$ the accuracy estimate (58) can be proven likewise, where the asymptotic expansion u_0^1 is defined similarly to (63), but with normalization multiplied by ε^{-1} in accordance with (46). In addition, one can conclude that for solution to problem (47) with the right-hand side $\sum_{i=1}^{\kappa} \hat{f}(x)\, \delta_1((x_2 - x_2^i)/\varepsilon)$ the accuracy estimate (58) for $L^2(\Omega)$-norms is valid. Thus, this case of weak diffusion and transport with zero average velocity is quite reminiscent of problem (31), where weak diffusion and transport are considered.

Note that the case when $\langle a_1 \rangle = \langle a_2 \rangle = 0$, $a_0(y) = 0$ in problem (24) was studied in detail using probabilistic methods in [6], Ch 2. In this work, theorems are proved on the strong in $L^2(\Omega)$ and pointwise convergence of the solution of the problem to the solution of the Dirichlet problem for a second-order equation with constant coefficients. Problem (59) was also considered there when $\langle a_0^1 \rangle = 0$.

Let us fix $\lambda \in \mathbb{C} \setminus \{0\}$ and consider problem (59) with the coefficient $a_0^1 = \lambda$, depending on the spectral parameter. Then expansion (63) means that the resolvent of the operator $-\Delta$ is defined on smooth compactly supported functions and is approximately a multiple of the unit operator $\varepsilon \lambda^{-1} I$. A similar remark is made for the operator of problem (47) when $A_1(x) = A_2(x) = 0$ and $A_0(x) > 0$ for $x \in \bar{\Omega}$. This means that the spectrum of order ε^{-1} for such operators is structured quite simply. It is strange that a spectrum of this order has not been found in publications, despite the fact that there are many publications where spectra of order ε^0 and ε^{-2} are considered for second-order equations. Note that the representation (63) for the solution to problem (59) was received and justified in [18], (1.51).

6. Asymptotic Expansions of Finite Order Type

Returning to the construction of asymptotic solution expansions of the previous section, starting with (37). Let us assume that $A^1_\beta = 0$ for β with $|\beta| = 1$. Then, it is natural to expect that homogenized second-order and higher-order equations may appear, since the second-order equations are well known and have been considered in numerous works, starting with fundamental works [1] and [3].

Homogenized second order equation. Let us choose $p = m - 2$ in (4), then to satisfy the first equality in (11) it is enough to choose $\{\beta_0\} = \{0\}$ and $N^0_0 = N(y)$ ($u^0_0(x)$ not yet defined). Next, the second equality (11) takes the form (37), where $C^0_0 = 1$ is chosen. To satisfy this equality, it is enough to choose $\{\beta_1\} = \{\beta : |\beta| = 1\} \cup \{0\}$ and $u^\beta_1 = D^\beta_x u^0_0(x)$ for $|\beta| = 1$ ($u^0_1(x)$ not yet defined). In the case, for N^1_β for $|\beta| = 1$ we obtain the equations

$$P(y, D_y)\, N^1_\beta(y) \;=\; -\,P^\beta N(y) \quad \text{for} \quad |\beta| = 1. \tag{64}$$

The equations are solvable since $\{P^\beta N\} = A^1_\beta = 0$ and $N^1_\beta = N_\beta(y) + C^1_\beta N(y)$ for $|\beta| = 1$, $N^1_0 = N(y)$. Consequently, the third equality in (11) takes the form

$$\sum_{\{\beta_2\}} [PN^2_{\beta_2}]u^{\beta_2}_2 + \sum_{|\alpha|=1}\sum_{|\beta|=1} [P^\beta(N_\alpha + C^1_\alpha N)]D^{\alpha+\beta}_x u^0_0$$

$$+ \sum_{|\beta|=1} [P^\beta N] D^\beta_x u^0_1 + \sum_{|\beta|=2} [P^\beta N] D^\beta_x u^0_0/\beta! \;=\; a\, f,$$

since $\beta! = 1$ for $|\beta| = 1$. Applying the weighted average operator $\{\,\cdot\,\}$ to this equality, we may obtain for u^0_0 a second-order equation with constant coefficients

$$\sum_{|\beta|=2} A^2_\beta\, D^\beta_x\, u^0_0(x) \;=\; \{a\}\, f(x), \tag{65}$$

where $A^2_\beta = \hat{A}^2_\beta + \hat{A}^{11}_{\alpha\beta_1}$, $\hat{A}^2_\beta = \{P^\beta N\}/\beta!$ for $|\beta| = 2$, $\hat{A}^{11}_{\alpha\beta_1} = \{P^{\beta_1} N_\alpha\}$ for $|\alpha| = |\beta_1| = 1$.

Suppose that $A^2_\beta \neq 0$ for some β with $|\beta| = 2$. Multiplying the last equation by $M(y)$ and subtracting the resulting relation from the penultimate one, we have an equality similar to (41). From this equality, with an appropriate choice of $\{\beta_2\}$ and $u^{\beta_2}_2$, we obtain equations for $N^2_{\beta_2}$. These equations are solvable due to the choice of appropriate constants and further construction is carried out by induction. Under the additional condition that the solutions to equation (65) are sufficiently smooth, the constructed asymptotic expansion u^l_{m-2} satisfies equation (10) up to $O(\varepsilon^{l-1})$.

Homogenized finite order equation. In the case when $A^1_{\beta_1} = 0,\ A^2_{\beta_2} = 0$ for all corresponding multi-indices β_1, β_2, naturally consider the possibility of obtaining an equation with constant coefficients for u^0_0 of a higher order. A natural question arises: will this process stop and, if so, at what step (in the sense that $A^1_{\beta_1} = 0, \ldots, A^{q-1}_{\beta_{q-1}} = 0$ and $A^q_{\beta_q} \neq 0$ for some q). Further, under a natural assumption, it will be proven that this process will indeed stop.

Let us assume that the solutions to problem (1) satisfy the inequality

$$\varepsilon^{\sigma-m} \parallel u \parallel_{L^2(\Omega)} \leqslant C \parallel a\, f \parallel_{L^2(\Omega)} \tag{66}$$

with positive constants σ and C independent of ε. Let us also assume that $A^1_{\beta_1} = 0, \ldots, A^q_{\beta_q} = 0$ for all relevant multi-indexes. We fix some nonzero functions $u^0_0, \ldots, u^0_{q-1}$ from $C^\infty_0(\Omega)$. In accordance with the constructions described above, the asymptotic expansion

$$u^{q-1}_{m-q} = \varepsilon^{m-q} N^0_0(x/\varepsilon)\, u^0_0(x) + \varepsilon^{m+1-q} N^1_0(x/\varepsilon)\, u^0_1(x) + \ldots$$

satisfies the boundary conditions from (1) exactly and equation (10) up to $O(1)$.

Consequently, for $f = 0$ from (66) we have $u = 0$ and the inequality

$$\varepsilon^{\sigma-q} \parallel N^0_0\, u^0_0 + \varepsilon\, N^1_0\, u^0_1 + \ldots \parallel_{L^2(\Omega)} \leqslant C_1$$

holds with a constant C_1 independent of ε. For a sufficiently small ε, from the known properties of the norm we have the inequality

$$(1/2) \parallel N^0_0 u^0_0 \parallel_{L^2(\Omega)} \leqslant \parallel N^0_0 u^0_0 \parallel_{L^2(\Omega)} - \varepsilon\, C_2 \leqslant \parallel N^0_0 u^0_0 + \varepsilon\, N^0_1 u^0_1 + \ldots \parallel_{L^2(\Omega)}$$

with a constant C_2 independent of ε. So one can conclude that

$$\varepsilon^{\sigma-q} \parallel N^0_0\, u^0_0 \parallel_{L^2(\Omega)} \leqslant 2\, C_1,$$

which is impossible if $\sigma - q < 0$ and ε is small enough. Thus, $1 \leqslant q \leqslant \sigma$ and necessarily $A^k_{\beta_k} \neq 0$ for $k \leqslant \sigma + 1$, if inequality (66) is satisfied.

The case of the asymptotic right-hand side (for $d = 1$, which is being considered now) does not introduce fundamental difficulties into the constructions carried out. In this case, it is necessary to change only the corresponding multi-index sets $\{\beta_s\}$.

The case of asymptotic coefficients in problem (1) can make a fundamental difference in construction. Let the coefficients in the equation from (1) have the form $a_\alpha(x/\varepsilon) = a^0_\alpha(x/\varepsilon) + \varepsilon\, a^1_\alpha(x/\varepsilon) + \ldots$. In this case $P(y) = P_0(y) + \varepsilon\, P_1(y) + \ldots$. Taking into account this expansion in (9) and naturally assuming that $P_0(y)$ is elliptic and has a one-dimensional kernel, we obtain instead of (38) the equality

$$\sum_{\{\beta_1\}} [P_0 N^1_{\beta_1}] u^{\beta_1}_1 + C^0_0 \sum_{|\beta|=1} [P^\beta_0 N] D^\beta_x u^0_0 + C^0_0 [P_1 N] u^0_0 = 0.$$

Applying the weighted average operator $\{\,\cdot\,\}$ to this equality, we have

$$\sum_{|\beta|=1} A^1_\beta\, D^\beta_x\, u^0_0(x) \;+\; A^0_0\, u^0_0(x) \;=\; 0,$$

where the constants $A^1_\beta = \{P^\beta_0 N\}$ for $|\beta| = 1$, $A^0_0 = \{P_1 N\}$ and, as usual, $C^0_0 = 1$.

Now suppose that $A^1_\beta \neq 0$ for some β with $|\beta| = 1$. Then, choosing $p = m - 1$, we obtain for u^0_0 a first-order equation with constant coefficients and a zero-order term, which can be introduced by the coefficients $a^1_\alpha(y)$. If $A^1_\beta = 0$ for all β with $|\beta| = 1$, then the equation for u^0_0 has zero order for $A^0_0 \neq 0$ and $p = m - 1$, preserving all further constructions. Specific examples of such equations have already been considered and justified for problems (47) and (59).

Thus, there is some instability in the construction by coefficients in the sense that adding to $P_0(y)$ an operator of the form $\varepsilon\, P_1(y)$ can lead to a change in the order of the corresponding equation for u^0_0. For example, this equation can become a zero-order equation, which cannot happen in the case of an equation without asymptotic coefficients in accordance with (40). We emphasize that the order of the equation for u^0_0 is very important for justifying the resulting asymptotic expansions, since the smoothness of the corresponding solution u^0_0 depends on it.

By the way, considering problem (1) with the operator symbol for which has only terms of order m and constant coefficients, it is easy to write out the asymptotic coefficients giving equations for u^0_0 of orders $1, 2, \ldots, m$. Indeed, in this case we can assume that $N(y) = 1$ and $M(y) = 1$ and note that symbols (5) define the operators that necessarily contain differentiation for $|\beta| \leqslant m - 1$.

Thus, following [18] and [19], we can now give a definition of the weak limit equation for problem (1). *A homogenized equation for problem* (1) will be called a differential operator $\hat{P} = \hat{P}(D_x)$ of order $\hat{s}$ with constant coefficients and a constant $\hat{p}$ such that if solutions $u = u_\varepsilon(x)$ to problem (1) converges to $v = v(x)$ weakly in $L^2(\Omega)$ for $\varepsilon \to 0$, then $v(x)$ satisfies the following equation

$$\hat{P}(D_x)\, v(x) \;=\; \hat{p}\, f \quad \text{for} \quad x \in \Omega$$

in the sense of distributions (from the space $\mathcal{D}'(\Omega)$ dual to $C^\infty_0(\Omega) = C^\infty_0(\Omega; \mathbb{C})$).

In fact, in all cases considered, expansion (4) begins with the oscillating function $N^0_0 = N^0_0(x/\varepsilon)$ multiplied by ε^p. So, it is natural to replace convergence in $L^2(\Omega)$ in the last definition with convergence in $L^2(\Omega, \varepsilon^p N^0_0(x/\varepsilon))$ (with weight $\varepsilon^p N^0_0(x/\varepsilon)$). By definition, $u_\varepsilon(x)$ converges weakly with the weight to $v(x)$ for $\varepsilon \to 0$, if

$$\int_\Omega u_\varepsilon(x)\,\varepsilon^p N_0^0(x/\varepsilon)\,\varphi(x)\,dx \to \int_\Omega v(x)\,\varphi(x)\,dx \quad \text{for every} \quad \varphi \in L^2(\Omega).$$

It is well known that to satisfy the last relation (for some subsequence $\varepsilon' \to 0$) it is sufficient that the inequality $\| u_\varepsilon(x)\,\varepsilon^p N_0^0(x/\varepsilon) \|_{L^2(\Omega)} \leqslant C$ with a constant independent of ε be satisfied. Convergence in $\mathcal{D}'(\Omega, \varepsilon^p N_0^0(x/\varepsilon))$ can also be defined.

Therefore, *a homogenized equation with respect to* $L^2(\Omega, \varepsilon^p N_0^0(x/\varepsilon))$ is defined according to [18], if solutions $u_\varepsilon(x)$ to problem (1) converges to $v(x)$ weakly in $L^2(\Omega, \varepsilon^p N_0^0(x/\varepsilon))$ for $\varepsilon \to 0$, then v satisfies the equation $\hat{P}(D_x)\,v = \hat{p}\,f$ in $\mathcal{D}'(\Omega)$.

This is proven in [18], Theorem 3.1, that if solutions $u_\varepsilon(x)$ to problem (1) converges to $v(x)$ weakly in $L^2(\Omega, \varepsilon^p N_0^0(x/\varepsilon))$ for some subsequence $\varepsilon' \to 0$, then v satisfies the homogenized equation in $\mathcal{D}'(\Omega)$ under the only condition that the inequality $\varepsilon^\kappa \| u_\varepsilon(x) \|_{L^2(\Omega)} \leqslant C$ with a constant independent of ε is satisfied for some fixed κ. Here the homogenized equation is defined according to general principles. As already noted, this equation can have an arbitrary order less than or equal to the order of the equation of problem (1). Hence, a sufficient condition for the convergence of solutions to problem (1) to the solution of the homogenized equation is the fulfillment of the inequality $\| \varepsilon^p\, u_\varepsilon(x) \|_{L^2(\Omega)} \leqslant C$ with a constant independent of ε, since the function $N_0^0(x/\varepsilon)$ is bounded, for example, in the $C^0(\bar{\Omega})$-norm.

The idea of proving such convergence is quite simple. First, the asymptotic expansion (4) for solutions to problem (1) with the adjoint operator must be constructed. Then, the functions $u_s^{\beta_s}$ are replaced by arbitrary ones from $C_0^\infty(\Omega)$ and the passage to the limit in the integral identity is performed for $\varepsilon \to 0$, as was already done in the transition from (55) to (56). Further details can be found in [18], where the case of $d \geqslant 2$ is also considered. Of course, information only about the homogenized equation may not be enough to obtain accuracy estimates; boundary conditions also need to be determined. But sometimes such boundary conditions appear naturally, for example, for zero-order equations.

The situation in the case of $d \geqslant 2$ is even more complicated, but it is treated similarly to the case $d = 1$ just considered. General expansions in the case $d = 2$ are given in [18]. Of course, in the case the asymptotic expansions begins like this

$$u^l_{p_1,p_2} = \varepsilon^{p_1} N_1(x/\varepsilon)\,u_1^0(x) + \varepsilon^{p_2} N_2(x/\varepsilon)\,u_2^0(x) + \ldots \tag{67}$$

and the solvability conditions for the corresponding cell problems will lead to a system of equations with constant coefficients for u_1^0 and u_2^0. Such systems can have an arbitrary but finite order if inequality (66) is satisfied for solutions to problem (1). Specific examples of these constructions will be given below.

Nondivergent second order equations. Let $n \geqslant 2$ be given a bounded domain $\Omega \subset \mathbb{R}^n$ with a smooth boundary and a smooth function $f \in C_0^\infty(\Omega)$. Let us define the function u as a solution to the problem for the second-order equation

$$a_{lk}(x/\varepsilon)\,\partial_h\partial_k\,u = a(x/\varepsilon)\,f \quad \text{in} \quad \Omega, \qquad u = 0 \quad \text{on} \quad \partial\Omega, \tag{68}$$

It is assumed that the coefficients $a_{lk}(y)$ and $a(y)$ are smooth 1-periodic (real-valued) functions satisfying the ellipticity condition (25) for $h, k = 1, \dots, n$.

It is known [6], [49] that for a fixed ε there is a unique solution to problem (68). From the ellipticity it also follows that 1-periodic solutions of the equations

$$PN(y) = a_{lk}(y)\,\partial_{y_h}\partial_{y_k}N(y) = 0, \quad P^*M(y) = \partial_{y_k}\partial_{y_h}\big(a_{lk}(y)\,M(y)\big) = 0,$$

$N = 1$ and $M = M(y) \geqslant \sigma > 0$ are uniquely defined and therefore $d = 1$ (for example, according to [6], [49]). Therefore, we can assume that $\langle\, M\,\rangle = 1$.

The symbol of the differential operator of problem (68) is $P(y,\xi) = a_{lk}(y)\,\xi_h\,\xi_k$. Consequently, the equalities $P^\beta(y, D_y)\,N(y) = 0$ are satisfied for $|\beta| \leqslant 1$. Taking this observation into account, we choose $p = 0$ in (4). Then to satisfy the first and second equalities in (11) it is enough to choose $\{\beta_0\} = \{\beta_1\} = \{0\}$ and $N_0^0 = N_0^1 = 1$ ($u_0^0(x)$ and $u_1^0(x)$ are not yet defined). Directly from the definitions we obtain

$$P^\beta(y, D_y) = a_\beta(y)\beta! \quad \text{for} \quad |\beta| = 2.$$

Thus, the third equality in (11) is of the following form

$$\sum_{\{\beta_2\}} \big[PN^2_{\beta_2}(y)\big]u_2^{\beta_2}(x) \,+\, a_{lk}(y)\,\partial_h\partial_k\,u_0^0(x) \,=\, a(y)\,f(x).$$

Applying the weighted average operator $\{\,\cdot\,\}$ to this equality, we have

$$\hat{a}_{lk}\,\partial_h\partial_k u_0^0(x) \,=\, \{a\}\,f(x) \qquad \text{in} \qquad \Omega, \tag{69}$$

where $\hat{a}_{lk} = \{a_{lk}\} = \langle\, a_{lk}\,M\,\rangle$ for $l, k = 1, \dots, n$.

Subtracting the last equality from the penultimate one, we get

$$\sum_{\{\beta_2\}} \big[PN^2_{\beta_2}(y)\big]u_2^{\beta_2}(x) + \big[a_{hk}(y) - \hat{a}_{hk}\big]\partial_h\partial_k u_0^0(x) = \big[a(y) - \{a\}\big]f(x).$$

To satisfy the equality, it is enough to choose $\{\beta_2\} = \{h, k \,:\, h, k = 1, \dots, n\} \cup \{1\} \cup \{0\}$, $u_2^1 = f(x)$, $u_2^{hk} = \partial_h\partial_k u_0^0(x)$ and $N_0^2 = 1$ ($u_2^0(x)$ not yet defined). Then

$$P\,N_1^2 = a(y) - \{a\}, \qquad P\,N_{hk}^2 = \hat{a}_{lk} - a_{lk}(y) \quad \text{for} \quad l, k = 1, \dots, n,$$

are the equations for $N^2_{\beta_2}$, which are solvable due to the choice of the constants.

The constant matrix $\{\hat{a}_{lk}\}^n_{l,k=1}$ is elliptic by (25). Consequently, the homogeneous Dirichlet problem for the equation (69) has a uniquely defined smooth solution $u^0_0 \in C^\infty(\overline{\Omega})$ (for example, according to [6], [49]). Continuing the construction by induction, we obtain for the functions $u^0_1, \ldots, u^0_s$ problems of the similar types

$$\hat{a}_{hk}\,\partial_h\partial_k u^0_j(x) = F_j\big(f, u^0_0, \ldots, u^0_{j-1}\big) \quad \text{in} \quad \Omega, \quad u^0_j(x) = 0 \quad \text{on} \quad \partial\Omega,$$

for $j = 0, \ldots, s$, where $F_j(f, u^0_0, \ldots, u^0_{j-1})$ denotes a linear combination of finite order derivatives of $f, u^0_0, \ldots, u^0_{j-1}$ for $j = 0, \ldots, s$. Consequently, the problems have uniquely defined smooth solutions $u^0_0, \ldots, u^0_s$. Thus, an asymptotic solution expansion for problem (68) has been constructed in the following form

$$u^s_0 = u^0_0(x) + \varepsilon\, u^0_1(x) + \varepsilon^2 u^0_2(x) + \varepsilon^2 N^2_1(x/\varepsilon) f(x) + \varepsilon^2 N^2_{hk}(x/\varepsilon)\,\partial_h\partial_k u^0_0(x) + \ldots.$$

This asymptotic solution expansion satisfies the relations

$$P_\varepsilon\,(u - u^s_0) = O(\varepsilon^{\,s+1-2}) \quad \text{in} \quad \Omega, \qquad u - u^s_0 = O(\varepsilon^2) \quad \text{on} \quad \partial\Omega.$$

A consequence of these relations for u^3_0 and A. D. Alexandrov's estimate from [49] for problem (68) is the following statement.

Theorem 2. *Let u be a solution to problem (68) and u^0_0 be a solution to the homogeneous Dirichlet problem for equation (69). Then*

$$\|\, u - u^0_0 \,\|_{C^0(\overline{\Omega})} \leqslant C\,\varepsilon, \qquad \|\, u - u^3_0 \,\|_{C^0(\overline{\Omega})} \leqslant C\,\varepsilon^2, \tag{70}$$

where the constant C is independent of ε for $0 < \varepsilon \leqslant \varepsilon_0$ with an appropriate ε_0.

Modern proofs of such accuracy estimates is given in [50], where one can find a contemporary review of Chinese papers on this topic. Note that the accuracy estimates (70) were proven independently in [18], Remark 2.2.

Diffusion and transfer by microneedles. The accuracy estimate (58) indicates normal transfer and weak filtration in the problem (47), which is actually associated with non-zero transfer and potentiality coefficients on average. Here we will reduce this coefficient. Let in (24) we have

$$af = \hat{a}(x/\varepsilon) f(x) + \varepsilon\,\partial_k\big(\hat{a}^k(x/\varepsilon)\hat{f}(x)\big)$$

$$= \hat{a}(y) f(x) + \big(\partial_{y_k}\hat{a}^k(y)\big)\hat{f}(x) + \varepsilon\,\hat{a}^k(y)\partial_k\hat{f}(x) \quad \text{for} \quad y = x/\varepsilon, \tag{71}$$

coefficients are of general types, $a_k = a_k^0(x,y) + \varepsilon\, a_k^1(x,y)$ and $a_0 = \varepsilon^2 a_0^2(x,y)$. Then the problem with asymptotic coefficient is of the following form

$$-\,\partial_l\big(a_{lk}(x,x/\varepsilon)\partial_k\, u\big) \,+\, \varepsilon^{-1}a_k(x,x/\varepsilon)\,\partial_k\, u \,+\, a_0^2(x,x/\varepsilon)\, u \;=$$

$$= \hat{a}(x/\varepsilon)f(x) + \varepsilon\,\partial_k\big(\hat{a}^k(x/\varepsilon)\hat{f}(x)\big) \quad \text{in} \quad \Omega, \quad u=0 \quad \text{on} \quad \partial\Omega, \tag{72}$$

where the coefficients $a_{lk}(x,y) = a_{kl}(x,y), \ldots, a_0^2(x,y)$ are smooth and 1-periodic in y for $l,k = 1,2$. Thus, we renormalized problem (24) (so that $p=0$ in (4)) and selected general asymptotic data in order to obtain a second order equation for u_0^0.

As before, the unique solution to this problem exists for a fixed ε by Schauder theorem [21]. In order to obtain an energy estimate that is exactly dependent on ε, we assume that $\partial_{y_k} a_k^0(x,y) = 0$, $\partial_k\, a_k^0(x,y) = 0$, $\partial_{y_k} a_k^1(x,y) = 0$ and inequality (25) is valid for all $(x,y) \in \Omega \times Y$. In addition, analogues of inequalities (26) are satisfied in the form $\partial_k\, a_k^1(x,y) \leqslant \sigma$ and $\sigma \leqslant a_0^2(x,y)$ for a positive σ.

Then, we can repeat the proof of (27) and get for solution to (47) the estimate

$$\|\,\nabla u\,\|_{L^2(\Omega)^2} + \|\,u\,\|_{L^2(\Omega)} \leqslant C\,\|\,\hat{a}\,f\,\|_{L^2(\Omega)} + \varepsilon\, C\,\|\,(\hat{a}^1 + \hat{a}^2)\,\hat{f}\,\|_{L^2(\Omega)} \tag{73}$$

and relevant analogue of estimate (32) with the constants C independent of ε.

In the situation under consideration, the operator of problem (72) is of the form $P(y) = P_0(y) + \varepsilon\, P_1(y) + \varepsilon^2 P_2(y)$, where the symbols for operators $P_0(y)$ and $P_1(y)$ were defined after inequality (49) and $p_2 = a_0^2(y)$. Here it is assumed that $\langle a_k^0 \rangle = 0$ for $k = 1,2$ otherwise we will get a first order equation, which is considered for (47). Note also that we cannot choose $a_0 = \varepsilon\, a_0^1(x,y)$, where $a_0^1(x,y)$ satisfies (26), since this will lead to a zero-order equation, which is considered for (59). Without (26), energy estimates similar to (73), which exactly depend on ε, may not be fulfilled.

To construct an asymptotic solution expansion for this problem, we note that $N=1, M=1$ and use, for example, equality (37), where we should also add $[P_1 N(y)]\, u_0^0$, which is zero by the definition of $P_1 = P_1(y)$. To satisfy this equality, as before (64), it is enough to choose $\{\beta_1\} = \{\beta : |\beta| = 1\} \cup \{0\}$ and

$$u_1^\beta = D_x^\beta u_0^0(x) \quad \text{for} \quad |\beta| = 1$$

($u_1^0(x)$ not yet defined). In the case, for N_β^1 for $|\beta| = 1$ we obtain equations (64).

The equations are solvable since $A_\beta^1 = 0$ and $N_\beta^1 = N_\beta(y) + C_\beta^1 N(y)$

for $|\beta| = 1$, $N_0^1 = 1$. Consequently, the third equality in (11) takes the form

$$\sum_{\{\beta_2\}} [P_0 N_{\beta_2}^2] u_2^{\beta_2} + \sum_{|\alpha|=1} \sum_{|\beta|=1} [P_0^{\beta} N_{\alpha}^1] D_x^{\alpha+\beta} u_0^0 + \sum_{|\beta|=1} [P_0^{\beta} N] D_x^{\beta} u_1^0$$

$$+ \sum_{|\beta|=2} [P_0^{\beta} N] D_x^{\beta} u_0^0/\beta! + \sum_{|\beta|=1} [P_1^{\beta} N] D_x^{\beta} u_0^0 + [P_2 N(y)]\, u_0^0 = a\, f.$$

Applying the average operator $\{\,\cdot\,\} = \langle\,\cdot\,\rangle$ to this equality, we may obtain for u_0^0 a second-order equation with lower order terms. This equation is of the form

$$\sum_{|\beta|=2} A_{\beta}^2\, D_x^{\beta}\, u_0^0(x) + \sum_{|\beta|=1} A_{\beta}^1\, D_x^{\beta}\, u_0^0(x) + A_0\, u_0^0(x) = \langle a \rangle\, f(x), \quad (74)$$

where $A_{\beta}^2 = \hat{A}_{\beta}^2 + \hat{A}_{\alpha\beta_1}^{11}$, $\hat{A}_{\beta}^2 = \langle P_0^{\beta} N \rangle / \beta!$ for $|\beta| = 2$, $\hat{A}_{\alpha\beta_1}^{11} = \langle P_0^{\beta_1} N_{\alpha} \rangle$ for $|\alpha| = |\beta_1| = 1$, $A_{\beta}^1 = \langle P_1^{\beta} N \rangle$ for $|\beta| = 1$, and $A_0 = \langle a_0^2 \rangle$ is a strictly positive function of $x \in \bar{\Omega}$.

Let us clarify the form of the other coefficients in equation (74). By definition, the symbol of the operator $P_1(y)$ is $p_1 = a_k^1(i\,\xi) - \partial_l a_{lk}(i\,\xi)$ and therefore

$$p_1^{(1,0)} = i\, a_1^1 - i\,\partial_l\, a_{l1}, \qquad p_1^{(0,1)} = i\, a_2^1 - i\,\partial_l\, a_{l2}.$$

Hence, the right-hand side in equation (64) is purely imaginary and so the solution to the equation is the same $N_{\beta}^1 = i\,\hat{N}_{\beta}^1$. Thus, the first order term in equation (74) is $A_k\,\partial_k\, u_0^0$, where $A_k = \langle a_k^1 \rangle - \partial_l\, \langle a_{lk} \rangle$, since $D_k = i^{-1}\partial_k$ for $k = 1, 2$.

Next, the symbol of the operator $P_0(y)$ is $p_1 = a_{lk}\,\xi_l\,\xi_k + b_k(i\,\xi)$ and therefore

$$p_1^{(1,0)} = p_1^1(y,\xi) = 2\,a_{l1}\,\xi_l + i\,b_1, \qquad p_1^{(0,1)} = p_1^2(y,\xi) = 2\,a_{l2}\,\xi_l + i\,b_2,$$

$p_1^{(2,0)} = 2\,a_{11}$, $p_1^{(0,2)} = 2\,a_{22}$ and $p_1^{(1,1)} = 2\,a_{22}$, where $b_k = a_k^0 - \partial_{y_l}\, a_{lk}$.

Let us present the solution to equation (64) as $N_s^1 = i\,\hat{N}_s^1$ for $s = 1, 2$. Then

$$-\,\partial_{y_l}\big(a_{lk}\,\partial_{y_k}\hat{N}_s^1\big) + a_k^0\,\partial_{y_k}\hat{N}_s^1 = \partial_{y_l} a_{ls} - a_s^0 \quad (75)$$

is a cell problem for $s = 1, 2$. Multiplying the equation by $\hat{N}_j^1$ and integrating, we get

$$\big\langle a_{lk}\,(\partial_{y_k}\hat{N}_s^1 + \delta_{ks})\partial_{y_l}\hat{N}_j^1 \big\rangle = -\big\langle a_k^0\,(\partial_{y_k}\hat{N}_s^1)\hat{N}_j^1 \big\rangle - \big\langle a_s^0\,\hat{N}_j^1 \big\rangle$$

for $s, j = 1, 2$, where δ_{ks} denotes the Kronecker symbol. Moreover, we have

$$A_{sj} = \big\langle a_{lk}\,\big(\partial_{y_k}\hat{N}_s^1 + \delta_{ks}\big)\big(\partial_{y_l}\hat{N}_j^1 + \delta_{lj}\big)\big\rangle \quad (76)$$

$$= -\left\langle a^0_k \left(\partial_{y_k} \hat{N}^1_s\right) \hat{N}^1_j \right\rangle - \left\langle a^0_s \hat{N}^1_j \right\rangle + \left\langle a_{jk} \partial_{y_k} \hat{N}^1_s \right\rangle + \left\langle a_{js} \right\rangle$$

for $s, j = 1, 2$. Further, in natural notation for $A^{11}_{sj} = \left\langle p^s_0(y, D_y) N^1_j \right\rangle$ we have

$$A^{11}_{sj} = \left\langle 2\, a_{ls} \partial_{y_l} \hat{N}^1_j \right\rangle - \left\langle a^0_s \hat{N}^1_j \right\rangle + \left\langle (\partial_{y_l} a_{ls}) \hat{N}^1_j \right\rangle = \left\langle a_{ls} \partial_{y_l} \hat{N}^1_j \right\rangle - \left\langle a^0_s \hat{N}^1_j \right\rangle,$$

since

$$\langle (\partial_{y_l} a_{ls}) \hat{N}^1_j \rangle + \langle a_{ls} \partial_{y_l} \hat{N}^1_j \rangle = \langle \partial_{y_l} (a_{ls} \hat{N}^1_j) \rangle = -\langle (a_{ls} \hat{N}^1_j) \partial_{y_l} 1 \rangle = 0$$

due to the Leibniz formula and integration by parts.

The left side of equality (76) is symmetrical with respect to s and j. Whereas the first term on the right side is skew-symmetric, because we have

$$\left\langle a^0_k \left(\partial_{y_k} \hat{N}^1_s\right) \hat{N}^1_j \right\rangle = -\left\langle \hat{N}^1_s\, \partial_{y_k} \left(a^0_k \hat{N}^1_j\right) \right\rangle = -\left\langle \hat{N}^1_s\, a^0_k \left(\partial_{y_k} \hat{N}^1_j\right) \right\rangle$$

by assumptions. Consequently, equation (74) can be written in the form

$$-A_{lk}\, \partial_l\, \partial_k\, u^0_0(x) + A_k\, \partial_k\, u^0_0(x) + A_0\, u^0_0(x) = \langle a \rangle\, f(x) \qquad (77)$$

with an elliptic symmetric matrix $\{A_{lk}\}_{l,k=1,2}$, which meets (25), according to (76).

Consequently, for smooth data, an asymptotic solution expansion for problem (72) can be constructed in the following form

$$u^s_0 = u^0_0 + \varepsilon\, u^0_1 + \varepsilon\, N^1_k(x, x/\varepsilon) \partial_k u^0_0 + \varepsilon^2 N^0_0(x, x/\varepsilon) f + \varepsilon^2 N^2_{lk}(x, x/\varepsilon) \partial_l \partial_k u^0_0 + \ldots. \quad (78)$$

This asymptotic solution expansion satisfies the relations

$$P_\varepsilon\, (\, u - u^s_0\,) = O(\varepsilon^{\,s+1-2}) \quad \text{in} \quad \Omega, \qquad u - u^s_0 = O(\varepsilon) \quad \text{on} \quad \partial\Omega.$$

These relations are not very suitable for deriving an estimate of the accuracy of the resulting asymptotic expansion, since the derivatives of the function u^0_0 in (78) may not satisfy the homogeneous Dirichlet boundary conditions.

Therefore, it is convenient to cut off these derivatives near the boundary $\partial\Omega$, which will be illustrated for problem (72), where the right-hand side of the equation is the periodic δ-functions as for problem (31). Let e^ε be a smooth function on $\overline{\Omega}$ that vanishes for $\mathrm{dist}(x, \partial\Omega) \le \varepsilon$ and equals to 1 for $\mathrm{dist}(x, \partial\Omega) \ge 2\varepsilon$, where $\mathrm{dist}(x, \partial\Omega)$ is the distance from the point x to the boundary $\partial\Omega$. The functions exist for small enough ε ($\varepsilon \le \varepsilon_0$ for some positive ε_0). In addition, we have

$$\varepsilon^\kappa\, |\, \nabla^\kappa_x\, e^\varepsilon(x)\, | \leqslant C \quad \text{in} \quad \bar{\Omega} \quad \text{for a fixed integer} \quad \kappa \qquad (79)$$

with a constant C independent of ε. The supports of the functions $e^\varepsilon - 1$ and $\nabla^s e^\varepsilon$ are contained in the boundary strip $\Omega_{2\varepsilon} \subset \Omega$ of width 2ε (see, for example, [7],4.1).

Let us denote by v_0^2 a function of the form (78), where the terms containing the derivatives $\partial_k u_0^0$ and $\partial_l \partial_k u_0^0$ are multiplied by $e^\varepsilon(x)$. The measure $|\,\Omega_{2\varepsilon}|$ of the strip can be estimated as follows $|\,\Omega_{2\varepsilon}| \leqslant C\,\varepsilon$ with a constant independent of ε. Thus, it is known [7] and can be directly verified that the following inequality

$$\|\, \nabla u_0^2 - \nabla v_0^2 \,\|^2_{L^2(\Omega)^2} + \|\, u_0^2 - v_0^2 \,\|^2_{L^2(\Omega)} \leqslant C\,\varepsilon \tag{80}$$

and an analogue for $L^{2,\infty}(\Omega)$-norms are valid with the constant C independent of ε. Hence, the expansions u_0^2 and v_0^2 are quite close and $v_0^2 = 0$ on the boundary $\partial\Omega$.

Now we can consider problem (72) with periodic δ-functions as for problem (31). Indeed, if in the notation for (31) the right-hand side of the problem equation (72) is the sum of δ-functions, then estimate (73) for $L^{2,\infty}(\Omega)$-norms is satisfied.

Thus, it follows directly from (73) and (80) that for solution u to problem (72) with the right-hand side $\sum_{i=1}^{\kappa} \delta_2((x - \hat{x}^i)/\varepsilon)$ the following accuracy estimate

$$\|\nabla u - \nabla u_0^2 \,\|^2_{L^{2,\infty}(\Omega)^2} + \|u - u_0^2 \,\|^2_{L^{2,\infty}(\Omega)} \leqslant \varepsilon\, C \tag{81}$$

is valid, where the constant does not depend on ε and u_0^2 is defined by the asymptotic representation (78). In the representation $u_0^0 = u_0^0(x)$ is a solution to equation (77) with homogeneous Dirichlet boundary conditions according to (1), $N_0^0 = N_0^0(x, x/\varepsilon)$ is a solution to problem (34) in which $a_0 = 0$, $a_k = a_k^0(x, y)$ and the right hand side is $\delta_2(y - \hat{y}) - 1$. In addition, for example, $N_s^0 = N_s^0(x, x/\varepsilon)$ is defined as a 1-periodic solution of the equation (75) for $s = 1, 2$.

Similarly, repeating the above arguments, one can conclude that for solution to problem (72) with the right-hand side $\sum_{i=1}^{\kappa} \hat{f}(x)\, \delta_1((x_2 - x_2^i)/\varepsilon)$ the accuracy estimate (81) for $L^2(\Omega)$-norms is valid with a constant independent of ε. Accordingly, in representation (78), for example, $N_0^0 = N_0^0(x, x/\varepsilon)$ is a solution to problem (34) in which $a_0 = 0$, $a_k = a_k^0(x, y)$ and the right hand side is $\delta_1(y_2 - \hat{y}_2) - 1$.

Thus, for a diffusion coefficient of the unity order and a large transfer coefficient with zero average, homogenized diffusion is modeled with guaranteed accuracy by a second-order elliptic equation with coefficients independent of the small parameter. The advantage of this approach is the simplicity of the numerical solution of equation (77) with homogeneous boundary conditions and cell problems (34) and (75). For example, the solutions may be represented as convergent series and partial sums of these series may be used to approximate the solution to problem (72). Whereas a direct numerical solution of the original problem (72) with guaranteed accuracy requires significant computer resources for small ε.

7. Elliptic Finite Order Equations of Divergent Form

In this section, the algorithm for constructing asymptotic expansions from the initial sections will be applied to elliptic equations of divergent form with periodic rapidly oscillating coefficients. For the constructed asymptotic expansions, accuracy estimates similar to those of Theorem 1 will be proven. Thus, this expansion will be justified with guaranteed accuracy in the small parameter ε.

Let for $n \geqslant 2$ be given an integer $r > 0$, bounded domain $\Omega \subset \mathbb{R}^n$ with smooth (class C^∞) boundary and smooth function $f \in C_0^\infty(\Omega)$. Let us define the function $u = u(x)$ as a solution to the following Dirichlet problem for the order $2r$-order partial differential equation of divergent form

$$\sum_{|\alpha|=|\gamma|=r} D_x^\alpha \left(a_{\alpha\gamma}(x/\varepsilon)\, D_x^\gamma u(x)\right) = a(x/\varepsilon)\, f(x) \quad \text{in} \quad \Omega,$$
$$\partial_\nu^s u(x) = 0 \quad \text{on} \quad \partial\Omega \quad \text{for} \quad s = 0, \ldots, r-1, \tag{82}$$

where ε is a small positive parameter and the notation introduced for problem (1) is used. It is assumed that the coefficients $a_{\alpha\gamma}(y)$ and $a(y)$ of equation (82) are 1-periodic (real) functions of class $C^\infty(\mathbb{R}^n)$. For the differential operator P_ε of problem (82) it is also assumed that the following ellipticity condition

$$\sum_{|\alpha|,|\gamma|=r} a_{\alpha\gamma}(y)\, \xi^\alpha \xi^\gamma \geqslant \nu\, |\, \xi\, |^{2r} \qquad \text{for} \quad y \in \mathbb{R}^n \quad \text{and} \quad \xi \in \mathbb{R}^n, \tag{83}$$

is satisfied, where ν is a given positive constant.

It is known [21] that for a fixed ε there is a unique solution to problem (82) and

$$\|\, u\, \|_{H_0^r(\Omega)} \leqslant C\, \|\, a\, f\, \|_{H^{-r}(\Omega)}, \tag{84}$$

where the constant C is independent of ε.

In accordance with the constructions from the initial sections, to determine the asymptotic behavior of the solution to problem (82), it is necessary first of all to obtain the symbol of the operator P_ε. Let us use the following composition formula for this: if A, B are differential operators with symbols $a(x, \xi)$ and $b(x, \xi)$, then the composition of these operators $C = A \cdot B$ has the symbol

$$c(x,\xi) = \sum_{\varkappa} \big(\partial_\xi^\varkappa a(x,\xi) \big) D_x^\varkappa b(x,\xi)/\varkappa!,$$

for example, according to [23]. From this formula for $A = D_x^\alpha$ and $B = a_{\alpha\gamma}^\varepsilon D_x^\gamma$ we conclude that the operator $C = A \cdot B = D_x^\alpha \big(a_{\alpha\gamma}^\varepsilon D_x^\gamma \cdot \big)$ has the symbol

$$c_\varepsilon(x,\xi) = \sum_{|\varkappa|\leqslant r} \big(\partial_\xi^\varkappa \xi^\alpha \big) D_x^\varkappa a_{\alpha\gamma}^\varepsilon \xi^\gamma/\varkappa! \,.$$

Direct calculation verifies the following equality

$$\partial_\xi^\varkappa \xi^\alpha = \varkappa! \, C_\alpha^\varkappa \xi^{\alpha-\varkappa}, \tag{85}$$

where $C_\alpha^\varkappa = \alpha!/\big(\varkappa! \, (\alpha-\varkappa)!\big)$ for a multi-index $(\alpha - \varkappa)$ not containing negative components and $C_\alpha^\varkappa = 0$ otherwise. Thus, we get

$$c_\varepsilon(x,\xi) = \sum_{|\varkappa|\leqslant r} C_\alpha^\varkappa \big(D_x^\varkappa a_{\alpha\gamma}(x/\varepsilon) \big) \xi^{\alpha-\varkappa+\gamma},$$

and therefore the symbol of the operator P_ε of problem (82) has the form

$$p_\varepsilon(x/\varepsilon,\xi) = \sum_{|\alpha|,|\gamma|=r} \sum_{|\varkappa|\leqslant r} C_\alpha^\varkappa \big(D_x^\varkappa a_{\alpha\gamma}(x/\varepsilon) \big) \xi^{\alpha-\varkappa+\gamma} = \tag{86}$$

$$= \sum_{|\alpha|,|\gamma|=r} \sum_{|\varkappa|\leqslant r} \varepsilon^{-|\varkappa|} C_\alpha^\varkappa \big(D_y^\varkappa a_{\alpha\gamma}(y) \big) \xi^{\alpha-\varkappa+\gamma} \quad \text{for} \quad y = x/\varepsilon$$

in accordance with the differentiation formula (7).

Similarly, we find that the symbol of the operator Q_ε, formally conjugate to the operator P_ε of problem (82), has the form

$$q_\varepsilon(x/\varepsilon,\xi) = \sum_{|\alpha|,|\gamma|=r} \sum_{|\varkappa|\leqslant r} \varepsilon^{-|\varkappa|} C_\gamma^\varkappa (D_y^\varkappa a_{\gamma\alpha}(y)) \xi^{\gamma-\varkappa+\alpha} \quad \text{for} \quad y = x/\varepsilon, \tag{87}$$

which coincides with symbol (86) if the coefficients are obviously symmetric.

We will look for an expansion of the solution to problem (82) in the form of an asymptotic sum (4), in which we need to choose $p = 0$ as will be clear from further constructions. For the coefficients of this sum, as before, we obtain equalities similar to (11) with the numbers $s = 0, \ldots, l$. Let $P = P(y, D_y)$ denote the operator on T^n obtained from symbol (86) for $\varepsilon = 1$. From (83) it follows that the equation $P(y, D_y)\, N(y) = 0$ on T^n has a unique solution $N(y) = 1$ satisfying the conditions (13). For a similar reason, $M(y) = 1$ and therefore $d = 1$. In accordance with (5), the operator $P^\beta = P^\beta(y, D_y)$ has the symbol

$$\partial_\xi^\beta p_1(y,\xi) = \sum_{|\alpha|,|\gamma|=r} \sum_{|\varkappa|\leqslant r} C_\alpha^\varkappa \left(D_y^\varkappa a_{\alpha\gamma}(y)\right) \partial_\xi^\beta \left(\xi^{\alpha-\varkappa+\gamma}\right) \tag{88}$$

$$= \sum_{|\alpha|,|\gamma|=r} \sum_{|\varkappa|\leqslant r} C_\alpha^\varkappa\, C_{\alpha-\varkappa+\gamma}^\beta \left(D_y^\varkappa a_{\alpha\gamma}(y)\right) \xi^{\alpha-\varkappa+\gamma-\beta} \beta!.$$

By virtue of equality (88), the operator $P^\beta(y, D_y)$ for $|\beta| < s$ necessarily contains differentiation and therefore

$$P^\beta(y, D_y) N(y) = 0 \qquad \text{for} \qquad |\beta| \leqslant s-1. \tag{89}$$

Consequently, equalities (11) with numbers $s = 0, \ldots, r-1$, will be satisfied if we choose $\{\beta_s\} = \{0\}$ and $N_0^s = 1$ ($u_s^0(x)$ not yet defined). Thus, taking into account (88) and (89), equality (11) with number r takes the form

$$\sum_{\{\beta_r\}} [PN_{\beta_r}^r]\, u_r^{\beta_r} + \sum_{|\beta|=r} [P^\beta N_0^0]\, D_x^\beta u_0^0/\beta! = 0\,, \tag{90}$$

where $P^\beta N_0^0 = \sum_{|\varkappa|=r} D_y^\varkappa a_{\varkappa\beta}(y)\beta!$ for $|\beta| = r$, since $C_\alpha^\varkappa = 0$ for $\varkappa \neq \alpha$ and $C_\alpha^\varkappa = 1$ with $\varkappa = \alpha$ for $|\alpha| = |\varkappa| = r$. Hence, we choose $\{\beta_r\} = \{\beta : |\beta| = r\} \cup \{0\}$, $u_r^\beta = D_x^\beta u_0^0(x)$ for $|\beta| = r$ and $N_0^r = 1$ ($u_r^0(x)$ not yet defined). Consequently, for N_β^r we obtain the following equations

$$P(y, D_y)\, N_\beta^r(y) = -\sum_{|\varkappa|=r} D_y^\varkappa a_{\varkappa\beta}(y) \quad \text{for} \quad |\beta| = r. \tag{91}$$

Let's denote by 1_k the multiindex with the kth component one and the rest zeros. For a multi-index α with $|\alpha| > 0$ and a smooth function $K(y)$ on T^n we get

$$\langle D_y^\alpha K \rangle = \langle D_y^\alpha K, 1 \rangle = \langle D_y^{\alpha-1_k} K, D_y^{1_k} 1 \rangle = 0, \tag{92}$$

where for definiteness $\alpha_k > 0$ for some k. From these equalities it follows that equation (91) is solvable, since $M(y) = 1$ in accordance with (15). Thus, equality (90) is satisfied due to (91) and the choice of $\{\beta_r\}$.

Now the equality numbered $r+1$ takes the form

$$\sum_{\{\beta_{r+1}\}} [PN_{\beta_{r+1}}^{r+1}]\, u_{r+1}^{\beta_{r+1}} + \sum_{\{\beta_r\}} \sum_{|\beta|=1} [P^\beta N_{\beta_r}^r]\, D_x^\beta u_r^{\beta_r}/\beta! + \ldots$$

$$+ \sum_{\{\beta_1\}} \sum_{|\beta|=r} [P^\beta N_{\beta_1}^1]\, D_x^\beta u_1^{\beta_1}/\beta! + \sum_{|\beta|=r+1} [P^\beta N_0^0]\, D_x^\beta u_0^0/\beta! = 0\,.$$

Consequently, taking (88) and (89) into account, we obtain

$$\sum_{\{\beta_{r+1}\}} [PN^{r+1}_{\beta_{r+1}}] u^{\beta_{r+1}}_{r+1} + \sum_{|\alpha|=r} \sum_{|\beta|=1} [P^\beta N^r_\alpha] D^\beta_x u^\alpha_r/\beta! \tag{93}$$

$$+ \sum_{|\beta|=r} [P^\beta N^1_0] D^\beta_x u^0_1/\beta! + \sum_{|\beta|=r+1} [P^\beta N^0_0] D^\beta_x u^0_0/\beta! = 0\,.$$

In addition, from (88) it follows that

$$P^\beta N^0_0(y) = \sum_{|\alpha|,|\gamma|=r} \sum_{\varkappa=\alpha+\gamma-\beta} C^\varkappa_\alpha D^\varkappa_y a_{\alpha\gamma}(y)\beta! \quad \text{for} \quad |\beta| = r+1.$$

Therefore, to satisfy equality (93), it is enough to choose

$$\{\beta_{r+1}\} = \{\beta : |\beta| = r+1\} \cup \{\beta = \beta^1 + \beta_r : |\beta^1| = 1, |\beta_r| = r\} \cup \{\beta : |\beta| = r\} \cup \{0\},$$

$u^\beta_{r+1} = D^\beta_x u^0_0(x)$ for $|\beta| = r+1$ $u^\beta_{r+1} = D^{\beta^1}_x u^{\beta_r}_r(x) = D^{\beta^1+\beta_r}_x u^0_0(x)$ for $\beta = \beta^1 + \beta_r$ with $|\beta^1| = 1$, $|\beta_r| = r$, $u^\beta_{r+1} = D^\beta_x u^0_1(x)$ for $|\beta| = r$ and $N^{r+1}_0 = 1$ ($u^0_{r+1}(x)$ not yet defined). Then for N^{r+1}_β we obtain the equations

$$PN^{r+1}_\beta(y) = -\sum_{|\alpha|,|\gamma|=r} \sum_{\varkappa=\alpha+\gamma-\beta} C^\varkappa_\alpha D^\varkappa_y a_{\alpha\gamma}(y) \quad \text{for} \quad |\beta| = r+1,$$

$$PN^{r+1}_\beta(y) = -P^{\beta^1} N^r_{\beta_r}(y) \quad \text{for} \ \beta = \beta^1 + \beta_r \ \text{with} \ |\beta^1| = 1, |\beta_r| = r, \tag{94}$$

$$PN^{r+1}_\beta(y) = -\sum_{|\varkappa|=r} D^\varkappa_y a_{\varkappa\beta}(y) \quad \text{for} \quad |\beta| = r.$$

In accordance with (87) for the operator $Q^\beta(y, D_y) = (P^\beta(y, D_y))^*$ with $|\beta| \leqslant r-1$ equalities (89) are satisfied. Consequently, we have the equalities

$$\langle P^\beta N^r_\alpha, M \rangle = \langle N^r_\alpha, Q^\beta M \rangle = 0 \quad \text{for} \quad |\beta| \leqslant r-1.$$

From these equalities and (92) it follows that equations (94) are solvable and have sufficiently smooth solutions if $r+1 < 2r$. Thus, equality (93) is valid due to the choice of $\{\beta_{r+1}\}$ and the fulfillment of equations (94).

Equalities (11) with numbers $r+s$ for $s = 2, \ldots, r-1$ are satisfied in a similar way. Further, equality (11) with number r takes the form

$$\sum_{\{\beta_{2r}\}} [PN^{2r}_{\beta_{2r}}] u^{\beta_{2r}}_{2r} + \sum_{\{\beta_{2r-1}\}} \sum_{|\beta|=1} [P^\beta N^{2r-1}_{\beta_{2r-1}}] D^\beta_x u^{\beta_{2r-1}}_{2r-1}/\beta! + \ldots +$$

$$+\sum_{\{\beta_r\}}\sum_{|\beta|=r}\big[P^\beta N^r_{\beta_r}\big]D^\beta_x u^{\beta_r}_r/\beta! + \ldots + \tag{95}$$

$$+\sum_{|\beta|=2r-1}\big[P^\beta N^1_0\big]D^\beta_x u^0_1/\beta! + \sum_{|\beta|=2r}\big[P^\beta N^0_0\big]D^\beta_x u^0_0/\beta! = a\,f\,.$$

Applying to the equality the average operator $\{\cdot\} = \langle\,\cdot\,\rangle$, defined by (15), we get

$$\sum_{\{\beta_r\}}\sum_{|\beta|=r}\langle P^\beta N^r_{\beta_r}\rangle D^\beta_x u^{\beta_r}_r/\beta! + \sum_{|\beta|=2r}\langle P^\beta N^0_0\rangle D^\beta_x u^0_0/\beta! = \langle\, a\,\rangle f, \tag{96}$$

where equalities (89) and (92) are taken into account. By definition, $\{\beta_r\} = \{\alpha : |\alpha| = r\} \cup \{0\}$, $u^\alpha_r = D^\alpha_x u^0_0(x)$ for $|\alpha| = r$ and $\langle P^\beta N^r_0\rangle = 0$ for $|\beta| = r$ in accordance with (92). Therefore, equality (96) can be rewritten as

$$\sum_{|\alpha|=r}\sum_{|\beta|=r}\langle P^\beta N^r_\alpha\rangle D^{\alpha+\beta}_x u^0_0/\beta! + \sum_{|\beta|=2r}\langle P^\beta N^0_0\rangle D^\beta_x u^0_0/\beta! = \langle\, a\,\rangle f. \tag{97}$$

From (88) it follows that $P^\beta N^0_0 = a_{\alpha\gamma}(y)\beta!$ for $\beta = \alpha+\gamma$ with $|\alpha| = |\gamma| = r$ and $P^\beta N^0_0 = 0$ for other β with $|\beta| = 2r$. In accordance with (87) and (88), the symbol of operator $Q^\beta(y, D_y) = (P^\beta(y, D_y))^*$ for $|\beta| = r$ has the form

$$q^\beta(y,\xi) = \sum_{|\alpha|,|\gamma|=r}\sum_{|\varkappa|\leqslant r} C^\varkappa_\gamma\, C^\beta_{\gamma-\varkappa+\alpha}\,\big(\,D^\varkappa_y\, a_{\gamma\alpha}(y)\,\big)\,\xi^{\gamma-\varkappa+\alpha-\beta}\beta! =$$

$$= \sum_{|\alpha|,|\gamma|=r}\sum_{|\varkappa|\leqslant r-1} C^\varkappa_\gamma\, C^\beta_{\gamma-\varkappa+\alpha}\,\big(\,D^\varkappa_y\, a_{\gamma\alpha}(y)\,\big)\,\xi^{\gamma-\varkappa+\alpha-\beta}\beta! +$$

$$+\sum_{|\varkappa|=r}\big(\,D^\varkappa_y\, a_{\varkappa\beta}(y)\,\big)\,\beta! \equiv q^\beta_1(y,\xi) + q^\beta_2(y),$$

where the sum over $|\varkappa|\leqslant r$ is presented as two sums over $|\varkappa|\leqslant r-1$ and $|\varkappa| = r$ and properties of binomial coefficients defined by (85) are used. The operator $Q^\beta_1 = q^\beta_1(y, D_y)$ corresponding to the symbol $q^\beta_1(y,\xi)$ contains differentiation and hence

$$\langle\, P^\beta N^r_\alpha\,\rangle = \langle\, N^r_\alpha, Q^\beta 1\,\rangle = \langle\, N^r_\alpha, Q^\beta_1 1\,\rangle + \langle\, N^r_\alpha, Q^\beta_2 1\,\rangle$$

$$= \Big\langle\, N^r_\alpha, \sum_{|\varkappa|=r}\big(D^\varkappa_y a_{\varkappa\beta}\big)\,\beta!\,\Big\rangle = \sum_{|\varkappa|=r}\langle\, N^r_\alpha, D^\varkappa_y a_{\varkappa\beta}\,\rangle\beta! = \sum_{|\varkappa|=r}\langle\, a_{\beta\varkappa} D^\varkappa_y N^r_\alpha\,\rangle\beta!$$

for $|\beta| = |\alpha| = r$. Thus, we get the following equation with constant coefficients

$$\sum_{|\alpha|=|\beta|=r} A_{\alpha\beta}\, D^{\alpha+\beta}_x\, u^0_0(x) = \langle\, a\,\rangle\, f(x), \tag{98}$$

for u_0^0 according to (97), where the notation is introduced

$$A_{\alpha\beta} = \left\langle a_{\alpha\beta} + \sum_{|\varkappa|=r} a_{\alpha\varkappa} D_y^{\varkappa} N_\beta^r \right\rangle \quad \text{for} \quad |\alpha| = |\beta| = r$$

and $N_\beta^r = N_\beta^r(y)$ are 1-periodic solutions of equations (91).

It is known [6] that the coefficients $A_{\alpha\beta}$ of equation (98) satisfy ellipticity condition (83). Thus, the homogeneous problem Dirichlet for equation (98) has a uniquely defined smooth solution $u_0^0 \in C^\infty(\overline{\Omega})$ (see, for example, [21] and [50]). Continuing by induction, we get for the functions $u_1^0, \ldots, u_l^0$ problems of the similar form

$$\sum_{|\alpha|=|\beta|=r} A_{\alpha\beta} D_x^{\alpha+\beta} u_j^0 = F_j(f, u_0^0, \ldots, u_{j-1}^0) \quad \text{in} \quad \Omega,$$

$$\partial_\nu^s u_j^0 = 0 \quad \text{on} \quad \partial\Omega \quad \text{for} \quad s = 0, \ldots, r-1 \quad \text{and} \quad j = 0, \ldots, l, \tag{99}$$

where $F_j(f, u_0^0, \ldots, u_{j-1}^0)$ denotes some linear combination of finite order derivatives of $f, u_0^0, \ldots, u_{j-1}^0$ for $j = 0, \ldots, l$. Consequently, problems (99) have uniquely defined smooth solutions $u_0^0, \ldots, u_l^0$. Thus, an asymptotic expansion of the solution to problem (82) has been constructed in the following form

$$\begin{aligned} u_0^l &= u_0^0(x) + \varepsilon\, u_1^0(x) + \ldots + \varepsilon^r u_r^0(x) + \varepsilon^r \sum_{|\beta|=r} N_\beta^r(x/\varepsilon) D_x^\beta u_0^0(x) \\ &+ \varepsilon^{r+1} \sum_{|\beta|=r+1} N_\beta^{r+1}(x/\varepsilon) D_x^\beta u_0^0(x) + \varepsilon^{r+1} \sum_{|\beta|=r} N_\beta^{r+1}(x/\varepsilon)\, D_x^\beta\, u_1^0(x) + \\ &+ \varepsilon^{r+1} \sum_{|\alpha|=1} \sum_{|\beta|=r} N_{\alpha+\beta}^{r+1}(x/\varepsilon) D_x^{\alpha+\beta} u_0^0(x) + \ldots + \varepsilon^{2r} N_0^{2r}(x/\varepsilon) f(x) + \ldots . \end{aligned} \tag{100}$$

This asymptotic expansion satisfies the relations

$$P_\varepsilon\,(u - u_0^l) = O(\varepsilon^{\,l+1-2r}) \quad \text{in} \quad \Omega, \tag{101}$$

$$u - u_0^l = O(\varepsilon^r), \ldots, \partial_\nu^{r-1}(u - u_0^l) = O(\varepsilon) \quad \text{on} \quad \partial\Omega.$$

From these relations the following statement will be deduced.

Theorem 3. *Let u be a solution to problem (82), u_0^0 be a solution to the homogeneous Dirichlet problem for equation (98) and u_0^r is defined by asymptotic equality (100) for $l = r$. Then*

$$\| u - u_0^0 \|_{L^2(\Omega)}^2 \leqslant C\,\varepsilon, \qquad \| u - u_0^r \|_{H_0^r(\Omega)}^2 \leqslant C\,\varepsilon,$$

where the constant C is independent of ε for $0 < \varepsilon \leqslant \varepsilon_0$ with an appropriate ε_0.

PROOF. Let $l = 2r$ in equality (100). In this equality all terms except those containing derivatives of $u_0^0, \ldots, u_r^0$ satisfy the homogeneous Dirichlet boundary conditions on $\partial\Omega$. It is useful to correct such terms as well. To do this, we use a function e^ε that is smooth in Ω and satisfies (79). As already noted, the supports of the functions $e^\varepsilon - 1$ and $\nabla^s e^\varepsilon$ are contained in the boundary strip $\Omega_{2\varepsilon} \subset \Omega$ of width 2ε (see, for example, [7],4.1). Moreover, the measure $|\,\Omega_{2\varepsilon}|$ of the strip can be estimated as follows $|\,\Omega_{2\varepsilon}| \leqslant C\,\varepsilon$ with a constant independent of ε. Thus, for an function $N \in H^r(T^n)$, the boundedness of the domain Ω implies the inequalities

$$\big\|(D_y^\alpha N)(x/\varepsilon)\big\|^2_{L^2(\Omega)} + \varepsilon^{-1}\big\|(D_y^\alpha N)(x/\varepsilon)\big\|^2_{L^2(\Omega_{2\varepsilon})} \leqslant C \quad \text{for} \quad |\alpha| \leqslant r, \tag{102}$$

where the constant C is independent of ε (see, for example, [52]).

Multiplying the terms containing the derivatives of $u_0^0, \ldots, u_r^0$ in asymptotic equality (100) for $l = 2r$ by the function e^ε, we determine the corrected asymptotics for solution to problem (82) by equality

$$v_a = u_0^{2r} + \varepsilon^r (e^\varepsilon - 1)\Big(\sum_{|\beta|=r} N_\beta^r(x/\varepsilon)\, D_x^\beta\, u_0^0(x) \tag{103}$$

$$+\, \varepsilon \sum_{|\alpha|=1}\sum_{|\beta|=r} N_{\alpha+\beta}^{r+1}(x/\varepsilon) D_x^{\alpha+\beta} u_0^0(x) + \ldots + \varepsilon^r \sum_{|\beta|=r} N_\beta^r(x/\varepsilon) D_x^\beta u_r^0(x)\Big).$$

The function v_a satisfies the boundary conditions from (82) exactly and

$$P_\varepsilon\,(\,u - v_a\,) = -\,P_\varepsilon\Big(\varepsilon^r (e^\varepsilon - 1)\Big(\sum_{|\beta|=r} N_\beta^r(x/\varepsilon)\, D_x^\beta\, u_0^0(x) + \ldots \tag{104}$$

$$+\,\varepsilon^r \sum_{|\beta|=r} N_\beta^r(x/\varepsilon) D_x^\beta u_r^0(x)\Big)\Big) + \varepsilon\, r(x,\varepsilon) \equiv -\sum_{|\alpha|=r} D_x^\alpha\, a_\alpha(x,\varepsilon) + \varepsilon\, r(x,\varepsilon),$$

where $|r(x,\varepsilon)| \leqslant C$ due to the relations (101), and we introduced the notation

$$a_\alpha(x,\varepsilon) = \sum_{|\gamma|=r} a_{\alpha\gamma}(x/\varepsilon) D_x^\gamma\Big(\varepsilon^r (e^\varepsilon - 1)\Big(\sum_{|\beta|=r} N_\beta^r(x/\varepsilon)\, D_x^\beta\, u_0^0(x) + \ldots$$

$$+\, \varepsilon^r \sum_{|\beta|=r} N_\beta^r(x/\varepsilon)\, D_x^\beta\, u_r^0(x)\Big)\Big) \quad \text{for} \quad |\alpha| = r.$$

From Leibniz formula for differentiation and properties of the function e^ε it follows that $|a_\alpha(x,\varepsilon)| \leqslant C$ for $|\alpha| = r$ with a constant C independent of

ε. In addition, the functions $a_\alpha(x,\varepsilon)$ are nonzero only in the border strip of width $2\,\varepsilon$ and therefore

$$\| \, a_\alpha(x,\varepsilon) \, \|^2_{L^2(\Omega)} \leqslant C\,\varepsilon \qquad \text{for} \qquad |\alpha| = r, \tag{105}$$

where the constant C is independent of ε in accordance with estimate (102).

Thus, taking into account inequalities (84) and (105), relation (104) and the known properties of the norm of the space $H^{-r}(\Omega)$, we have the estimate

$$\| \, u \, - \, v_a \, \|^2_{H^r_0(\Omega)} \leqslant C\,\varepsilon$$

with a constant C independent of ε. From this estimate and relations (100), (103) for the functions u_0^{2r} and v_a the statements of Theorem 3 follow directly. □

Construction of the asymptotics of the solution to problem (82) for an elliptic divergence equation with lower terms of the following form

$$\sum_{|\alpha|=|\gamma|\leqslant r} D^\alpha_x \left(a_{\alpha\gamma}(x/\varepsilon)\, D^\gamma_x\, u(x)\right) \; = \; a(x/\varepsilon)\, f(x) \quad \text{in} \quad \Omega \tag{106}$$

repeats the construction of the asymptotics for problem (82), if we note that the case of asymptotic coefficients occurs. Indeed, the operator for the problem can be represented in the form $P_\varepsilon = P_0 + \varepsilon^2 P_2 + \cdots + \varepsilon^{2r} P_{2r}$, where

$$P_{2s} u = \varepsilon^{-2r} \sum_{|\alpha|=|\gamma|\leqslant r-s} (\varepsilon D_x)^\alpha \left(a_{\alpha\gamma}(x/\varepsilon)\, (\varepsilon D_x)^\gamma\, u(x)\right) \quad \text{for} \quad s = 0, \ldots, r.$$

The symbol of operator P_{2s} has the form (86) with r replaced by $r - s$. Consequently, equalities (11) with numbers $s = 0, \ldots, 2r-1$ can be satisfied using solvable problems on cells. In turn, applying the average operator to relation (11) with the number $2r$ leads to the homogenized equation

$$\sum_{|\alpha|=|\beta|\leqslant r} A_{\alpha\beta}\, D^{\alpha+\beta}_x\, u^0_0(x) \; = \; \langle\, a \,\rangle\, f(x),$$

where the coefficients $A_{\alpha\beta}$ for $|\alpha| = |\beta| = r$ coincide with the coefficients in (98), and the remaining coefficients are calculated in a similar way. In turn, if for the homogeneous Dirichlet problem for equation (106) the energy estimate (84) is satisfied, then Theorem 3 is also true for such a problem, which can be proven directly.

Energy estimate (84) is satisfied, in particular, for the homogeneous Dirichlet problem for the following equation

$$\sum_{|\alpha|=|\gamma|=r} (-1)^r \partial^\alpha_x \left(a_{\alpha\gamma}(x/\varepsilon)\, \partial^\gamma_x\, u(x)\right) + a_0(x/\varepsilon)\, u(x) = a(x/\varepsilon)\, f(x) \quad \text{in} \quad \Omega, \tag{107}$$

where the function $a_0(y)$ satisfies inequality (26) and therefore is strictly positive. Here D_x is replaced by ∂_x to ensure that the relevant quadratic form is positive definite. In turn, the replacement is not made when the convenient decomposition formula (6) is used. In the case, we have $P_\varepsilon = P_0 + \varepsilon^{2r} P_{2r}$, where $P_{2r} = \varepsilon^{-2r} a_0(x/\varepsilon)$. Thus, repeating the above constructions, we get for u_0^0 homogenized equation (98) to the left side of which the additional term $\langle a_0 \rangle\, u_0^0$ is added.

Note that if one multiply the coefficient $a_0(x/\varepsilon)$ by ε^{-s} for $s = 1, 2, \ldots, 2\,r$ in (107), then the answer will be completely different, namely a zero-order homogenized equation for $u_0^0 = u_0^0(x)$, as was considered for problem (59), or an expansion of Taylor type for $s = 2r$. In particular, this means that the spectrum of order ε^{-s} for $s = 1, 2, \ldots, 2\,r - 1$ has a fairly simple structure.

In the latter case, the perturbations of the operator from (107) were singular through multiplication by negative powers ε. On the other hand, it is not difficult to verify that the homogenization process for equation (107) is stable in terms of coefficients. This observation can be used when considering, for example, equation (107) with coefficients that are approximated by periodic functions. Of course, in this case one can only expect the convergence of the solutions of the Dirichlet problem for equation (107) to the solution of the homogenized problem, if there is no control over the accuracy of the approximation of the coefficients. A similar remark is actual for equation (106), if the energy estimate (84) is satisfied, for example, the corresponding quadratic form is positive definite.

Note also that if one replace the coefficient $a_0(x/\varepsilon)$ with a differential operator of order s, for example, with constant coefficients multiplied by ε^{-s} for $s = 1, \ldots, 2\,r - 1$ in (107), then $u_0^0 = u_0^0(x)$ can be a solution to the equation of order s.

The interested reader may also consider the Direchlet problem for equation (107), in the case where the sum has the form $\sum_{|\alpha|,\,|\gamma| \leqslant r}$. In this case, the relevant operator can be represented in the form $P_\varepsilon = P_0 + \varepsilon\, P_1 + \varepsilon^2 P_2 + \varepsilon^3 P_3 + \cdots + \varepsilon^{2r} P_{2r}$ and the corresponding constructions are more cumbersome. One can also check that Theorem 3 holds under much weaker data regularity conditions, for example, if the coefficients are elements of the space $L^\infty(Y)$.

Elliptic equations with constant coefficients. It is directly verified that if the coefficients of the equation of problem (82) are constant, then these coefficients coincide with the coefficients of the homogenized equation (98). This remark also holds true in a more general situation.

Let $n \geqslant 2$ be given an even integer $m > 0$, bounded domain $\Omega \subset \mathbb{R}^n$ with smooth boundary and a smooth complex-valued function $f \in C_0^\infty(\Omega; \mathbb{C})$. Let us define the function $u = u(x)$ as a solution to the following Dirichlet problem for the m-order partial differential equation with constant coefficients

$$\sum_{|\alpha|\leqslant m} a_\alpha D_x^\alpha\, u(x) = a(x/\varepsilon)\, f(x) \quad \text{in} \quad \Omega, \quad \partial_\nu^s\, u(x) = 0 \quad \text{on} \quad \partial\Omega, \tag{108}$$

where the assumptions for problem (1) are used, in particular, the ellipticity (3). Assumed also that a unique solution to problem (108) exists and meets the inequality

$$\| \, u \, \|_{H_0^m(\Omega)} \leqslant C \, \| \, a \, f \, \|_{L^2(\Omega)} \tag{109}$$

with a constant C independent of ε. Thus, the number $a_{(0,\ldots,0)}$ is not an eigenvalue for problem (108) and the conditions that are sufficient to satisfy this inequality are well known (see, for example, [21] and [51]).

Problem equation (108) can be rewritten as

$$\varepsilon^{-m}\Big(\sum_{|\alpha|=m} a_\alpha(\varepsilon D_x)^\alpha + \varepsilon \sum_{|\alpha|=m-1} a_\alpha(\varepsilon D_x)^\alpha + \ldots + \varepsilon^m a_{(0,\ldots,0)} \Big)\, u = a(x/\varepsilon) f.$$

Therefore, in problem (108) the case of asymptotic coefficients occurs. The operator P_ε of this problem can be represented as $P_\varepsilon = P_0 + \varepsilon\, P_1 + \cdots + \varepsilon^m P_m$, where

$$P_0 = \varepsilon^{-m} \sum_{|\alpha|=m} a_\alpha\, (\varepsilon D_x)^\alpha, \; \ldots, \; P_m = \varepsilon^m\, a_{(0,\ldots,0)}\,.$$

From the ellipticity of the operator P_0 it follows that $M = N = 1$ and therefore $d = 1$. For $|\beta| \leqslant m - 1$ the operator P_0^β necessarily contains differentiation and therefore for such β the equalities $P_0^\beta(D_y)N(y) = 0$ are satisfied. For operators $P_1, \ldots, P_{m-1}$ the following similar equalities

$$P_l^\beta(D_y)\, N(y) = 0 \quad \text{for} \quad |\beta| \leqslant m - 1 - l \quad \text{and} \quad l = 0, \ldots, m-1.$$

also hold. Given this observation, one can take $p = 0$ in (4). Then, to satisfy relations (11) with numbers $1, \ldots, m-1$ it is enough to choose $\{\beta_0\} = \ldots = \{\beta_{m-1}\} = \{0\}$ and $N_0^0 = \ldots = N_0^{m-1} = 1$ ($u_0^0(x), \ldots, u_{m-1}^0(x)$ not yet defined). Thus, relation (11) with the number m has the form

$$\sum_{\{\beta_m\}} \big[PN_{\beta_m}^m(y)\big] u_m^{\beta_m}(x) + \sum_{|\alpha|\leqslant m} a_\alpha D_x^\alpha\, u_0^0(x) = a(y)\, f(x).$$

Applying the average operator $\langle \cdot \rangle$ to this equality and remembering the boundary conditions, we obtain for u_0^0 the following problem

$$\sum_{|\alpha|\leqslant m} a_\alpha D_x^\alpha\, u_0^0(x) = \langle a \rangle f(x) \quad \text{in} \quad \Omega, \quad \partial_\nu^s\, u_0^0(x) = 0 \quad \text{on} \quad \partial\Omega, \tag{110}$$

which is similar to the original problem (108). Continuing the construction, as before, we obtain problems of a similar type for the functions $u_1^0, \ldots, u_l^0$. Consequently, these problems have uniquely defined smooth solutions $u_0^0, \ldots, u_l^0 \in C^\infty(\overline{\Omega})$ (see, for example, [21] and [51]). Thus, an asymptotic expansion of the solution to problem (108) has been constructed in the following form

$$u_0^l = u_0^0 + \ldots + \varepsilon^m\, u_m^0 + \varepsilon^m N_0^m(x/\varepsilon)\, f + \varepsilon^{m+1} \sum_{|\beta| \leqslant m} N_\beta^{m+1}(x/\varepsilon) D_x^\beta f + \ldots, \tag{111}$$

which satisfies the relation $P_\varepsilon\,(\,u - u_0^l\,) = O(\varepsilon^{\,l+1-m})$ in Ω and boundary conditions exactly, since in (111) only the function $f(x)$ and its derivatives are multiplied by oscillating functions. Therefore, we have proven the following statement.

Theorem 4. *Let u be a solution to problem (108), v be a solution to problem (110), and u_0^l defined by equality (111). Then*

$$\|\, u - u_0^0 \,\|_{L^2(\Omega)} \leqslant C\,\varepsilon, \qquad \|\, u - u_0^l \,\|_{H_0^m(\Omega)} \leqslant C\,\varepsilon^{l+1-m},$$

where the constant C is independent of ε for $0 < \varepsilon \leqslant \varepsilon_0$ with an appropriate ε_0.

The presented theorem confirms the naturalness of the formulated general principles for constructing asymptotic expansions. In accordance with (110), to obtain a homogenized problem for an elliptic equation with constant coefficients, one should calculate the average of the right-hand side over rapidly oscillating variables.

Krylov-Bogolyubov-Mitropolsky averaging method. The simplest elliptic operator is the differentiation operator on the circle. Let u be a solution to the problem with the initial condition for this operator in the form

$$u_x' = U(x/\varepsilon, u), \qquad u\big|_{x=0} = \varrho, \tag{112}$$

where ϱ is some constant and $U(y,u)$ is a 1-periodic in y and smooth function.

In accordance with the above recipe for the main term $v = u_0^0(x)$ of the asymptotic solution expansion for problem (112) we have the following problem

$$v_x' = \int_0^1 U(y,v)\,dy, \qquad v\big|_{x=0} = \varrho. \tag{113}$$

More precisely, one should first replace u with u_0^l in (112) and expand $U(y, u_0^l)$ into a Taylor series in the second variable. Then for the main term of this expansion we obtain problem (113), just as for problem (108) problem (110) was obtained.

Let us change the variable $t = x/\varepsilon$ in (112). Then u satisfies the relations

$$u'_t = \varepsilon U(t, u), \qquad u\big|_{x=0} = \varrho. \tag{114}$$

Thus, equalities (113) determine (up to change $t = x/\varepsilon$) the *averaged problem* of the Krylov-Bogolyubov-Mitropolsky method for problem (114). The theorems justifying the transition from problem (114) to problem (113) are well known and form the basis of this method, for example, in accordance with [20], [54].

The interested reader may well consider instead of (112) the generalized problem

$$u'_x + \varepsilon^{-1}A(x/\varepsilon)\, u = U(x/\varepsilon, u), \qquad u\big|_{x=0} = \varrho,$$

where the notation from (112) is used and $A = A(t)$ is a smooth 1-periodic function.

After changing the variables $t = x/\varepsilon$, this problem takes the following form

$$u'_t + A(t)\, u = \varepsilon U(t, u), \qquad u\big|_{x=0} = \varrho. \tag{115}$$

It is known that there is a 1-periodic function $Q = Q(t)$ such that after replacing the unknown function $u = Q\,\hat{u}$ in problem (115), the function $A = A(t)$ can be considered as a constant (for example, according to [55], Floquet theory). Thus, it is possible to indicate a countable set of such A for which the Krylov-Bogolyubov-Mitropolsky method is applicable to problem (115), otherwise one can use asymptotic expansions of Taylor type that were presented in section 3.

Likewise, the interested reader may consider a system of equations of the form (115), where u is a vector function and $A = A(t)$ is a smooth 1-periodic matrix function. Some details on this can be found in [56]. In this case, partly the Krylov-Bogolyubov-Mitropolsky method and partly asymptotic expansions of Taylor type can be applied. Strange, but this approach has not been found in the literature. In particular, the following problem can be reduced to such a system

$$u''_{tt} + B(t)\, u'_t + A(t)\, u = \varepsilon U(t, u), \quad u'_t\big|_{x=0} = \rho, \quad u\big|_{x=0} = \varrho, \tag{116}$$

where the notation from (115) is used, $B = B(t)$ is a smooth 1-periodic function and ρ is some constant. With a particular choice of A and B in (116), this problem is a problem of perturbation of a harmonic oscillator. Thus, the interested reader can establish a complete picture of all possible trajectories of this oscillator.

Respectively, it is also possible to consider the case of problems (115) and (116), when $u = u(t)$ takes values in a Banach space and the operators A and B, for example, are constant with respect to t and have a not very complex spectrum.

Let us recall that a function $\hat{a}(x) = a(\omega_1 x, \dots, \omega_n x)$ for $x \in \mathbb{R}$ is called *quasiperiodic* with the frequency vector $\omega = (\omega_1, \dots, \omega_n) \in \mathbb{R}^n$ if the function $a = a(y_1, \dots, y_n)$ is the smooth and 1-periodic in $y = (y_1, \dots, y_n) \in \mathbb{R}^n$ function. Such a function is almost periodic if $(m, \omega) = m_1\omega_1 + \dots + m_1\omega_1 \neq 0$ for all $m \in \mathbb{Z}^n \setminus \{0\}$ (the exact definition of almost periodicity is given, for example, in [57]). The last condition is called *non-resonance* of the frequency vector ω.

In turn, for a complete description of almost periodic functions it is necessary, in a sense, to consider non-resonant frequency vectors and periodic functions on an infinite-dimensional torus, which is called the *Bohr compact*, according to [57].

The Krylov-Bogolyubov-Mitropolsky method (in the coordinate $t = x/\varepsilon$) was also used for the following multi-frequency problem

$$u'_x = U(\theta_1 + \omega_1 x/\varepsilon, \dots, \theta_n + \omega_n x/\varepsilon, u), \qquad u\big|_{x=0} = \varrho, \tag{117}$$

where the notation from (112) is used, $\omega = (\omega_1, \dots, \omega_n)$ is a non-resonant frequency vector, $\theta = (\theta_1, \dots, \theta_n) \in T^n$ are free parameters and $U(y_1, \dots, y_n, u)$ is a 1-periodic in $y \in \mathbb{R}^n$ and smooth function.

However, when using this method, the following condition is usually necessary

$$\sigma\, |m|^{-\nu} \leqslant |(m, \omega)| = |\, m_1\omega_1 + \dots + m_1\omega_1| \quad \text{for all} \quad m \in \mathbb{Z}^n \setminus \{0\}, \tag{118}$$

where σ is positive and ν is non-negative given constants. The condition is called *strong non-resonant* of the frequency vector and was first introduced by Kolmogorov.

Before considering problem (117), let's explore a simpler equation with smooth $a = a(y)$ and $f = f(x)$ of this form

$$u'_x = a(\theta_1 + \omega_1 x/\varepsilon, \dots, \theta_n + \omega_n x/\varepsilon) f(x) \quad \text{for} \quad x \in \mathbb{R}. \tag{119}$$

Let's also represent the asymptotic solution expansion for the equation in the form

$$u_0^l = N_0(y)\, u_0(x) + \varepsilon\, N_1(y)\, u_1(x) + \dots \quad \text{for} \quad y = \theta + \frac{\omega\, x}{\varepsilon},$$

with smooth functions $N_0(y), N_1(y), \dots$ and $u_0(x), u_1(x), \dots$ as in (4). Therefore, an analogue of expansion (10) is written in the form

$$\varepsilon^{-1}\big(P_\omega N_0(y)\big)u_0 + N_0(y)(u_0)'_x + \big(P_\omega N_1(y)\big)u_1 + \dots = a(y)f(x) \text{ for } y = \theta + \frac{\omega\, x}{\varepsilon},$$

by Leibniz formula, where $P_\omega = \omega_1\partial_1 + \ldots + \omega_n\partial_n$. Hence, equation (14) looks like

$$P_\omega\, w(y) = g(y) \qquad \text{on} \qquad T^n \tag{120}$$

with smooth function $g(y)$. The function can be represented as the Fourier series $g = \sum_{m\in\mathbb{Z}^n} g_m \exp(i\,2\pi\,(m,y))$, where $|\,g_m| = O(|m|^{-\kappa})$ as $|m| \to \infty$ for every fixed positive κ. Hence, $w(y)$ exists if and only if $\langle g\rangle = \int_0^1 \ldots \int_0^1 g(y)\,dy = 0$ and

$$w = \sum_{m\in\mathbb{Z}^n\setminus\{0\}} w_m \exp(i\,2\pi\,(m,y)) = \sum_{m\in\mathbb{Z}^n\setminus\{0\}} g_m\big(i\,2\pi\,(m,\omega)\big)^{-1}\exp(i\,2\pi\,(m,y)),$$

where $|\,w_m| = O(|m|^{-\kappa})$ as $|m| \to \infty$ for every fixed positive κ due to (118).

Thus, the function $w = w(y)$ is smooth, $N = M = 1$ in notation (13) and

$$v'_x = \int_0^1 \ldots \int_0^1 U(y_1,\ldots,y_n,v)\,dy_1\ldots dy_n, \qquad v\big|_{x=0} = \varrho$$

is an averaged problem for problem (117). Remark that this result is uniform with respect to the free parameters θ, since the integral does not depend on them. The justification for this case is discussed, for example, in [54] and [58].

Note that when finding a solution $w(y)$ to (120) one must necessarily divide the Fourier coefficients of the function $g(y)$ by $i\,2\pi\,(m,\omega)$ for $m \in \mathbb{Z}^n \setminus \{0\}$. This necessity is called the *small divisor problem* in the theory of multi-frequency systems of the form (117) and KAM theory, where equations (120) occur. Condition (118) means that such division does not greatly change the growth rate of the Fourier coefficients of the solution. But, if $g(y)$ is not regular, then $w(y)$ may not be defined as a function of a certain class. This phenomenon is known as 'loss of smoothness' affected by the solution operator P_ω^{-1} and is the main culprit why small divisor problems are technically so involved (according to [59]).

Sometimes, such a loss of smoothness can become catastrophic, since there are frequency vectors ω for which condition (118) is not satisfied. This means that a small divisor problem will lead to the fact that the solution to problem (120) will not even be a distribution, since the Fourier series $\sum w_m \exp(i\,2\pi\,(m,y))$ defines a distribution iff there are constants C and c such that $|w_m| \leqslant C\,|m|^c$ for all $m \in \mathbb{Z}^n \setminus \{0\}$. In turn, this means that there are elliptic equations with quasiperiodic and almost periodic coefficients for which a homogenized equation is not defined, in contrast to what is written in some papers and books. Indeed, homogenized coefficients are determined through solutions of cell problems, such as problem (75). But, if the solution is not a distribution, then the coefficients (76) will not be definite.

The above remarks highlight the problem of regularity in homogenization, which was already noted when considering expansions of Taylor type. In fact, for quasi-periodic functions this problem is even more complicated, since such a function $\hat{a}(x) = a(\omega_1 x, \ldots, \omega_n x)$ is defined only if $a = a(y_1, \ldots, y_n)$ is regular. Indeed, this function is a trace of the function a of variables $y_1, \ldots, y_n$ on the straight line $y_1 = \omega_1 x, \ldots, y_n = \omega_n x$ with the direction vector ω and origin at 0. The exact conditions for the existence of traces are well known, for example, in the theory of Sobolev functions. Thus, sufficient regularity and condition (118) are necessary for constructing and justifying the asymptotic solution expansions even for problem (119). This approach is easy to implement if this necessity is met.

Quasiperiodic functions and condition (118) for the case of several variables are defined, for example, in [60], where the homogenization of second-order linear divergent equations with such coefficients was considered. However, there are still papers that ignore condition (118), which generates incorrect homogenization results.

The research of problem (117) in the case when $(m, \omega) = m_1\omega_1 + \cdots + m_1\omega_1 = 0$ for some $m \in \mathbb{Z}^n \setminus \{0\}$ (that is, the vector ω is resonant) is also of interest, for example, in celestial mechanics, in accordance with [54]. In this case, after replacing the coordinates, one can assume that $\omega = (\hat{\omega}, 0) = (\hat{\omega}_1, \ldots, \hat{\omega}_s, 0, \ldots, 0)$, where $\hat{\omega}$ is a non-resonant frequency vector of dimension s. Assuming that condition (118) is satisfied for the frequency vector $\hat{\omega}$ and $\hat{m} \in \mathbb{Z}^s \setminus \{0\}$, it is also possible to obtain and justify the averaged problem for problem (117). Such an averaged problem will be infinite-dimensional with $(n - i)$ parameters, namely $\theta_{i+1}, \ldots, \theta_n$. This corresponds to the existence of $(n - i)$-dimensional invariant tori in problem (117), the perturbation of which is described by the averaged problem. Further details can be found in [61], where the case of the resonance frequency vector ω was considered.

In the last case, for the quasiperiodic function $\hat{a}(x) = a(\theta_1 + \omega_1 x, \ldots, \theta_n + \omega_n x)$ with free parameters $\theta \in T^n$ the following equality

$$\lim_{T\to\infty} \frac{1}{2T} \int_{-T}^{T} \hat{a}(x)\, dx = \int_0^1 \cdots \int_0^1 a(y_1, \ldots, y_s, \theta_{s+1}, \ldots, \theta_n)\, dy_1 \ldots dy_s$$

is valid according to [61]. This explains how invariant tori arise after averaging, since after averaging over time, the dependence on some of the free parameters remains. This dependence determines the function $\breve{a} = \breve{a}(\theta_{s+1}, \ldots, \theta_n)$. If we consider the set of such functions as an algebra, then we can construct an analogue of the Hilbert space over this algebra and, in a sense, reduce problem (117) to problem (112) over this algebra. Further details and a suitable bibliography can be found in [61].

In conclusion, we note that condition (118) is exactly the definition of the *hypoellipticity* of the operator P_ω over the torus T^n, for example, according to [23]. Such operators preserve many properties of elliptic operators, in particular, the smoothness of the generators and the finite-dimensionality of the kernel and cokernel are valid as is known [23]. This makes it possible to apply for such operators the general principles of constructing asymptotic solution expansions, as was noted in [18]. As a result, one can generalize the asymptotic methods presented here to multi-frequency problems of the form (115) and (116) under the assumption that the relevant operators are hypoelliptic and the data are sufficiently smooth.

8. Spectral Asymptotics for the Mathieu Equation

Here the general principles of constructing asymptotic solution expansions will be applied to the spectral problem for the Mathieu equation. This equation is so famous that there is a large book dedicated exactly to this equation [62]. The main attention will be paid to the consideration of the spectrum of order ε^{-2}, since the spectrum of order ε^{-1} is structured simply, as already noted for problem (59), and the spectrum of order ε^0 is quite well studied in the available literature for more general equations with rapidly oscillating coefficients. It is natural to call such a spectrum under consideration high-frequency, since it is formed by sufficiently large eigenvalues of the form $\lambda\,\varepsilon^{-2}$. The constructions presented here can be considered as model ones, although this approach allows for natural generalizations.

We consider the spectral problem for the second-order Mathieu equation with a rapidly oscillating potential and periodic boundary conditions on the segment $[0, 2\pi]$: find the real eigenvalue λ_ε and the nonzero eigenfunction u_ε such that

$$\begin{aligned} &-\varepsilon^2(u_\varepsilon)'' + 2q\,u_\varepsilon \cos\left(\frac{2x}{\varepsilon}\right) = \lambda_\varepsilon u_\varepsilon \quad \text{for} \quad x \in (0, 2\pi), \\ &u_\varepsilon(0) = u_\varepsilon(2\pi), \quad u_\varepsilon'(0) = u_\varepsilon'(2\pi), \end{aligned} \tag{121}$$

where $\varepsilon = (N)^{-1}$ with some positive integer N. The goal is to study the eigenvalues and functions of this problem for sufficiently large N and, accordingly, for sufficiently small ε. The equation of problem (121) depends on a fixed real parameter q and is written in the *form traditionally accepted* in the theory of Mathieu functions defined by the problem under consideration for $\varepsilon = 1$ in accordance with [62], [63]. Note also that after considering problem (121) with periodic boundary conditions, from the resulting asymptotic solutions it will not be difficult to select solutions that satisfy the homogeneous Dirichlet (or Neumann) boundary conditions.

The equation of problem (121) can be divided into ε^2 and thus considered as a spectral problem for the one-dimensional Schrodinger equation with a very large oscillating potential (for sufficiently small ε) on a finite interval. In this equation one can make a change of variable $y = x/\varepsilon$ and thus consider the problem on an expanding interval of the length $2\pi N = 2\pi/\varepsilon$. In solid state physics it is customary to replace the large N by infinity. However, the correctness of such a replacement is not entirely clear, since real crystals for which such models are accepted have a finite length and actual exploration of infinite crystals would be difficult. For example, after constructing the asymptotics solution to problem (121) in [62], [63], examples of bound states were constructed in [66] for an equation with an oscillating potential only on the interval $[0, 2\pi]$. Such bound states decay exponentially outside the interval and have finite energy, which cannot be the case on an infinite interval.

Problems on the asymptotic expansion of eigenvalues and functions for a second-order equation with a small periodic potential, considered on an expanding interval, were first posed and studied in the book by Titchmarsh [67]. In fact, in this book the case of a spectrum of order ε^0 was considered, and a spectrum of order ε^{-1} was considered here for problem (59); these spectra are lost when N is replaced by ∞. In this sense, problems considered on an interval that asymptotically expands are more complex than problems on the entire line, since they depend on a large parameter N. In turn, such problems can be reduced to problems with a rapidly oscillating potential on a finite interval. Moreover, such problems are equivalent and, for a fixed N, have a discrete spectrum, the dependence of which on N is quite complex and will be studied here for the Mathieu equation.

For a fixed ε it is known [68] that there are countable sets of eigenvalues $\lambda_\varepsilon^0, \lambda_\varepsilon^1, \ldots, \lambda_\varepsilon^n, \ldots$ and orthonormal eigenfunctions $u_\varepsilon^0, u_\varepsilon^1, \ldots, u_\varepsilon^n, \ldots$, which are solutions problems (121), and considering the possible multiplicity, the inequalities

$$\lambda_\varepsilon^0 \leq \lambda_\varepsilon^1 \leq \ldots \leq \lambda_\varepsilon^n \leq \ldots$$

are satisfied and $\lambda_\varepsilon^n \to \infty$ for $n \to \infty$. Multiplicity here means the possible coincidence of eigenvalues $\lambda_\varepsilon^n, \lambda_\varepsilon^{n+1}, \ldots$ for different eigenfunctions $u_\varepsilon^n, u_\varepsilon^{n+1}, \ldots$. Thus, taking into account multiplicity means numbering the eigenvalues in accordance with the numbering of different eigenfunctions.

In order to construct asymptotic expansions of the eigenvalues and functions of problem (121), we consider the following spectral problem

$$\begin{gathered} -u'' + 2q\, u \cos(2y) = \lambda\, u \quad \text{for} \quad y \in (0, 2\pi), \\ u(0) = u(2\pi), \quad u'(0) = u'(2\pi), \end{gathered} \tag{122}$$

which coincides with problem (121) for $\varepsilon = 1$. For a fixed $q \neq 0$ it is known [62], [63] that there are countable sets of single eigenvalues $\Lambda = \{\lambda_1^n\}_{n=0}^{\infty}$ and orthonormal functions $u_1^0, \ldots, u_1^n, \ldots$, which are solutions to problem (122) and the inequalities

$$\lambda_1^0 < \lambda_1^1 < \cdots < \lambda_1^n < \ldots$$

are satisfied and $\lambda_1^n \to \infty$ for $n \to \infty$. Such functions, normalized in a relevant way, are usually called special *Mathieu functions*. Of course, problem (122) is exactly the cell problem (14) in the case under consideration.

Chapters in reference books on special functions and entire book are devoted to the study of Mathieu functions [62], [63], where special notations are used. According to the notations, the set of eigenvalues of problem (122) is divided into two sets $\Lambda = \{a_n\}_{n=0}^{\infty} \bigcup \{b_n\}_{n=1}^{\infty}$, and the eigenfunctions for $m = 0, 1, \ldots$ into four classes

$$\begin{aligned} u_1^n &= \sqrt{2}\, ce_{2m}(y) \quad \text{for} \quad a_{2m} \in \{a_n\}_{n=0}^{\infty} \quad \text{and} \quad n = 2m, \\ u_1^n &= \sqrt{2}\, ce_{2m+1}(y) \quad \text{for} \quad a_{2m+1} \in \{a_n\}_{n=0}^{\infty} \quad \text{and} \quad n = 2m+1, \\ u_1^n &= \sqrt{2}\, se_{2m+1}(y) \quad \text{for} \quad b_{2m+1} \in \{b_n\}_{n=1}^{\infty} \quad \text{and} \quad n = 2m+1, \\ u_1^n &= \sqrt{2}\, se_{2m+2}(y) \quad \text{for} \quad b_{2m+2} \in \{b_n\}_{n=1}^{\infty} \quad \text{and} \quad n = 2m+2, \end{aligned} \tag{123}$$

where the root appears from the orthonormalization condition $(2\pi)^{-1}\int_0^{2\pi} (u_1^n)^2 dy = 1$ and the traditionally accepted [62], [63] following conditions

$$ce_n(0) > 0, \; \int_0^{2\pi} ce_n(y)^2\, dy = \pi, \; se'_{n+1}(0) > 0, \; \int_0^{2\pi} se_{n+1}(y)^2\, dy = \pi$$

for Mathieu functions. Moreover, it is known [62] that $ce_n(0) \neq 0$ and $se'_{n+1}(0) \neq 0$ for every fixed q on which $a_n, ce_n(y)$ and $b_{n+1}, se_{n+1}(y)$ depend in accordance with the equation from (122). In addition, the eigenfunctions $ce_n(y)$ and $se_{n+1}(y)$ with different eigenvalues a_n and b_{n+1} are mutually orthogonal for $n = 0, 1, \ldots$.

It is also known [62], [63] that the eigenvalues a_n and b_{n+1} are real and distinct for $n = 0, 1, \ldots$ and satisfy the relations

$$\begin{aligned} a_0 < b_1 < a_1 < b_2 < a_2 < b_3 < a_3 < \ldots \quad &\text{for} \quad q > 0, \\ a_0 < a_1 < b_1 < b_2 < a_2 < a_3 < b_3 < b_4 < \ldots \quad &\text{for} \quad q < 0, \end{aligned} \tag{124}$$

and the value of a_0 is always negative for $q \neq 0$ and $a_0 = 0$ for $q = 0$.

For $n = 0, 1, \ldots$ the Mathieu functions $ce_n(y)$ on the interval $[0, \pi]$ also satisfy the following homogeneous Neumann boundary conditions

$$ce'_n(0) = ce'_n(\pi) = 0, \tag{125}$$

and the functions $se_{n+1}(y)$ satisfy the Dirichlet boundary conditions

$$se_{n+1}(0) = se_{n+1}(\pi) = 0. \tag{126}$$

By definition, for $n = 0, 1, \ldots$ the Mathieu functions $ce_n(y)$ and $se_{n+1}(y)$ are 2π periodic and therefore can be extended along the entire line. Such continuations are also denoted by $ce_n(y)$ and $se_{n+1}(y)$. For $n = 0, 1, \ldots$ the functions $ce_n(y)$ are even, and the functions $se_{n+1}(y)$ are odd with respect to the origin [62], [63].

In addition, for $m = 0, 1, \ldots$ the functions $ce_{2m}(y)$ and $se_{2m+2}(y)$ have period π, and the functions $ce_{2m+1}(y)$ and $se_{2m+1}(y)$ are aperiodic on the interval $[0, \pi]$, satisfying, for example, the boundary conditions

$$ce_{2m+1}(0) = -\,ce_{2m+1}(\pi), \quad ce'_{2m+1}(0) = -\,ce'_{2m+1}(\pi).$$

For $m = 0, 1, \ldots$ it is directly verified that the functions

$$\begin{aligned} u^n_{1\varepsilon} &= \sqrt{2}\, ce_{2m}(x/\varepsilon), \quad u^n_{1\varepsilon} = \sqrt{2}\, ce_{2m+1}(x/\varepsilon), \\ u^n_{1\varepsilon} &= \sqrt{2}\, se_{2m+1}(x/\varepsilon), \quad u^n_{1\varepsilon} = \sqrt{2}\, se_{2m+2}(x/\varepsilon) \end{aligned} \tag{127}$$

are eigenfunctions of problem (121) with eigenvalues a_{2m}, a_{2m+1}, b_{2m+1}, b_{2m+2}, respectively, where it is taken into account, for example, the following equality

$$(2\pi)^{-1}\int_0^{2\pi} (u^n_{1\varepsilon})^2\, dx = (\pi)^{-1}\int_0^{2\pi} (ce_{2m}(x/\varepsilon))^2\, dx = (\pi N)^{-1}\int_0^{2\pi N} (ce_{2m}(y))^2\, dy = 1$$

that guarantees the orthonormality of the functions defined in (127).

Thus, problem (121) under consideration is partially solved. However, functions (127) do not exhaust all eigenfunctions of the problem. Next, approximate eigenfunctions of problem (121) will be constructed, generating additional eigenfunctions of this problem, which branch from functions (127) and are different from these functions. To do this, additional notation and definitions will be needed.

The equation of problem (122) is of the second order and it is always possible to determine two basic solutions $\phi_1(y)$ and $\phi_2(y)$ of this equation. Such solutions are called *fundamental* if the following conditions are met

$$\phi_2(0) = \phi'_1(0) = 0, \qquad \phi'_2(0) = \phi_1(0) = 1. \tag{128}$$

C**-solutions.** For $\lambda = a_n$ in the equation of problem (122) with $n = 0, 1, \ldots$ as the first fundamental solution, taking into account (125), we define

$$\phi_1(y) = \phi^c_{1n}(y) = ce_n(y)(ce_n(0))^{-1}.$$

In this case, following [62], one can find a uniquely defined periodic function $f^c_n(y)$ on the interval $[0, 2\pi]$ such that $f^c_n(0) = f^c_n(2\pi) = 0$ and

$$\phi_2(y) = \phi^c_{2n}(y) = \alpha_n\, y\, \phi^c_{1n}(y) + \alpha_n\, f^c_n(y), \tag{129}$$

where $n = 0, 1, \ldots$ and the constant α_n is chosen from the condition of equalities (128), from which it follows that $\alpha_n = (1 + (f_n^c)'_y(0))^{-1}$.

The eigenvalue $\lambda = a_n$ for problem (122) for $n = 0, 1, \ldots$ is single and the function $\phi_{2n}^c(y)$ cannot be periodic with period 2π for $q \neq 0$ (as noted in [62]). For every fixed and nonzero q, it follows from equalities (128) and (129) that

$$\phi_{2n}^c(2\pi) = \alpha_n \, 2\pi$$

and therefore $\alpha_n \neq 0$ for $n = 0, 1, \ldots$. For the fundamental solution ϕ_{2n}^c traditionally [62] the following representation

$$\phi_2(y) = \phi_{2n}^c(y) = C_n \, y \, ce_n(y) + f_n(y)$$

is also used, where the constant $C_n = \alpha_n \, (ce_n(0))^{-1}$ and the periodic function $f_n = \alpha_n f_n^c(y)$ are uniquely determined by conditions (128) and representation (129). Note that the function $f_n = f_n(y)$ has a special notation and is sometimes called the third Mathieu function, while the first two are $ce_n = ce_n(y)$ and $se_n = se_n(y)$. As it turns out later, the function $f_n(y)$ is a solution to a cell problem for problem (121).

For small and non-zero q it is known that the constant C_n for $n = 0, 1, \ldots$ is a multiple of q^n in accordance with [62]. Therefore, the constant α_n for $n = 0, 1, \ldots$ from representation (129) is always positive for even n and has the sign of q for odd n and non-zero q due to the theorem on continuous dependence of solutions of ordinary differential equations on parameters [55].

***S*-solutions.** For $\lambda = b_n$ in the equation of problem (122) with $n = 1, 2, \ldots$ as the second fundamental solution, taking into account (126), we define

$$\phi_2(y) = \phi_{2n}^s(y) = se_n(y)(se'_n(0))^{-1}.$$

In the case of such a choice, following [62], one can find such periodic function $g_n^s(y)$ on the interval $[0, 2\pi]$ such that $(g_n^s)'_y(0) = (g_n^s)'_y(2\pi) = 0$ and

$$\phi_1(y) = \phi_{1n}^s(y) = \beta_n \, y \, \phi_{2n}^s(y) + \beta_n \, g_n^s(y), \tag{130}$$

where $n = 1, 2, \ldots$ and the constant β_n is chosen from the condition of fulfillment of equations (128), from which it follows that $\beta_n = g_n^s(0)^{-1}$.

The eigenvalue $\lambda = b_n$ for problem (122) with $n = 1, 2, \ldots$ is single and the function $\phi_{1n}^s(y)$ cannot be periodic with period 2π for $q \neq 0$ (as noted in [62]). For every fixed and nonzero q, it follows from equalities (128) and (129) that

$$(\phi_{1n}^s)'_y(2\pi) = \beta_n \, 2\pi$$

and therefore $\beta_n \neq 0$ for $n = 1, 2, \ldots$. For the fundamental solution ϕ_{1n}^s the representation is also traditionally used

$$\phi_1(y) = \phi_{1n}^s(y) = S_n \, y \, se_n(x) + g_n(y),$$

where the constant S_n for $n = 1, 2, \dots$ is a multiple of q^n for small and non-zero q in accordance with [62]. Thus, the constant β_n for $n = 1, 2, \dots$ from (130) is always positive for even n and has the sign of q for odd n and non-zero q when choosing $\lambda = b_n$ in the equation of problem (122). Again, as will be clarified later, the third Mathieu function $g_n(y)$ is a solution to the cell problem for problem (121).

Asymptotic expansions for the spectral problem. Following the general principles, we will define the expansion of the eigenfunctions u_ε of problem (121) as an asymptotic sum of the following form

$$u_\varepsilon^a(x) = N_0(x/\varepsilon)\, v_0(x) + \varepsilon\, N_1(x/\varepsilon)\, v_1(x) + \varepsilon^2 N_2(x/\varepsilon)\, v_2(x), \tag{131}$$

where, as before, the unknown periodic functions $v_0(x)$, $v_1(x)$, $v_2(x)$ and $N_0(y)$, $N_1(y)$, $N_2(y)$ with period 2π will be found from the condition of approximate fulfillment of the conditions of problem (121) for small ε.

Let's represent the expansion of the eigenvalues λ_ε of problem (121) in the form

$$\lambda_\varepsilon^a = \lambda_0 + \varepsilon\lambda_1 + \varepsilon^2\lambda_2 \tag{132}$$

with unknown values λ_0, λ_1, λ_2 determined during the construction process. Thus, the case of asymptotic coefficients will be realized obviously, so as to find the the asymptotic expansions of the eigensolutions of the spectral problem (121). Note that if problem (121) had the right-hand side $a(x/\varepsilon)f(x)$ and λ_ε was fixed, the answer would be different, as for example for (107).

Substituting expansions (131) and (132) instead of solutions in (121), we have

$$L^y N_0(y)v_0 + \varepsilon\, L^y N_1(y)v_1 + \varepsilon^2 L^y N_2(y)v_2 - \lambda_0 N_0(y)v_0 - \lambda_0\, \varepsilon\, N_1(y)v_1$$
$$-\lambda_0\varepsilon^2 N_2(y)v_2 - \lambda_1\varepsilon N_0(y)v_0 - \lambda_1\varepsilon^2 N_1(y)v_1 - \lambda_1\varepsilon^3 N_2(y)v_2 - \lambda_2\varepsilon^2 N_0(y)v_0$$
$$-\lambda_2\varepsilon^3 N_1(y)v_1 - 2\,\varepsilon\, N_0'(y)v_0'(x) - 2\,\varepsilon^2 N_1'(y)v_1'(x) - 2\,\varepsilon^3 N_2'(y)v_2'(x)$$
$$-\varepsilon^2 N_0(y)v_0''(x) - \varepsilon^3 N_1(y)v_1''(x) - \varepsilon^4 N_2(y)v_2''(x) - \lambda_2\varepsilon^4 N_2(y)v_2 = 0$$

for $y = x/\varepsilon$, where the usual Leibniz formula has been used and the notation

$$L^y N_0(y) = -(N_0(y))'' + 2\, q\cos(2y) N_0(y).$$

Let us equate the coefficients at the same powers of ε in the resulting relation to determine the unknown functions and values from (131) and (132). For ε^0, ε^1 and ε^2, respectively, we obtain the following equations

$$L^y N_0(y)v_0(x) - \lambda_0 N_0(y)v_0(x) = 0, \tag{133}$$

$$(L^y N_1(y) - \lambda_0 N_1(y))\, v_1(x) = 2N_0'(y)v_0'(x) + \lambda_1 N_0(y)v_0(x), \tag{134}$$

$$(L^y N_2(y) - \lambda_0 N_2(y))\, v_2 = 2N_1'(y)v_1' + N_0(y)v_0'' + \lambda_1 N_1(y)v_1 + \lambda_2 N_0(y)v_0. \tag{135}$$

C-asymptotics. Let us choose $\lambda_0 = a_n$ for $n = 1, 2, \ldots$. Then for arbitrary $v_0(x)$ the first relation (133) will be satisfied if we define

$$N_0(y) = N_n^c(y) = \sqrt{2}\, ce_n(y) = \sigma_n\, \phi_{1n}^c(y), \tag{136}$$

where $\sigma_n = \sqrt{2}\, ce_n(0)$ is chosen by the normalization $(2\pi)^{-1}\int_0^{2\pi}(N_n^c)^2 dy = 1$.

We further fix the choice of $\lambda_0 = a_n$ for $n = 0, 1, \ldots$ and consider relation (134) as an equation for $N_1(y)$ for every $x \in [0, 2\pi]$. In accordance with (15), such an equation has a solution if and only if the right-hand side is orthogonal to $N_n^c(y)$. The last condition is written in the form

$$v_0'\,(2\pi)^{-1}\int_0^{2\pi} 2\,(N_n^c)'(N_n^c)\, dy = -\lambda_1\, v_0\, (2\pi)^{-1}\int_0^{2\pi} (N_n^c)^2\, dy.$$

Integrating periodic functions by parts, for the left side we have

$$\int_0^{2\pi} (N_n^c)'(N_n^c)\, dy = -\int_0^{2\pi} (N_n^c)(N_n^c)'\, dy,$$

therefore this integral is equal to zero. Thus, equation (134) can be solvable for an arbitrary v_0 only if $\lambda_1 = 0$.

It is natural to choose $v_1(x) = v_0'(x)$ in (134). In this case, this relation holds for $n = 0, 1, \ldots$, if we define $N_1(y) = N_{1n}^c(y)$ as a 2π-periodic solution of the equation

$$L^y N_{1n}^c(y) - a_n N_{1n}^c(y) = 2\,(N_n^c(y))'. \tag{137}$$

This equation is solvable due to the last equality for the integrals, and one can try to find a solution to this equation by the method of varying constants. However, it is known [62] and can be directly verified that the periodic function

$$f_n^c(y) = \alpha_n^{-1}\phi_{2n}^c(y) - y\,\phi_{1n}^c(y)$$

for $n = 0, 1, \ldots$, given by equality (129), is a particular solution to the equation

$$L^y f_n^c(y) - a_n f_n^c(y) = 2\,(\phi_{1n}^c(y))'.$$

Comparing the equation with the previous one and using (136), we conclude that the general 2π-periodic solution to equation (137) is determined by the equality

$$N_1(y) = N_{1n}^c(y) = \sigma_n\left(\alpha_n^{-1}\phi_{2n}^c(y) - y\,\phi_{1n}^c(y)\right) + \gamma_n\, N_n^c(y), \tag{138}$$

where $\sigma_n = \sqrt{2}\, ce_n(0)$ and the constant γ_n is chosen from the orthogonality condition $(2\pi)^{-1}\int_0^{2\pi} N_{1n}^c(y) N_n^c(y)\, dy = 0$, which will be useful in the future.

Let us now consider relation (135) as an equation for $N_2(y)$. This equation has a solution if the right-hand side is orthogonal to $N_n^c(y)$. The last orthogonality condition for $n = 0, 1, \dots$ can be written as

$$\lambda_2\, v_0 + \left(2\,(2\pi)^{-1} \int_0^{2\pi} (N_{1n}^c(y))'\, N_n^c(y)\, dy \right) v_0'' + v_0'' = 0, \tag{139}$$

since $v_1(x) = v_0'(x)$ and $(2\pi)^{-1}\int_0^{2\pi} (N_n^c(y))^2 dy = 1$. Let us introduce the notation

$$I_n = \pi^{-1} \int_0^{2\pi} (N_{1n}^c(y))'\, N_n^c(y)\, dy = -\pi^{-1} \int_0^{2\pi} N_{1n}^c(y)\, (N_n^c(y))'\, dy$$
$$= -\sigma_n^2\, \alpha_n^{-1}\, \pi^{-1} \left(\int_0^{2\pi} \phi_{2n}^c(y)\, (\phi_{1n}^c(y))'\, dy - \alpha_n \int_0^{2\pi} y\, \phi_{1n}^c(y)\, (\phi_{1n}^c(y))'\, dy \right),$$

where we used (136), (138) and integration by parts for $n = 0, 1, \dots$.

For fundamental solutions, the Wronski identity

$$\phi_1(y)\phi_2'(y) - \phi_1'(y)\phi_2(y) = 1$$

is satisfied according to [62] and $(\phi_2\phi_1)' = \phi_1'\phi_2 + \phi_1\phi_2'$ by Leibniz formula. Hence $2\,\phi_2(y)\phi_1'(y) = (\phi_2(y)\phi_1(y))' - 1$. Taking into account (128), for $n = 0, 1, \dots$ we get

$$I_n = \sigma_n^2\, \alpha_n^{-1}\, (2\pi)^{-1} \left(\int_0^{2\pi} \left(1 - (\phi_{2n}^c \phi_{1n}^c)'\right) dy + \alpha_n \int_0^{2\pi} y\, \left((\phi_{1n}^c)^2\right)'\, dy \right) =$$
$$= \sigma_n^2\, \alpha_n^{-1} (2\pi)^{-1} \left((y - \phi_{2n}^c(y)\phi_{1n}^c(y)) \Big|_0^{2\pi} + \alpha_n\, y\, (\phi_{1n}^c(y))^2 \Big|_0^{2\pi} - \alpha_n \int_0^{2\pi} (\phi_{1n}^c)^2 dy \right) =$$
$$= \sigma_n^2\, \alpha_n^{-1} \left((1 - \alpha_n) + \alpha_n - \alpha_n\, \sigma_n^{-2} \right) = \sigma_n^2\, \alpha_n^{-1} - 1.$$

Consequently, for every fixed n, relation (139) is a spectral problem for determining λ_2 and 2π-periodic function $v_0(x)$ of the following form

$$-\left(\sigma_n^2\, \alpha_n^{-1}\right) v_0''(x) = \lambda_2\, v_0(x) \quad \text{for} \quad x \in (0, 2\pi). \tag{140}$$

It is known and directly verified that the function e^{ikx} for a nonzero integer k defines a 2π-periodic eigenfunction for problem (140) with an eigenvalue $k^2\, (\sigma_n^2\, \alpha_n^{-1})$ for every fixed n. It is convenient to renumber such eigenfunctions and values (taking into account multiplicity) as follows

$$v_0^1 = e^{ix},\ v_0^2 = e^{-ix},\ v_0^3 = e^{i2x},\ v_0^4 = e^{-i2x},\ v_0^5 = e^{i3x},\ v_0^6 = e^{-i3x}, \dots,$$

$$\lambda_{2n}^{1c} = \rho_n, \; \lambda_{2n}^{2c} = \rho_n, \; \lambda_{2n}^{3c} = 2^2\rho_n, \; \lambda_{2n}^{4c} = 2^2\rho_n, \; \lambda_{2n}^{5c} = 3^2\rho_n, \; \lambda_{2n}^{6c} = 3^2\rho_n, \ldots,$$

where the notation $\rho_n = \sigma_n^2 \, \alpha_n^{-1}$ is introduced for every fixed n.

Thus, for every fixed $n = 0, 1, \ldots$ there are given

$$v_0^l = e^{ikx}, \qquad \lambda_{2n}^{lc} = ((l+1)/2)^2 \left(\sigma_n^2 \, \alpha_n^{-1}\right) \qquad \text{for } \; l = 2k-1,$$
$$v_0^l = e^{-ikx}, \quad \lambda_{2n}^{lc} = (l/2)^2 \left(\sigma_n^2 \, \alpha_n^{-1}\right) \quad \text{for } \; l = 2k \; \text{ when } \; k = 1, 2, \ldots \tag{141}$$

as eigenfunctions and numbers for the homogenized equation (140) with periodic boundary conditions. The eigenfunctions can also be represented as linear combinations of $\cos(kx)$ and $\sin(kx)$ for $k = 1, 2, \ldots$, if real-valued solutions are interesting.

Let us define $v_2 = v_0''(x)$, fix some l and v_0^l and take into account the homogenized equation (140) in (135). Then relation (135) will be satisfied if $N_2(y) = N_{2n}^c(y)$ is defined as a periodic solution of the equation

$$L^y N_{2n}^c(y) - a_n N_{2n}^c(y) = 2(N_{1n}^c(y))' + \left(1 - \sigma_n^2 \, \alpha_n^{-1}\right) N_n^c(y),$$

which is solvable due to the above calculations, and one can find a solution to this equation such that $(2\pi)^{-1}\int_0^{2\pi} N_{2n}^c(y) N_n^c(y) \, dy = 0$.

Thus, for non-negative integer l and n, approximate solutions to problem (121) have been determined, which can be represented in the following form

$$\lambda_{ac}^{ln} = a_n + \varepsilon^2 \lambda_{2n}^{lc},$$
$$u_{ac}^{ln} = N_n^c\left(\frac{x}{\varepsilon}\right) v_0^l(x) + \varepsilon \, N_{1n}^c\left(\frac{x}{\varepsilon}\right) \left(v_0^l(x)\right)' + \varepsilon^2 N_{2n}^c\left(\frac{x}{\varepsilon}\right) \left(v_0^l(x)\right)''. \tag{142}$$

Remark that these approximate solutions are also consistent for $l = 0$, since in this case $v_0^0 = 1$ and $\lambda_2^0 = 0$ are a solution to problem (140), and equalities (142) define exact solutions to the problem (121) in accordance with (127).

***S*-asymptotics.** Let us choose $\lambda_0 = b_n$ for $n = 1, 2, \ldots$. Then for arbitrary $v_0(x)$ relation (133) will also be satisfied if we define

$$N_0(y) = N_n^s(y) = \sqrt{2}\, se_n(y) = \kappa_n \, \phi_{2n}^s(y), \tag{143}$$

where $\kappa_n = \sqrt{2}\, se_n'(0)$ is chosen by the normalization $(2\pi)^{-1}\int_0^{2\pi} (N_n^s)^2 dy = 1$.

Let us further fix the choice of $\lambda_0 = b_n$ for $n = 1, 2, \ldots$ and consider relation (134) as an equation for $N_1(y)$ for every $x \in [0, 2\pi]$. As previously verified that this equation can be solvable for an arbitrary v_0 only if $\lambda_1 =$

0. Next, choosing $v_1(x) = v_0'(x)$, we conclude that relation (134) will be satisfied for $n = 1, 2, \ldots$, if we define $N_1(y) = N_{1n}^s(y)$ as a 2π-periodic solution of the equation

$$L^y N_{1n}^s(y) - b_n N_{1n}^s(y) = 2\,(N_n^s(y))'. \tag{144}$$

This equation is solvable, which is verified as for equation (137), and the general 2π-periodic solution of equation (144) is determined by the equality

$$N_1(y) = N_{1n}^s(y) = \kappa_n \left(\beta_n^{-1}\phi_{1n}^s(y) - y\,\phi_{2n}^s(y)\right) + \varkappa_n\, N_n^s(y), \tag{145}$$

where $\kappa_n = \sqrt{2}\, se_n'(0)$ and the constant $\varkappa_n$ is chosen from the orthogonality condition $(2\pi)^{-1}\int_0^{2\pi} N_{1n}^s(y) N_n^s(y)\, dy = 0$, useful in the future. In comparison with (138), it is important here that the indices 1 and 2 have swapped places.

Consequently, considering relation (135) as an equation for $N_2(y)$, we obtain for $n = 1, 2, \ldots$ the following solvability condition

$$\lambda_2\, v_0 + J_n\, v_0'' + v_0'' = 0, \tag{146}$$

where

$$J_n = \pi^{-1}\int_0^{2\pi} (N_{1n}^s(y))'\, N_n^s(y)\, dy = -\pi^{-1}\int_0^{2\pi} N_{1n}^s(y)\,(N_n^s(y))'\, dy =$$

$$= -\kappa_n^2\, \beta_n^{-1}\, \pi^{-1}\left(\int_0^{2\pi} \phi_{1n}^s(y)\,(\phi_{2n}^s(y))'\, dy - \beta_n \int_0^{2\pi} y\,\phi_{2n}^s(y)\,(\phi_{2n}^s(y))'\, dy\right).$$

Here we also use (143), (145) and integration by parts.

From Wronski identity and Leibniz formula for fundamental solutions we obtain

$$2\,\phi_1(y)\phi_2'(y) = (\phi_2(y)\phi_1(y))' + 1.$$

What is important here is the change in sign compared to the previous case, since the indices 1 and 2 are not symmetric in Wronski identity. Next, taking into account (126), (128) and (130), we get $\phi_{1n}^s(0) = \phi_{1n}^s(2\pi) = 1$, $\phi_{2n}^s(0) = \phi_{2n}^s(2\pi) = 0$ for $n = 1, 2, \ldots$. Consequently, we conclude that

$$J_n = \kappa_n^2\, \beta_n^{-1}\,(2\pi)^{-1}\left(-\int_0^{2\pi} \left(1 + (\phi_{2n}^s\phi_{1n}^s)'\right) dy + \beta_n \int_0^{2\pi} y\, \left((\phi_{2n}^s)^2\right)'\, dy\right) =$$

$$= \kappa_n^2\, \beta_n^{-1}(2\pi)^{-1}\left(-\left(y + \phi_{2n}^s\phi_{1n}^s\right)\Big|_0^{2\pi} + \beta_n\, y\,(\phi_{2n}^s)^2\Big|_0^{2\pi} - \beta_n\int_0^{2\pi} (\phi_{2n}^s)^2 dy\right) =$$

$$= \kappa_n^2\, \beta_n^{-1}\left(-1 - \beta_n\, \kappa_n^{-2}\right) = -\,\kappa_n^2\, \beta_n^{-1} - 1.$$

Thus, for every fixed n, relation (146) is a spectral problem for determining λ_2 and 2π-periodic function $v_0(x)$ of the following form

$$\left(\kappa_n^2\,\beta_n^{-1}\right) v_0''(x) = \lambda_2\, v_0(x) \quad \text{for} \quad x \in (0, 2\pi). \tag{147}$$

Therefore, for every fixed $n = 0, 1, \ldots$ there are given

$$v_0^l = e^{i\,k\,x},\ \lambda_{2n}^{ls} = ((l+1)/2)^2\left(-\kappa_n^2\,\beta_n^{-1}\right) \text{ for} l = 2k-1,$$
$$v_0^l = e^{-\,i\,k\,x},\ \lambda_{2n}^{ls} = (l/2)^2\left(-\kappa_n^2\,\beta_n^{-1}\right) \text{ for} l = 2k \,\text{when} k = 1, 2, \ldots \tag{148}$$

as eigenfunctions and values for equation (147) with periodic boundary conditions.

Let us define $v_2 = v_0''(x)$, fix some l and v_0^l and take into account the homogenized equation (147) in (135). Then relation (135) will be satisfied if $N_2(y) = N_{2n}^s(y)$ is defined as a periodic solution of the equation

$$L^y N_{2n}^s(y) - b_n N_{2n}^s(y) = 2(N_{1n}^s(y))' + \left(1 + \kappa_n^2\,\beta_n^{-1}\right) N_n^s(y),$$

which is solvable due to the above calculations, and one can find a solution to this equation such that $(2\pi)^{-1}\int_0^{2\pi} N_{2n}^s(y) N_n^s(y)\, dy = 0.$

Thus, for non-negative integer l and n, approximate solutions to problem (121) have been determined, which can be represented in the following form

$$\lambda_{as}^{ln} = b_n + \varepsilon^2 \lambda_{2n}^{ls},$$
$$u_{as}^{ln} = N_n^s\left(\frac{x}{\varepsilon}\right) v_0^l(x) + \varepsilon\, N_{1n}^s\left(\frac{x}{\varepsilon}\right)\left(v_0^l(x)\right)' + \varepsilon^2 N_{2n}^s\left(\frac{x}{\varepsilon}\right)\left(v_0^l(x)\right)''. \tag{149}$$

A significant difference between expansions (142) and (149) is the difference in signs in the definitions of eigenvalues (141) and (148). As noted earlier, for example, for positive q in problem (121), the values of $\sigma_n^2\,\alpha_n^{-1}$ in (141) are always positive, and the values of $-\kappa_n^2\,\beta_n^{-1}$ in (148) are always negative. This means that the eigenvalues λ_{ac}^{ln} from (142) are strictly increasing, and the eigenvalues $\lambda_{as}^{l(n+1)}$ from (149) are strictly decreasing with increasing number l for fixed n and positive q, which is fully illustrated in **Fig. 2.** Taking into account the comments on representations (129) and (130), one can also present a similar figure for negative q, where relations (124) should also be taken into account.

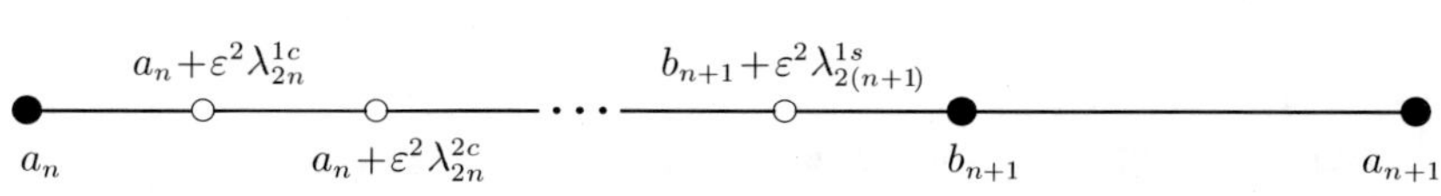

Figure 2. The eigenvalues λ_{ac}^{ln} and $\lambda_{as}^{l(n+1)}$ for $n, l = 0, 1, \ldots$ and $q > 0$.

This representation of approximate eigenvalues means that there will always be intervals on the spectral line that are free from these eigenvalues. Such intervals are called *gaps*. The appearance of gaps is well known for equations with periodic potentials on the entire line [4]. However, for equations on the entire line, the spectral intervals adjacent to the gaps are filled with points of the continuous spectrum. Then, as in the case considered here, such intervals are filled with eigenvalues, which nevertheless tend to fill the relevant interval $[a_n, b_{n+1}]$ when ε decreases.

Note that the coefficients of homognized problems (140) and (147) are defined by the eigenvalues of the monodromy matrix for equation in (121). Thus, the interested reader may well be able to construct similar asymptotic solution expansions for a problem similar to (121) with a regular $2\,\pi$-periodic potential. Indeed, in the above constructions only Wronski identity and Leibniz formula were used, while the monodromy matrix structure is well known for equations with a regular potential.

Asymptotic expansions (142) and (149) are actually modulation (superposition) of standard harmonics $e^{i\,k\,x}$, $e^{-i\,k\,x}$ and rapidly oscillating Mathieu functions with period $2\pi\varepsilon$. However, the expansions are not complete, since it is necessary to additionally prove their consistency on intersection intervals, which requires additional constructions. Nevertheless, for the initial (starting with a_0) expansions (142) established jointly with M. Basileva in [64], the following statement was proven.

Theorem 5. *For eigenvalues* λ^l_ε *and eigenfunctions* u^l_ε *of problem (121) there is a constant* C *independent of* ε *and* l *such that*

$$|\,\lambda^l_\varepsilon - \left(a_0 + \varepsilon^2 \lambda^{lc}_{20}\right)| \le C\,(\varepsilon\,l)^3, \qquad \|\,u^l_\varepsilon - v^l_\varepsilon\,\|_{L^2(0,2\pi)} \le C\,\varepsilon\,l^2$$

for $l^2 \ll \varepsilon^{-1}$ *and* $0 < \varepsilon \le \varepsilon_0$ *for some* ε_0, *where* λ^{lc}_{20} *and* v^l_ε *are defined by the eigenvalues and functions of homogenized problem (140) and* $\mathrm{ce}_0(x/\varepsilon)$.

What is meant here, for example, is that $l^2 = \varepsilon^{-1+\nu}$ with some positive ν, else the given accuracy estimates may not have an asymptotic meaning. In addition, in this theorem, the functions v^0_ε, $v^1_\varepsilon, \ldots$ are constructed sequentially by orthogonalizing the functions from (142), since initially the functions may not be orthonormal.

Mathieu functions of real fractional order. Let $q > 0$ in (121) and (122). Along with special Mathieu functions, it is also customary to define such functions

$$ce_{n+\beta} = ce_{n+\beta}(y), \quad se_{n+\beta} = se_{n+\beta}(y) \qquad \text{for} \quad 0 < \beta < 1$$

which are the eigenfunctions of the equation from (122) with an eigenvalue $a_{n+\beta}$ from the interval (a_n, b_{n+1}) without taking into account the boundary conditions. Here $n = 0, 1, \ldots$ and the notation for a_n and b_{n+1} from (123) is used and it is helpful to keep in mind Fig. 2. It is known [62], that the following limit relations

$$ce_{n+\beta}(y) \to ce_n(y), \quad se_{n+\beta}(y) \to 0 \qquad \text{for} \quad \beta \to 0,$$

$$ce_{n+\beta}(y) \to 0, \quad se_{n+\beta}(y) \to se_{n+1}(y) \qquad \text{for} \quad \beta \to 1$$

are satisfied uniformly with respect to y in a fixed compact interval for $n = 0, 1, \dots$.

Let's fix a fairly large simple integer s. It is known [62] that the functions

$$ce_{n+\beta}(y), \quad se_{n+\beta}(y) \qquad \text{for} \quad \beta = 1\, s^{-1}, 2\, s^{-1}, \dots, (s-1)\, s^{-1}$$

are periodic with exact period $s2\pi$. The period may be shorter if the s is not simple.

Returning to problem (121), we fix the large $N = \varepsilon^{-1}$. Therefore, the functions

$$ce_{n+\beta}(x/\varepsilon),\ se_{n+\beta}(x/\varepsilon) \text{ for } \beta = 1\,\varepsilon, 2\,\varepsilon, \dots, (N-1)\,\varepsilon \text{ and } n = 0, 1, \dots$$

are eigenfunctions of problem (121) with the eigenvalue $a_{n+\beta} \in (a_n, b_{n+1})$.

In turn, the functions $ce_{n+\beta}$ and $se_{n+\beta}$ are obtained in [62] as the half-sum and half-difference of Floquet solutions (or Bloch solutions in physical terminology). Consequently, returning to such solutions, we have

$$ce_{n+\beta}(y) + se_{n+\beta}(y) = e^{i\beta x} p_{n+\beta}(y),\ ce_{n+\beta}(y) - se_{n+\beta}(y) = e^{-i\beta x} p_{n+\beta}(y)$$

for $n = 0, 1, \dots$ and $0 < \beta < 1$, where $p_{n+\beta}(y)$ is periodic with period 2π.

Thus, returning to problem (121) again, one can conclude that

$$e^{ix} p_{n+\varepsilon}(x/\varepsilon),\ e^{-ix} p_{n+\varepsilon}(x/\varepsilon),\ \dots,\ e^{-i(N-1)x} p_{n+(N-1)\varepsilon}(x/\varepsilon) \tag{150}$$

are exactly all eigenfunctions of problem (121) with the eigenvalue $a_{n+\beta} \in (a_n, b_{n+1})$ for $\beta = 1\,\varepsilon, 2\,\varepsilon, \dots, (N-1)\,\varepsilon$ and $n = 0, 1, \dots$.

Let's consider the functions for $n = 0$, add to them $ce_0(x/\varepsilon), se_1(x/\varepsilon)$ from (127) and orthonormalize, starting with $ce_0(x/\varepsilon)$ and then with $se_1(x/\varepsilon)$. Since a finite number of functions are considered, such an orthonormalization can be agreed upon and considered to be coincident. Thus, from Theorem 5 it follows that expansions (142) and (149) can be consistent through orthonormalization and determine exactly the $2N$ of eigenfunctions with eigenvalues from the segment $[a_0, b_1]$. This was first noticed by A. Krylova for problems on networks [69]. Naturally, these arguments can be repeated for $n = 1, 2, \dots$ and in the case of $q < 0$.

This approach seems preferable compared to representation (150), since only 2π-periodicity is known about the function $p_{n+\beta}(x/\varepsilon)$, while much more is known about the functions from (142) and (149), for example, boundary conditions (125) and (126) (and if necessary, one can replace the exponentials with sines and cosines). In particular, one can find graphs of spectral segments $[a_n, b_{n+1}]$ and additional gap intervals (b_{n+1}, a_{n+1}) depending on the parameter value g in the equation from (121) with $n = 0, 1, \ldots,$ for example, according to [62], [63]. Initially, the spectral segments are narrow (and composed of negative eigenvalues) and the gap intervals are wide, and vice versa for increasing n (with all positive eigenvalues) and a fixed $g > 0$ (for $g < 0$ one should take into account (124)). In particular, the relation $a_r - b_r = O(\, g^r\, r^{1-r})$ is valid for $r \to \infty$ in accordance with [63], 20.2.26.

The above observations can be used to estimate the number of gaps for the following 2π-periodic problem for the d-dimensional Schrodinger operator

$$-\,\varepsilon^2 \Delta\, u_\varepsilon + 2q\, u_\varepsilon (\cos\,(2x_1/\varepsilon) + \ldots + \cos\,(2x_n/\varepsilon)) = \lambda_\varepsilon u_\varepsilon \quad \text{for} \quad x \in (0, 2\pi)^d,$$

which is naturally related to problem (121). Indeed, let λ_ε^s and $u_\varepsilon^s(x)$ be the eigenvalue and function to (121) for $s = 0, 1, \ldots.$ It is directly verified that the function $u_\varepsilon^{s_1}(x_1)\, u_\varepsilon^{s_2}(x_2) \ldots u_\varepsilon^{s_d}(x_d)$ will be an eigenfunction for the problem with the eigenvalue $\lambda_\varepsilon^{s_1} + \ldots + \lambda_\varepsilon^{s_d}$. Looking at the graphs and tables of (b_n, a_n) in [62], for example, with $q = 1$, one can notice that already such a 2-dimensional operator has only one gap (b_1, a_1), since if $\lambda \in (b_2, a_2)$ then there are $\lambda_{ac}^{l1}, \lambda_{as}^{l1}$ from (142) or (149) such that their sum approximates λ for small ε (it is also true for $d > 2$). For a similar reason, there will be only tree gaps at $q = 5$. The statement about the finiteness of gaps in the spectrum of the Schrodinger operator with a periodic potential is called the Bethe-Sommerfeld conjecture, details can be found in [70].

In addition, following [66] and extending the potential from the last problem by the zero outside $(0, 2\pi)^d$, one can construct the asymptotics of bound states for the relevant Schrodinger operator on $\mathbb{R}^d$ that oscillate inside the cube and decay exponentially outside the cube. Of course, this is only feasible for negative eigenvalues. For example, on the segment $[a_0, b_1]$ it can be constructed $N^d = \varepsilon^{-d}$ of such bound states if $q > 0$ is not too small. For $q < 0$ it can be constructed for $[a_0, a_1]$. The cube can be replaced with a parallelogram. Perhaps such bound states model free electrons in real crystals, which accelerate electricity when necessary.

Thus, the general principles presented here are also useful for spectral problems as problems for elliptic operators. These principles, set out on dozens of pages in [18], were rewritten in [71], where at the end of the introduction it was even noted when the rewritten results simply coincided with the results of [18]. It is also noted there that such principles have been encountered before, but it is not indicated where, although there are dozens of references in [71]. Strangely, in subsequent rewrites, for example, in review [72], the author forgot to note the sources of matching results. In fact, the first rewrite can be found in [73] also without citing sources. Of interest are the methods from [73], where problem (121) is considered for a more general operator. There, for example, equation (137) was obtained and it is written that: 'We represent the solution of (8) in the form $b_m(y)\varphi_m(y)$, where $b_m(y)$ is an unknown function'. In the notation adopted here, the latter means: 'We represent the solution of (137) in the form $b_n(y)\phi^c_{1n}(y)$'. By virtue of (138), this is possible if and only if $\phi^c_{2n}(y) = 0$. The latter means that in [73] it was proven that a second-order equation is a first-order equation and vice versa, since $\phi^c_{2n}(y)$ is fundamental solution. One can prove anything using these methods.

The asymptotic behavior of many quantities for equation (122) at small q is known [62]. Hence, choosing, for example, $q = \varepsilon^2 \hat{q}$ in the equations from (121) and (122), one can obtain the asymptotics of the spectral problem for the equation. But, it is easier to use general principles and get asymptotics, for example, as for equation (107) with a constant operator and $a_0 = 2\hat{q}\cos(2y) - \lambda$, where $\langle a_0 \rangle = -\lambda$. As noted, the justification for such asymptotics is known and is given, for example, in [74], naturally for eigenvalues and functions with not very large numbers.

9. Homogenized Models for Problems with Parameters

Artificial composites have shown usefulness and applications in many fields. The periodicity of such composites is natural and technological, since by specifying a periodicity cell one can obtain such a composite as a periodic continuation of this cell. As a rule, there are many such cells, and this is how a multi-scale composite continuous medium arises, characterized by the microscale parameter, for example, ε. It is not natural to develop artificial composites from materials with similar properties. But, if the properties of the components of the composite are very different, then this difference should be compared with the microscale parameter.

This creates the need to construct asymptotic solution expansions for problems with contrasting coefficients. Such problems for parabolic equations will be considered here. Asymptotic solution expansions for such problems were obtained and justified for the parabolic case in [27], [28], [75]–[78] and for the hyperbolic case in [29], [79]–[81], where the system of elasticity equations was considered.

In order to begin with, periodic media with small microscale ε are defined as follows. Let $n, s \geqslant 1$ be an integers and $F_1, \ldots, F_s$ are open 1-periodic subsets of $\mathbb{R}^n$ having pairwise disjoint closures ($\bar{F}_q \cap \bar{F}_p = \emptyset$ for $s \geqslant 2$), $F_0 = \mathbb{R}^n \setminus (\bar{F}_1 \cup \ldots \cup \bar{F}_s)$ and

$$F_k^\varepsilon = \varepsilon F_k = \{\varepsilon x : x = (x_1, \ldots, x_n) \in F_k\} \quad \text{for} \quad k = 0, 1, \ldots, s.$$

Here and below, 1-periodicity of sets means periodicity with period 1 for each of the independent variables $x_1, \ldots, x_n$. Thus, $Y = (0,1)^n$ is *a periodicity cell* and the sets $F_k = F_k^1$ are determined by the sets $Y_k = F_k \cap Y$ for $k = 0, 1, \ldots, s$. Assumed that the sets Y_k have positive Lebesgue measures in $\mathbb{R}^n$ for $k = 0, 1, \ldots, s$.

In addition, by identifying the opposite faces of the periodicity cell Y and using 1-periodicity, the sets Y_k can be considered as subsets of an n-dimensional torus. It is also assumed that for each $k = 0, 1, \ldots, s$ the subset Y_k is a torus submanifold with Lipschitz boundary $S_k = \partial Y_k$. By assumption, the subset Y_i for $i = 1, \ldots, s$ is connected, and subset Y_0 can have a finite number of connected components.

Let the bounded domain $\Omega \subset \mathbb{R}^n$ also be given. The sets F_k^ε for fixed ε define the ε-periodic media $\Omega_k^\varepsilon = F_k^\varepsilon \cap \Omega$ for $k = 0, 1, \ldots, s$, which correspond to various materials located in the domain Ω and are bounded by the boundary $\partial\Omega$. Schematic drawings illustrating the media are shown in **Fig. 3**.

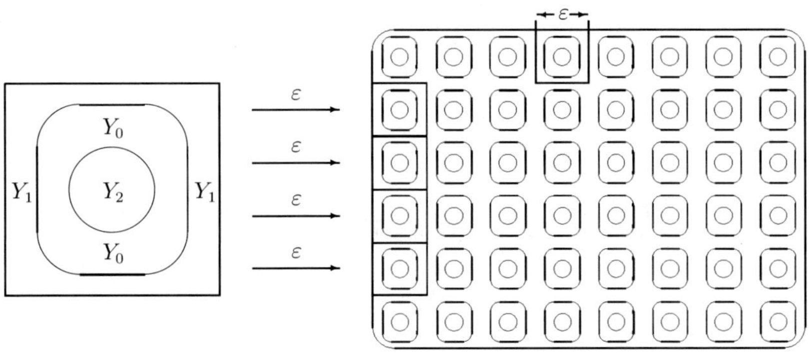

Figure 3. Models of periodic media by ε scaling of cells and periodization.

For such ε-periodic media, the real density coefficients and permeability matrices of these materials in the domain Ω will be determined by the equalities

$$\begin{gathered} m_\mu^\varepsilon = \mu\, m_0(x/\varepsilon), \;\; r_\mu^\varepsilon = \mu\, r_0(x/\varepsilon) \quad \text{in} \quad \Omega_0^\varepsilon, \\ A_\sigma^\varepsilon = \sigma A_0(x/\varepsilon), \;\; b^\varepsilon = b_0(x/\varepsilon) \quad \text{in} \quad \Omega_0^\varepsilon, \\ m_\mu^\varepsilon = m_i(x/\varepsilon), \;\; r_\mu^\varepsilon = r_i(x/\varepsilon), \;\; A_\sigma^\varepsilon = A_i(x/\varepsilon), \;\; b^\varepsilon = b_i(x/\varepsilon) \quad \text{in} \quad \Omega_i^\varepsilon \end{gathered} \tag{151}$$

for $i = 1, \ldots, s$, where μ and σ are positive parameters and ε is a small parameter. Further, for precision $0 < \mu \leq 1, 0 < \sigma \leq 1$ and $0 < \varepsilon \leq \varepsilon_0$ for some positive ε_0.

Suppose that all functions $m_k(y), r_k(y)$ and elements of the matrix $A_k(y)$, $b_k(y)$ from these equalities are defined as elements of the space $L^\infty(Y)$ for $k = 0, 1, \ldots, s$. In addition, the matrices $A_k = \{a_{lj}^k(y)\}_{l,j=1}^n$ are symmetric and elliptic (thus, inequalities (25) are satisfied), and $m_k(y)$ are separated from zero and therefore

$$\nu \leqslant m_k(y) \leqslant \gamma \quad \text{for almost every} \quad y \in Y,$$

where ν and γ are given positive constants and $k = 0, 1, \ldots, s$.

Let the functions $f \in C_0^\infty((0,T)\times\Omega)$, $u_0 \in C_0^\infty(\Omega)$, $\varrho(y) \in L^\infty(Y)$ and the vector function $g \in C_0^\infty((0,T)\times\Omega)^n$ be given, where positive T is fixed. Define $u = u(t,x)$ as a solution in the sense of distributions to the initial boundary value problem

$$\begin{gathered} m_\mu^\varepsilon\, u_t' \;-\; \operatorname{div} A_\sigma^\varepsilon\,(\,\nabla u \,+\, b^\varepsilon g\,) \;=\; r_\mu^\varepsilon\, f \qquad \text{in} \qquad \Omega\times(0,T), \\ u\big|_{t=0} = \varrho(x/\varepsilon)\, u_0 \;\; \text{in} \;\; \Omega, \quad u = 0 \;\; \text{on} \;\; \partial\Omega\times(0,T), \end{gathered} \tag{152}$$

For fixed parameters μ, σ and ε, problem (152) has a unique solution such that $u \in C^0([0,T]; L^2(\Omega)) \cap L^2(0,T; H_0^1(\Omega))$ and $u_t' \in L^2(0,T; H^{-1}(\Omega))$, for example, according to [21]. Let's try to construct an asymptotic expansion of the solution.

Note that here we use a vector representation of the equations, whereas previously we used a component-wise one as (2) and (24). This is due to the convenience of using the generalized Leibniz formula (6) for high-order equations and equations with pseudodifferential operators. In this case, representation (24) was used as a simple case of applying this approach. Here it is more convenient to use the usual Leibniz formula, which allows us to somewhat simplify the notation for explaining constructions. Naturally, it is possible to use formula (6) and a representation similar to (24), if necessary and convenient, but this does not change the results.

Regular asymptotic solution expansions. The equation of problem (152) is parabolic for fixed parameters μ, σ and ε. But, for very small σ this parabolicity can degenerate. Therefore, first we consider the simpler case $\sigma = 1$ for warming up.

Let's introduce the notation $U = \hat{u}$, $F = \hat{f}$ and $G = \hat{g}$ for the Laplace transforms of solutions and data from problem (152). For example, by definition we have

$$\widehat{u(t)} = \int_0^\infty e^{-z\,t}\, u(t)\, dt = U(z)$$

where $x \in \Omega$ is considered as a parameter and the complex parameter z usually belongs to the complex half-plane $\mathbb{C}_0 = \{\, z \in \mathbb{C} \,:\, z = z_1 + iz_2,\; z_1 > 0\,\}$.

Applying the Laplace transform to problem (152), we obtain for $U \in H_0^1(\Omega)$ the following equivalent problem with a complex parameter

$$m_\mu^\varepsilon\, z U \;-\; \operatorname{div} A_\sigma^\varepsilon\,(\,\nabla U \,+\, b^\varepsilon G\,) = m_\mu^\varepsilon\, \varrho^\varepsilon u_0 + r_\mu^\varepsilon\, F \quad \text{in} \quad \Omega, \tag{153}$$

for $z \in \mathbb{C}_0$ and $\varrho^\varepsilon = \varrho(x/\varepsilon)$, where, as usual, the function $\varrho(y)$ on Y are continued as 1-periodic functions on $\mathbb{R}^n$. Further details about such equivalence can be found in [26]. It should be emphasized that this equivalence preserves the norms of suitable spaces and is therefore an isometric isomorphism [26]. It is also well known that

$$\widehat{u'_t(t)} = \int_0^\infty e^{-z\,t}\, u'_t(t)\, dt = z\,U - u\big|_{t=0}.$$

For a fixed $z \in \mathbb{C}_0$, problem (153) is quite similar to the homogeneous Dirichlet problem for a second-order divergent equation, such as (72) for $a_k = 0$ or (107) for $r = 1$. Hence, one can choose an asymptotic solution expansion in the form

$$U_a = V(x) + \varepsilon\, V_1(x, x/\varepsilon) + \varepsilon^2 V_2(x, x/\varepsilon) + \varepsilon^3 V_3(x, x/\varepsilon), \tag{154}$$

since $N = M = 1$ in the notation from (13), where the functions $V_1(x, y)$, $V_2(x, y)$, $V_3(x, y)$ are, as usual, 1-periodic in the variable $y \in \mathbb{R}^n$.

To determine the terms of the asymptotic expansion (154), the two formulas

$$\begin{aligned} \nabla(W(x, x/\varepsilon)) &= \left(\varepsilon^{-1}\nabla_y W(x, y) + \nabla_x W(x, y)\right)\big|_{y=x/\varepsilon}, \\ \operatorname{div}(W(x, x/\varepsilon)) &= \left(\varepsilon^{-1}\operatorname{div}_y W(x, y) + \operatorname{div}_x W(x, y)\right)\big|_{y=x/\varepsilon} \end{aligned} \tag{155}$$

for functions of the form $W(x, x/\varepsilon)$ will be used in what follows. Naturally, as will become clear later, these formulas are special cases of expansion (8). But, in the case under consideration, these formulas are more convenient and provide another version of the principles for constructing asymptotic solution expansions.

Let's substitute the asymptotic expansion (154) instead of U into equation (153) and use (155). Then, keeping only the terms essential for what follows, we get

$$\begin{aligned} &m^1_\mu\, z\, V - \varepsilon^{-1}\operatorname{div}_y(A^1_1 \nabla_y V_1) - \varepsilon^{-1}\operatorname{div}_y A^1_1(\,\nabla_x V + b^1 G\,) \\ &- \operatorname{div}_y(A^1_1 \nabla_y V_2) - \operatorname{div}_x A^1_1(\,\nabla_x V + b^1 G\,) - \operatorname{div}_x(A^1_1 \nabla_y V_1) \\ &- \operatorname{div}_y(A^1_1 \nabla_x V_1) - \varepsilon\operatorname{div}_x(A^1_1 \nabla_x V_1) - \varepsilon\operatorname{div}_x(A^1_1 \nabla_y V_2) \\ &- \varepsilon\operatorname{div}_y(A^1_1 \nabla_x V_2) - \varepsilon\operatorname{div}_y(A^1_1 \nabla_y V_3) = m^1_\mu\, \varrho^1 u_0 + r^1_\mu\, F \end{aligned} \tag{156}$$

for $y = x/\varepsilon$ and $x \in \Omega$, where, for example, by definition $m^1_\mu = m_i(y)$ for $y \in Y_i$ with $i = 1, \ldots s$, $m^1_\mu = \mu\, m_0(y)$ for $y \in Y_0$ and $\varrho^1 = \varrho(y)$. Here, in fact, $\varepsilon = 1$ was chosen in (151). Recall that the case $\sigma = 1$ is considered.

Equating the coefficients of ε^{-1} in the asymptotic equality (156) we have

$$-\operatorname{div}_y(A_1^1(y)\nabla_y V_1(x,y)) = \operatorname{div}_y(A_1^1(y)\nabla_x V(x) + A_1^1(y)\, b^1(y)G(x)). \tag{157}$$

The equality will be fulfilled if we choose $V_1 = N_a(y)\nabla_x V(x) + N_b(y)G(x)$ and define the vector functions $N_a(y)$ and $N_b(y)$ as 1-periodic solutions of the problems

$$-\operatorname{div}_y(A_1^1\nabla_y N_a) = \operatorname{div}_y A_1^1, \quad -\operatorname{div}_y(A_1^1\nabla_y N_b) = \operatorname{div}_y A_1^1\, b^1 \quad \text{in} \quad Y. \tag{158}$$

It is known [7] that solutions to the problems exist and are defined up to constants, which will be fixed by the conditions $\int_Y N_a(y)\, dy = 0$ and $\int_Y N_b(y)\, dy = 0$. Remark that these problems are solvable, since $M = 1$ and, for example, $\int_Y \operatorname{div}_y A_1^1\, dy = 0$, which was already noted in (92) due to integration by parts.

Thus, in equality (157) it was possible to separate the variables x and y and the equality is satisfied for all x and y, in particular for $y = x/\varepsilon$. With this choice of V_1, the coefficients of ε^{-1} in the asymptotic equality (156) are equal to zero.

Equating the coefficients of ε^0 in the asymptotic equality (156) we have

$$-\operatorname{div}_y(A_1^1\nabla_y V_2) = \operatorname{div}_x\big((A_1^1 + A_1^1\nabla_y N_a)\nabla_x V + (A_1^1 b^1 + A_1^1\nabla_y N_b)G\big)$$

$$+\operatorname{div}_y(A_1^1 N_a \nabla_x^2 V + A_1^1 N_b \nabla_x G) + m_\mu^1 \varrho^1\, u_0 + r_\mu^1\, F - m_\mu^1\, z\, V, \tag{159}$$

where it was taken into account in the obvious notation that $V_1 = N_a\nabla_x V + N_b\, G$.

By analogy with (157), we consider this equality as a 1-periodic problem for $V_2 = V_2(x,y)$ for a fixed $x \in \Omega$. It is known [7] that this problem is correct if and only if the right-hand side is orthogonal to 1 in $L^2(Y)$ and therefore

$$m\, z\, V - \operatorname{div}_x A(\nabla_x V) = \operatorname{div}_x B\, G + m_\varrho u_0 + r F, \tag{160}$$

where constants and matrices with constant components are defined by the equalities

$$m = \int_Y m_\mu^1(y)\, dy = \mu \int_{Y_0} m_0(y)\, dy + \sum_{i=1}^{s} \int_{Y_i} m_i(y)\, dy \equiv \mu\, m^0 + m^s,$$

$$r = \int_Y r_\mu^1(y) dy \equiv \mu\, r^0 + r^s, \quad m_\varrho = \int_Y m_\mu^1(y)\varrho^1(y) dy \equiv \mu\, m_\varrho^0 + m_\varrho^s, \tag{161}$$

$$A = \int_Y \big(A_1^1(y) + A_1^1(y)\nabla_y N_a(y)\big) dy, \quad B = \int_Y \big(A_1^1(y) b^1(y) + A_1^1(y)\nabla_y N_b(y)\big) dy.$$

Equality (160) defines a homogenized problem with constant coefficients for problem (152). Indeed, applying the inverse Laplace transformation to this equality, we have for $v = v(t,x)$ the following homogenized initial boundary value problem

$$\begin{aligned} & m\, v_t' \;-\; \mathrm{div} A\,(\nabla v) \;=\; r\, f + \mathrm{div} B\, g \quad \text{in} \quad \Omega \times (0,T), \\ & v\big|_{t=0} = m^{-1} m_\varrho u_0 \quad \text{in} \quad \Omega, \quad v = 0 \quad \text{on} \quad \partial\Omega \times (0,T), \end{aligned} \tag{162}$$

where the Laplace transform formula for the derivative with respect to t is used.

It is known [7] that the matrix A is elliptic (this actually follows from representation (76)) and so a unique solution to problem (162) exists and is smooth enough. Hence, two terms in the asymptotic expansion (154) are defined. To determine the third term $V_2(x,y)$, we subtract from (159) equality (160) for the Laplace transform $\hat{v} = V$ of the solution to problem (162). Then, in natural notation, we get

$$\begin{aligned} -\,\mathrm{div}_y\big(A_1^1 \nabla_y V_2\big) = \Big(\mathrm{div}_y A_1^1 N_a + \widetilde{A}\Big) \nabla_x^2 V + \Big(\mathrm{div}_y A_1^1 N_b + \widetilde{B}\Big) \nabla_x G \\ + \widetilde{m}_\varrho\, u_0 + \widetilde{r}\, F \;-\; \widetilde{m}\, z\, V \quad \text{in} \quad Y, \end{aligned} \tag{163}$$

where the following 1-periodic matrix functions and 1-periodic functions

$$\widetilde{A} = A_1^1 + A_1^1 \nabla_y N_a - A, \qquad \widetilde{B} = A_1^1 b^1 + A_1^1 \nabla_y N_b - B,$$

$$\widetilde{m}_\varrho = m_\mu^1 \varrho^1 - m_\varrho, \quad \widetilde{r} = r_\mu^1 - r, \quad \widetilde{m} = m_\mu^1 - m$$

are such that all their components are orthogonal to 1 in the space $L^2(Y)$.

Thus, equality (163) and therefore equality (159) are satisfied if we choose

$$V_2 = N_a^1(y) \nabla_x^2 V + N_b^1(y) \nabla_x G + N_\varrho(y)\, u_0 + N_r(y) F - N_m(y)\, z\, V, \tag{164}$$

and define the functions N_r and N_m as 1-periodic solutions of the problems

$$-\,\mathrm{div}_y(A_1^1 \nabla_y N_r) = \widetilde{r}, \quad \mathrm{div}_y(A_1^1 \nabla_y N_m) = \widetilde{m} \quad \text{in} \quad Y$$

and 1-periodic functions $N_\varrho(y)$, $N_\varphi(y)$ and matrix functions $N_a^1(y)$, $N_b^1(y)$ in (164) one can define as solutions to similar problems. By construction, solutions to these problems exist and are defined up to constants that can be fixed as before. Through this choice of V_2, all coefficients for ε^0 in asymptotic equality (156) are zero.

Thus, the asymptotic equality (156) is satisfied up to $O(\varepsilon)$ even for $V_3 = 0$ in definition (154). Therefore, the asymptotic expansion $U_a(x, x/\varepsilon)$

also approximates the equation of problem (153) up to $O(\varepsilon)$. Following [27], [78], [80] and using the inverse Laplace transform of $U_a(x, x/\varepsilon)$, one can check using energy methods that the last statement implies, for example, the following accuracy estimate

$$\| u - v \|^2_{C^0([0,T];L^2(\Omega\setminus\Omega_0^\varepsilon))} + \mu \| u - v \|^2_{C^0([0,T];L^2(\Omega_0^\varepsilon))} \leq C\,\varepsilon, \tag{165}$$

where u is a solution to problem (152), v is a solution to the homogenized problem (162) and the constant C is independent of μ and ε for $0 < \mu \leq 1$, $0 < \varepsilon \leq \varepsilon_0$ with some ε_0. Note that the appearance of the parameter μ in the presented estimate is natural, since this estimate is uniform in μ and the left side of (165) is a component of the energy norm, which is simply to estimate due to energy equality. Whereas the use of non-energy methods for problems with parameters is quite complex.

In addition, for the case of small μ one can derive the following accuracy estimate

$$\| u - v_0 \|^2_{C^0([0,T];L^2(\Omega\setminus\Omega_0^\varepsilon))} + \mu \| u - v_0 \|^2_{C^0([0,T];L^2(\Omega_0^\varepsilon))} \leq C\,(\varepsilon + \mu^2), \tag{166}$$

where u is a solution to problem (152), v_0 is a solution to the homogenized problem (162) for $\mu = 0$ and the constant C is independent of μ and ε for $0 < \mu \leq \mu_0$, $0 < \varepsilon \leq \varepsilon_0$ with corresponding μ_0 and ε_0. Thus, to determine v_0, one can simply choose $\mu = 0$ in the definition of homogenized coefficients (161). This means that the parameter μ in the case under consideration is not singular. In addition, using simple asymptotic expansions, one can significantly improve the exponent of the small parameter μ in estimate (166) in accordance with [78] and [80].

One can also define $V_3 = V_3(x, x/\varepsilon)$ in the asymptotic approximation (154) so that $U_a(x, x/\varepsilon)$ approximates the equation of problem (153) up to $O(\varepsilon^2)$. However, this does not lead to an improvement in the exponent of ε in estimates (165) and (166), since the term $V_1 = N_a \nabla_x V + N_b G$ in expansion (154) may not satisfy homogeneous boundary conditions on the boundary $\partial\Omega$. In fact, to check the given accuracy estimates, it is enough to choose $V_2 = 0$ and $V_3 = 0$ in expansion (154).

However, in the latter case it would not be possible to establish a homogenized equation without knowing the exact structure of V_2. In addition, it is inconvenient to directly use $V_2 = V_2(x, x/\varepsilon)$ when checking such estimates, since after applying the inverse Laplace transform to (154), the delta-function $(N_\varrho - m^{-1} m_\varrho N_m)\,\delta(t)\,u_0$ will arise. This indicates the appearance of boundary layer effects near the boundary $t = 0$, which are introduced by fast oscillations of the initial condition and are zero if $\varrho = 1$. In turn, such effects near the boundary $\partial\Omega$ prevent the improvement of the exponent of the parameter ε in the accuracy estimates (165) and (166).

This boundary layer effect can be corrected. Indeed, to exclude such delta functions from this equality, one can represent the asymptotic approximation of the solution to problem (152) in the form $v_a = u_a + \varepsilon^2 w'_t$, where u_a denotes the inverse Laplace transform U_a from (154) for $V_3 = 0$, and w satisfies the initial conditions

$$w\big|_{t=0} = -(N_\varrho - m^{-1} m_\varrho N_m)\, u_0 \quad \text{in} \quad \Omega$$

and the Laplace transform of w'_t satisfies the equation of problem (153) up to $O(\varepsilon)$. Such w can be easily determined by renormalizing the time variable and $w = 0$ only if $\varrho = 1$. However, such a refinement does not improve the accuracy estimates (165) and (166), again due to boundary effects near the boundary $\partial\Omega$. Sometimes, for domains of a special type, such effects can be eliminated by adding boundary layer terms to the expansions; then it makes sense to use the terms $V_2 = V_2(x, x/\varepsilon)$, $V_3 = V_3(x, x/\varepsilon), \ldots$ in the asymptotic solution expansion (154).

Thus, when modeling diffusion (filtration, heat) processes in ε-periodic media, instead of solving problem (152) (for example, numerically), one can solve homogenized problem (162) and this approach will give results with guaranteed accuracy for small ε. Naturally, the numerical solution of problem (152) is much more complicated than the solution of problem (162), since a very fine computational grid is needed to take into account the geometry of a very large number of cells in the domain Ω. Note that the construction of initial terms of asymptotic solution expansions for problem (152) is known [6], [7] without using the Laplace transform, at least for $\mu = 1$ and $u_0 = 0$. But the approach presented here will be natural and useful later for more complex problems with parameters.

Asymptotic expansions in perforated domains, Dirichlet conditions. Assuming in (151) formally $\mu = \infty$ and $\sigma = \infty$, one can remark that diffusion in the domain Ω_0^ε is rigidly fixed (or, for example, a membrane is fixed). Thus, problem (151) will turn (formally) into a Dirichlet problem in the domain $\Omega \setminus \Omega_0^\varepsilon$, since the solution to this problem must be continuous in some sense. Forgetting about formalities, let's consider such a problem rewritten in the following form

$$m_\mu^\varepsilon u'_t - \operatorname{div}(A_\sigma^\varepsilon \nabla u) = \operatorname{div}(A_\sigma^\varepsilon b^\varepsilon g) + r_\mu^\varepsilon f \quad \text{in} \quad \Omega_{1s}^\varepsilon \times (0, T),$$

$$u\big|_{t=0} = \varrho^\varepsilon u_0 \ \text{ in } \ \Omega_{1s}^\varepsilon, \quad u = 0 \ \text{ on } \ \partial\Omega_{1s}^\varepsilon \times (0, T), \tag{167}$$

where the dependence of the coefficients on μ and σ is formal, since one can assume that $u = 0$ on $\bar{\Omega}_0^\varepsilon$. Here $\Omega_{1s}^\varepsilon = \Omega \setminus \bar{\Omega}_0^\varepsilon = \Omega_1^\varepsilon \cup \ldots \cup \Omega_s^\varepsilon$ by definition.

For a fixed ε, the solution $u \in C^0([0, T]; L^2(\Omega_{1s}^\varepsilon)) \cap L^2(0, T; H_0^1(\Omega_{1s}^\varepsilon))$ to this problem exists and is unique, for example, according to [21]. To construct an asymptotic expansion of the solution, one can note that there is an expansion of Taylor type case, for example, as for problem (24). Indeed, using (30), for the right side of the equation from (167) one can obtain

$$\operatorname{div}\left(A_\sigma^\varepsilon\, b^\varepsilon g\right) = \varepsilon^{-1}\left(\operatorname{div}_y A_\sigma^1\, b^1\right) g + \left(A_\sigma^1\, b^1\right)\nabla_x\, g \quad \text{for} \quad y = x/\varepsilon\,.$$

Consequently, similarly as for the asymptotic expansion (29), we can write

$$u_1^l = \varepsilon\, N_g(x/\varepsilon)\, g(t,x) + \varepsilon^2 N_{\beta_1}^1(x/\varepsilon)\, u_1^{\beta_1}(t,x) + \varepsilon^3 N_{\beta_2}^2(x/\varepsilon)\, u_2^{\beta_2}(t,x) + \ldots,$$

where, for example, the vector function $N_g(y)$ is 1-periodic solutions to the problem

$$-\operatorname{div}_y\left(A_\sigma^1\, \nabla_y N_g\right) = \operatorname{div}_y\left(A_\sigma^1\, b^1\right) \quad \text{in} \quad Y_{1s}, \qquad N_g = 0 \quad \text{on} \quad \bar{Y}_0, \tag{168}$$

where $Y_{1s} = Y_1 \cup \ldots \cup Y_s$. The equation of the problem can be considered on several sets $Y_1, \ldots, Y_s$ as submanifolds of the torus. But this does not affect the result, since a unique solution exists and can be extended to the entire space.

Thus, continuing by induction, we can conclude that u_1^l satisfies the equation of problem (167) up to $O(\varepsilon^{l-1})$, for example, in $C^0([0,T]; L^2(\Omega))$-norm. Note that $u_1^l\big|_{t=0} = 0$ due to the conditions imposed on f and g. Remark also that if equation (153) for the Laplace transform image is used to construct the asymptotic solution expansion, then a delta function with respect to time will again appear in such an expansion. This indicates boundary layer effects near $t = 0$.

Indeed, using the linearity of the equation from (167), we can assume that $f = 0$ and $g = 0$ in this equation. Let's define

$$v_0^l = N_0(t/\varepsilon^2, x/\varepsilon) u_0(x) + \varepsilon^1 N_{\beta_1}^1(t/\varepsilon^2, x/\varepsilon) u_1^{\beta_1}(x) + \varepsilon^2 N_{\beta_2}^2(t/\varepsilon^2, x/\varepsilon) u_2^{\beta_2}(x) + \ldots.$$

Substituting this expansion into (167) and equating the power coefficients, we conclude that the 1-periodic in y function $N_0(\tau, y)$ is a solution of the problem

$$\begin{aligned} &m_\mu^1\,(N_0)'_\tau - \operatorname{div}_y\left(A_\sigma^1\, \nabla_y N_0\right) = 0 \quad \text{in} \quad Y_{1s} \times (0,\infty), \\ &N_0\big|_{\tau=0} = \varrho(y) \ \text{ in } \ Y_{1s}, \quad N_0 = 0 \quad \text{on} \quad \bar{Y}_0 \times (0,\infty). \end{aligned} \tag{169}$$

A suitable solution $N_0 \in L^\infty(0,\infty; L^2(Y)) \cap L^2(0,\infty; H_0^1(Y_{1s}))$ to the problem exists and is unique, which follows from the energy estimate, for example, according to [22]. Here and below Y_{1s} is also considered as a submanifold of a torus, else one need to write $H_{per}^1(Y_{1s})$ for 1-periodic functions. It is also known that such a solution decreases exponentially for $\tau \to \infty$. In addition, solutions to problem (169) with an exponentially decreasing right-hand side also decrease exponentially. This actually follows from the hypoellipticity of the heat conductivity operator.

Thus, continuing by induction, we can conclude that v_0^l satisfies the equation of problem (167) up to $O(\varepsilon^{l-2})$, for example, in $L^1(0,\infty;L^2(\Omega))$-norm. Hence, from the energy estimate for problem (167) we can derive the following accuracy estimate

$$\| u - u_1^l - v_0^l \|_{L^\infty(0,\infty;L^2(\Omega))} + \| \nabla(u - u_1^l - v_0^l) \|_{L^2(0,\infty;L^2(\Omega)^n)} \le C\,\varepsilon^{l-2}, \tag{170}$$

where u is a solution to problem (167) extended by zero on Ω_0^ε, u_1^l and v_0^l are defined above and the constant C is independent of ε when $0 < \varepsilon \le \varepsilon_0$ for some ε_0.

Remark that the construction of initial terms of asymptotic solution expansions for stationary version of problem (152) is known, at least for $A_\mu^\varepsilon = E$, $r_\mu^\varepsilon = 1$ and $g = 0$. But, studies of such a stationary version were concentrated on the case when $s = 1$, Y_0 is compactly embedded in the cell Y (like Y_2 in Fig. 3) and the size (for example, diameter) d_ε of this set is much smaller than ε. Consequently, the coefficients of the cell problem (168) depend on the additional parameter d_ε and it is also necessary to consider the asymptotics of the solution to the cell problem. Usually, the limiting properties of solutions to the problem are only studied. Further details and references can be found, for example, in [9].

Remark also that the known and most studied analogue of the problem considered is the stationary Stokes system, in which the velocity satisfies homogeneous Dirichlet boundary conditions, and the pressure is determined up to a constant. Namely, the presence of this constant (from the kernel of the Stokes operator) leads to one homogenized equation, which is called Darcy's law, well known in filtration theory. Thus, for the Stokes system, mixed constructions with asymptotic expansions of Taylor type for velocity are used. How such asymptotic expansions are constructed for non-stationary problems can be found in [82]–[84].

Asymptotic expansions in perforated domains, Neumann conditions. Assuming in (151) formally $\mu = 0$ and $\sigma = 0$, one can remark that diffusion in the domain Ω_0^ε is completely unbound. Thus, problem (151) will turn (formally) into a Neumann problem in the domain $\Omega \setminus \Omega_0^\varepsilon$. Anyway, let's consider the problem

$$\begin{gathered} m_\mu^\varepsilon u_t' - \operatorname{div} A_\sigma^\varepsilon(\nabla u + b^\varepsilon g) = r_\mu^\varepsilon f \quad \text{in} \quad \Omega_{1s}^\varepsilon \times (0,T), \\ u\big|_{t=0} = \varrho^\varepsilon u_0 \ \text{ in } \ \Omega_{1s}^\varepsilon, \quad u = 0 \ \text{ on } \ (\partial\Omega \cap \partial\Omega_{1s}^\varepsilon) \times (0,T), \\ \Upsilon_\varepsilon \cdot (A_\sigma^\varepsilon \nabla u + A_\sigma^\varepsilon b^\varepsilon g) = 0 \quad \text{on} \quad (\partial\Omega_i^\varepsilon \setminus \partial\Omega) \times (0,T), \end{gathered} \tag{171}$$

where $i = 1, \ldots, s$, Υ_ε denotes the outer normal to the locally Lipschitz boundary $\partial\Omega_i^\varepsilon \setminus \partial\Omega$ and the dot $\cdot$ means the scalar product of vectors. Here one should be careful with regularity, since the boundaries $\partial\Omega_i^\varepsilon$ and $\partial\Omega$ can be tangent and therefore lose Lipschitz property. However, in the case, we can continue u and g by zero and assume that the Neumann condition is satisfied on ∂F_i^ε for $i = 1, \dots, s$.

By definition, sets Ω_i^ε for $i = 1, \dots, s$ have pairwise disjoint closures. Hence, problem (171) is divided into s independent problems, and it is enough to consider only one of the problems, simply assuming that $s = 1$. In addition, the set Ω_1^ε itself can crumble (like Ω_2^ε in Fig. 3). In this case, the set will not be a domain (a connected bounded set); to avoid this, we assume that F_1^ε is connected and Ω_1^ε is a domain for ε under consideration. Note also that the dependence of the coefficients on μ and σ is formal in (171), since the solution is considered only on the set Ω_1^ε.

For a fixed ε, the unique solution $u \in C^0([0,T]; L^2(\Omega_1^\varepsilon)) \cap L^2(0,T; H^1(\Omega_1^\varepsilon))$ to the problem exists, for example, according to [21]. Applying the Laplace transform to (171), one get for $U \in H^1(\Omega_1^\varepsilon)$ the equivalent problem with a complex parameter similar to (153) in Ω_1^ε, to which we should add Neumann boundary conditions.

Next, one can choose an asymptotic solution expansion in the form (154) with $V_3(x,y) = 0$, since $N = M = 1$ as will become clear later. Consequently, we have asymptotic equality (156) for $y = x/\varepsilon$ and $x \in \Omega_1^\varepsilon$. Similarly, keeping only the essential terms, for the Neumann conditions, we obtain

$$\Upsilon \cdot (A_1 \nabla_y V_1 + A_1 \nabla_x V + A_1 b_1 G + \varepsilon A_1 \nabla_y V_2 + \varepsilon A_1 \nabla_x V_1) = 0 \qquad (172)$$

for $y = x/\varepsilon$ and $x \in \partial F_1^\varepsilon$, where Υ is the outer normal to the manifold ∂Y_1.

Consequently, the coefficients for ε^{-1} in (156) and for ε^0 in (172) are zero, if we choose $V_1 = N_a^1(y)\nabla_x V(x) + N_b^1(y) G(x)$ and define the vector functions $N_a^1(y)$ and $N_b^1(y)$ as 1-periodic solutions of the problems

$$\begin{aligned} &-\operatorname{div}_y(A_1 \nabla_y N_a^1) = \operatorname{div}_y A_1 \ \text{ in } Y_1, \quad -\Upsilon \cdot (A_1 \nabla_y N_a^1) = \Upsilon \cdot A_1 \ \text{ on } \partial Y_1, \\ &-\operatorname{div}_y(A_1 \nabla_y N_b^1) = \operatorname{div}_y A_1\, b_1 \ \text{ in } Y_1, \quad -\Upsilon \cdot (A_1 \nabla_y N_b^1) = \Upsilon \cdot A_1\, b_1 \ \text{ on } \partial Y_1. \end{aligned} \qquad (173)$$

In what follows, the following statement will be useful.

Theorem 6. *Let* $M \in L^2(Y_1)$, $W \in H^{-1/2}(\partial Y_1)$ *and* $D \in L^2(Y_1)^n$ *be such that*

$$\int_{Y_1} M(y)\, dy = \int_{\partial Y_1} W(\tau)\, d\tau,$$

where Y_1 and ∂Y_1 are considered as a submanifold of the torus. Then there exists a unique solution $N \in H^1(Y_1)/\mathbb{R}$ of the following 1-periodic Neumann problem

$$-\operatorname{div}_y(A_1\nabla_y N) = M + \operatorname{div}_y D \text{ in } Y_1, \quad -\Upsilon\cdot(A_1\nabla_y N) = W + \Upsilon\cdot D \quad \text{on } \partial Y_1.$$

The statement is not so simple. Precisely, it is known in the case of bounded coefficients and the smooth boundary ∂Y_1. For Lipschitz coefficients and the boundary, it was proved in [85]. A review of results for problems with Lipschitz coefficients and only such ones is given in [86]. At last, for bounded coefficients and the Lipschitz boundary this statement was proved in [32]. More precisely, it was proved in [32] for domains with a Lipschitz boundary, but the proofs are local and therefore applicable to manifolds. By the way, from Temam results [87] and the statement follows the Stokes formula for integration by parts for domains with Lipschitz boundaries. The formula was proved in [87] under the assumption that there is a lifting operator from the boundary. Theorem 6 makes it possible to define the operator, as noted in [78].

Thus, solutions to problems (173) exist and are defined up to constants, which will be fixed as before. Next, we get relation (159) and it follows from (173), that

$$-\Upsilon\cdot(A_1\nabla_y V_2) = \Upsilon\cdot\big(A_1 N_a^1\nabla_x^2 V + A_1 N_b^1\nabla_x G\big)$$

for $y = x/\varepsilon$ and $x \in \partial F_1^\varepsilon$. By analogy with (157), we consider equation (159) with such boundary conditions as a 1-periodic problem for $V_2(x,y)$ for a fixed $x \in \Omega_1^\varepsilon$. By Theorem 6, the problem is correct if and only if equality (160) is satisfied, where constants and matrices with constant components are calculated using formulas (161) in which integration over Y is replaced by integration over Y_1.

Consequently, in the case under consideration, the homogenized problem is (162). It is known [7] that the matrix A is elliptic (this again follows from representation (76) in fact) and so a unique solution to problem (162) exists and is smooth enough. Hence, one can continue to determine all terms in (154) as before.

Following [27], [78], [80] again, one can verify the following accuracy estimate

$$\big\| u - v \big\|^2_{C^0([0,T];L^2(\Omega_1^\varepsilon))} \le C\,\varepsilon,$$

where u is a solution to problem (152), v is a solution to the homogenized problem (162) and the constant C is independent of ε for $0 < \varepsilon \le \varepsilon_0$ with some ε_0.

Note that for $s > 1$ there will be s Neumann problems and, accordingly, s relevant homogenized problems for which accuracy estimates can be proven. Now we will explain how such problems will interact for small σ.

Asymptotic expansions for problems with parameters. Let us further assume that σ is a small parameter in problem (152). In this case it is lost parabolicity of this problem on the part Ω_0^ε of the domain Ω.

The domain is composed of $s+1$ sets Ω_0^ε and $\Omega_1^\varepsilon, \ldots, \Omega_s^\varepsilon$, which are separated by common boundaries $\partial\Omega_i^\varepsilon \setminus \partial\Omega$ for $i = 1, \ldots, s$. Therefore, following [7], problem (152) can be rewritten in equivalent form as the following initial boundary value problem with conjugation conditions at the interface in the form

$$\mu\, m_0^\varepsilon\, u_t' \;-\; \sigma \operatorname{div} A_0^\varepsilon\,(\,\nabla u \,+\, b_0^\varepsilon\, g\,) \;=\; \mu\, r_0^\varepsilon\, f \quad \text{in} \quad \Omega_0^\varepsilon \times (0,T),$$

$$m_i^\varepsilon\, u_t' \;-\; \operatorname{div} A_i^\varepsilon\,(\,\nabla u \,+\, b_i^\varepsilon\, g\,) \;=\; r_i^\varepsilon\, f \quad \text{in} \quad \Omega_i^\varepsilon \times (0,T),$$

$$u\big|_{t=0} = \varrho^\varepsilon u_0 \quad \text{in} \quad \Omega, \qquad u = 0 \quad \text{on} \quad \partial\Omega \times (0,T),$$

$$\Upsilon_\varepsilon \cdot \big(A_i^\varepsilon \nabla u + A_i^\varepsilon\, b_i^\varepsilon\, g\big) = \sigma\, \Upsilon_\varepsilon \cdot \big(A_0^\varepsilon \nabla u + A_0^\varepsilon\, b_0^\varepsilon\, g\big) \;\text{ on }\; (\partial\Omega_i^\varepsilon \setminus \partial\Omega) \times (0,T) \tag{174}$$

for $i = 1, \ldots, s$ and $u \in H_0^1(\Omega)$ for almost all $t \in (0,T)$. The last conjugation conditions mean the continuity of flows at the internal boundary (the contact surface of the media under consideration). In such conditions, for example, the function $\Upsilon_\varepsilon \cdot (A_i^\varepsilon \nabla u)$ at point $x_0 \in \partial\Omega_i^\varepsilon$ is calculated as the limit for $x \to x_0$ of the function $\Upsilon_\varepsilon \cdot (A_i^\varepsilon \nabla u)$, which is considered at internal points $x \in \Omega_i^\varepsilon$, $i = 1, \ldots, s$. Here we introduce also the notations $m_0^\varepsilon = m_0(x/\varepsilon)$, $A_0^\varepsilon = A_0(x/\varepsilon), \ldots, r_s^\varepsilon = r_s(x/\varepsilon)$.

This form for problem (152) precisely explains that media with small density and permeability coefficients are considered on the part Ω_0^ε for small μ and σ. It is clear that one can multiply the equations of problem (174) by σ^{-1} and consider the case of large density and permeability coefficients on the parts Ω_i^ε, $i = 1, \ldots, s$ for small σ and $\mu = \sigma$. Accordingly, multiplying these equations by $\sigma^{-1/2}$ one can consider the case of small coefficients on Ω_0^ε and large coefficients on Ω_i^ε, $i = 1, \ldots, s$ for small σ and $\mu = \sigma$. Other options are possible when multiplying by $\mu^{-\kappa}$ for a positive κ. Of course, such multiplications do not affect the final form of the asymptotic solution expansions for problem (152).

Applying the Laplace transform to problem(174) in the notation from(153), we obtain for $U \in H_0^1(\Omega)$ the following equivalent problem with a complex parameter

$$\mu\, m_0^\varepsilon\, z\, U - \sigma \operatorname{div} A_0^\varepsilon\,(\,\nabla U + b_0^\varepsilon\, G\,) = \mu\, m_0^\varepsilon\, \varrho^\varepsilon u_0 + \mu\, r_0^\varepsilon\, F \;\text{ in }\; \Omega_0^\varepsilon,$$

$$m_i^\varepsilon\, z\, U - \operatorname{div} A_i^\varepsilon\,(\,\nabla U + b_i^\varepsilon\, G\,) = m_i^\varepsilon\, \varrho^\varepsilon u_0 + r_i^\varepsilon\, F \quad \text{in} \quad \Omega_i^\varepsilon,$$

$$\Upsilon_\varepsilon \cdot \big(A_i^\varepsilon \nabla U + A_i^\varepsilon\, b_i^\varepsilon\, G\big) = \sigma\, \Upsilon_\varepsilon \cdot \big(A_0^\varepsilon \nabla U + A_0^\varepsilon\, b_0^\varepsilon\, G\big) \;\text{ on }\; (\partial\Omega_i^\varepsilon \setminus \partial\Omega). \tag{175}$$

for $i = 1, \ldots, s$. Thus, for a fixed $z \in \mathbb{C}_0$ there are $1 + s$ elliptic equations.

According to the general principles, to construct asymptotic solution expansions for the equations, it is necessary to describe the kernels of operators on 1-periodic functions for $\varepsilon = 1$. For the first equation in (175), the kernel is trivial, for example, when $\sigma = \varepsilon^2$ and the expansion of Taylor type may be expected. For the rest of the equations, the kernels are formed by functions that are multiples of 1. Hence, we have a s-dimensional case. But, all the equations are considered on one domain, so the generators of the kernels can be specified as solutions to the equations

$$\operatorname{div}_y A_0\big(\nabla_y E_i\big) = 0 \quad \text{in } Y_0, \quad E_i = \delta_{ij} \quad \text{on } \bar{Y}_j \quad \text{for } j = 1, \ldots, s, \tag{176}$$

which do not affect the first equation in any way. Here, $\delta_{ii} = 1$ and $\delta_{ij} = 0$ for $i \neq j$.

Thus, using (67), one can choose an asymptotic solution expansion in the form

$$\begin{gathered} U_a = E_1(x/\varepsilon)V^1(x) + \ldots + E_s(x/\varepsilon)V^s(x) + V_0(x, x/\varepsilon) \\ + \varepsilon V_1^1(x, x/\varepsilon) + \ldots + \varepsilon V_1^s(x, x/\varepsilon) + \varepsilon^2 V_2^1(x, x/\varepsilon) + \ldots + \varepsilon^2 V_2^s(x, x/\varepsilon), \end{gathered} \tag{177}$$

where by definition $V_0(x, x/\varepsilon) = 0$ on Ω_{1s}^ε and $V_1^i(x, x/\varepsilon) = V_2^i(x, x/\varepsilon) = 0$ on $\Omega_{1s}^\varepsilon \setminus \Omega_i^\varepsilon$ for $i = 1, \ldots, s$. Here $V_0(x, x/\varepsilon)$ appears as the main term in the expansion of Taylor type on Ω_0^ε exactly as for the Dirichlet problem in a perforated domain.

Let's substitute the asymptotic expansion (177) instead of U into $(1+i)$-equation from (175). Using (155) and keeping only the terms essential for what follows, we get

$$\begin{gathered} m_i\, z\, V^i - \varepsilon^{-1}\operatorname{div}_y(A_i \nabla_y V_1^i) - \varepsilon^{-1}\operatorname{div}_y A_i(\,\nabla_x V^i + b_i G\,) \\ - \operatorname{div}_y(A_i \nabla_y V_2^i) - \operatorname{div}_x A_i(\,\nabla_x V^i + b_i G\,) - \operatorname{div}_x(A_i \nabla_y V_1^i) \\ - \operatorname{div}_y(A_i \nabla_x V_1^i) = m_i \varrho\, u_0 + r_i F \quad \text{for} \quad y = x/\varepsilon, \;\; x \in \Omega_i^\varepsilon, \end{gathered} \tag{178}$$

where $E_i(x/\varepsilon) = 1$ and $i = 1, \ldots, s$. Here it is useful to keep in mind equality (156). Similarly, retaining only the essential terms, from conjugation conditions, we have

$$\begin{gathered} \Upsilon \cdot \big(A_i \nabla_y V_1^i + A_i \nabla_x V^i + A_i b_i G + \varepsilon A_i \nabla_y V_2^i + \varepsilon A_i \nabla_x V_1^i\big) \\ = \varepsilon\, \Upsilon \cdot (\vartheta A_0 \nabla_y V_0 + \vartheta A_0 \nabla_y E_1 V^1 + \ldots + \vartheta A_0 \nabla_y E_s V^s + \varepsilon\, \vartheta A_0 \nabla_x V_0) \end{gathered} \tag{179}$$

for $y = x/\varepsilon$ and $x \in \partial F_i^\varepsilon$, where $\vartheta = \sigma/\varepsilon^2$ is considered as a parameter. Here it is useful to keep in mind equality (172). If the parameter ϑ is not convenient for consideration, one can choose $\sigma = \varepsilon^2$ and therefore take $\vartheta = 1$ to begin with. In the following, the convenience of using the parameter ϑ will be explained.

Consequently, the coefficients for ε^{-1} in (178) and for ε^0 in (179) are zero, if we choose $V_1^i = N_a^i(y)\nabla_x V^i(x) + N_b^i(y)G(x)$ and define the vector functions $N_a^i(y)$ and $N_b^i(y)$ as 1-periodic solutions of the problems

$$-\operatorname{div}_y(A_i\nabla_y N_a^i) = \operatorname{div}_y A_i \text{ in } Y_i, \quad -\Upsilon\cdot\big(A_i\nabla_y N_a^i\big) = \Upsilon\cdot A_i \text{ on } \partial Y_i,$$

$$-\operatorname{div}_y(A_i\nabla_y N_b^i) = \operatorname{div}_y A_i\, b_i \text{ in } Y_i, \quad -\Upsilon\cdot(A_i\nabla_y N_b^i) = \Upsilon\cdot A_i\, b_i \text{ on } \partial Y_i \tag{180}$$

for $i = 1, \ldots, s$. By virtue of Theorem 6, solutions to the problems exist and are defined up to constants, which will be fixed as before. In addition, one can define these solutions by zero on $\overline{Y_{1s}} \setminus Y_i$ for $i = 1, \ldots, s$.

Let us continue the vector functions on Y_0 as 1-periodic solutions of problems

$$\begin{aligned} &-\operatorname{div}_y(A_0\nabla_y N_a^i) = 0 \text{ in } Y_0, \quad N_a^i = E_i N_a^i \text{ on } \overline{Y_{1s}}, \\ &-\operatorname{div}_y(A_0\nabla_y N_b^i) = 0 \text{ in } Y_0, \quad N_b^i = E_i N_b^i \text{ on } \overline{Y_{1s}}. \end{aligned} \tag{181}$$

Equating the coefficients of ε^0 in the asymptotic equality (178) we have

$$\begin{aligned} -\operatorname{div}_y(A_i\nabla_y V_2^i) &= \operatorname{div}_x\big((A_i + A_i\nabla_y N_a^i)\nabla_x V^i + (A_i b_i + A_i\nabla_y N_b^i)G\big) \\ &+ \operatorname{div}_y(A_i N_a^i \nabla_x^2 V^i + A_i N_b^i \nabla_x G) + m_i \varrho\, u_0 + r_i\, F - m_i\, z\, V^i \end{aligned} \tag{182}$$

for $i = 1, \ldots, s$, where it was taken into account that $V_1^i = N_a^i\nabla_x V^i + N_b^i G$.

Accordingly, the coefficient for ε^1 in (179) can be written in the form

$$\begin{aligned} -\Upsilon\cdot\big(A_i\nabla_y V_2^i\big) &= \Upsilon\cdot\big(A_i N_a^i \nabla_x^2 V^i + A_i N_b^i \nabla_x G\big) \\ -\Upsilon\cdot(\vartheta A_0\nabla_y V_0) &- \Upsilon\cdot\big(\vartheta A_0\nabla_y E_1 V^1 + \ldots + \vartheta A_0 \nabla_y E_s V^s\big) \end{aligned}$$

for $y = x/\varepsilon$, $x \in \partial F_i^\varepsilon$ and $i = 1, \ldots, s$. Let us consider this equality as a boundary condition for equation (182), which defines the 1-periodic Neumann problem for $V_2^i = V_2^i(x, y)$. By Theorem 6, this problem is correct if and only if

$$\begin{aligned} \hat{m}_i\,(z\,V^i - e_i u_0) - \operatorname{div}_x \hat{A}_i(\nabla_x V^i) &= \vartheta \int_{\partial Y_i} \Upsilon\cdot A_0 \nabla_y V_0\, d\tau \\ + \operatorname{div}_x \hat{B}_i G + \hat{r}_i F - \vartheta\, a_i^1\, V^1 &- \ldots - \vartheta\, a_i^s\, V^s \end{aligned} \tag{183}$$

for $i = 1, \ldots, s$, where, using (176) and integrating in parts, constants and matrices with constant components are calculated using the following formulas

$$\hat{m}_i = \int_{Y_i} m_i(y)dy, \ \ \hat{m}^i_\varrho = \int_{Y_i} m_i(y)\varrho(y)dy, \ \ \hat{r}_i = \int_{Y_i} r_i(y)dy, \ \ e_i = \hat{m}_i^{-1}\hat{m}^i_\varrho$$

$$\hat{A}_i = \int_{Y_i}\Big(A_i(y)+A_i(y)\nabla_y N^i_a(y)\Big)dy, \ \ \hat{B}_i = \int_{Y_i}\Big(A_i(y)b_i(y)+A_i(y)\nabla_y N^i_b(y)\Big)dy,$$

$$a^j_i = \int_{Y_0}(\nabla_y E_j, A_0\nabla_y E_i)dy = -\int_{\partial Y_i}\Upsilon\cdot A_0\nabla_y E_j\, d\tau \quad \text{for} \quad i,j = 1,\ldots,s.$$

Let's substitute equality (177) instead of U into the first equation from (175) and take into account (155). So, keeping only the terms essential for what follows, we get

$$\mu\, m_0\, z\, V_0 - \vartheta \mathrm{div}_y(A_0\nabla_y V_0) = \mu\, r_0 F + \mu\, m_0(\varrho - e^s E^s)u_0$$
$$- \mu\, m_0 E_1(z\, V^1 - e_1 u_0) - \ldots - \mu\, m_0 E_s(z\, V^s - e_s u_0) \tag{184}$$

for $y = x/\varepsilon$, $x \in \Omega^\varepsilon_0$ and $e^s E^s = e_1 E_1 + \ldots + e_s E_s$. Here $\vartheta = \sigma/\varepsilon^2$ is considered as a parameter. Thus, as for the Dirichlet problem in a perforated region, we can separate the variables. Consequently, equality (184) will be satisfied for every fixed $z \in \mathbb{C}_0$ if we choose $V_0 = V_0(z,x,y)$ in the following form

$$V_0 = P_r(z,y)F(z,x) + P_\varrho(z,y)\, u_0(x)$$
$$- Q_1(z,y)(z\, V^1(z,x) - e_1 u_0) - \ldots - Q_s(z,y)(z\, V^s(z,x) - e_s u_0), \tag{185}$$

where P_r, P_ϱ, Q_1,..., Q_s are defined as 1-periodic solutions of the problems

$$\mu\, m_0(zP_\varrho + e^s E^s - \varrho) - \vartheta \mathrm{div}_y(A_0\nabla_y P_\varrho) = 0 \ \text{in} \ Y_0, \quad P_\varrho = 0 \ \text{on} \ \overline{Y}_{1s},$$

$$\mu\, m_0(z\, P_r - m_0^{-1} r_0) - \vartheta \mathrm{div}_y(A_0\nabla_y P_r) = 0 \ \text{in} \ Y_0, \quad P_r = 0 \ \text{on} \ \overline{Y}_{1s},$$

$$\mu\, m_0(z\, Q_1 - E_1) - \vartheta \mathrm{div}_y(A_0\nabla_y Q_1) = 0 \ \text{in} \ Y_0, \quad Q_1 = 0 \ \text{on} \ \overline{Y}_{1s}, \ldots,$$
$$\mu\, m_0(z\, Q_s - E_s) - \vartheta \mathrm{div}_y(A_0\nabla_y Q_s) = 0 \ \text{in} \ Y_0, \quad Q_s = 0 \ \text{on} \ \overline{Y}_{1s}, \tag{186}$$

Note that the inverse Laplace transform of such problems determines equivalent (isometrically isomorphic) nonstationary 1-periodic cell problems of the form

$$\mu\, m_0(p_\varrho)'_t - \vartheta\, \mathrm{div}_y(A_0\nabla_y\, p_\varrho) = 0 \quad \text{in} \quad Y_0 \times (0,\infty),$$

$$p_\varrho\big|_{t=0} = \varrho - e^s E^s \ \text{in} \ Y_0, \ \ p_\varrho = 0 \ \text{on} \ \overline{Y}_{1s} \times (0,\infty),$$

$$\mu\, m_0(p_r)'_t \;-\; \vartheta\, \mathrm{div}_y(A_0 \nabla_y\, p_r) = 0 \quad \text{in} \quad Y_0 \times (0,\infty),$$
$$p_r\big|_{t=0} = m_0^{-1} r_0 \;\text{ in } Y_0, \quad p_r = 0 \;\text{ on } \overline{Y}_{1s} \times (0,\infty), \tag{187}$$
$$\mu\, m_0(q_i)'_t \;-\; \vartheta\, \mathrm{div}_y(A_0 \nabla_y\, q_i) = 0 \quad \text{in} \quad Y_0 \times (0,\infty),$$
$$q_i\big|_{t=0} = E_i \;\text{ in } Y_0, \;\; q_i = 0 \;\text{ on } \overline{Y}_{1s} \times (0,\infty) \text{ for } i = 1, \ldots s,$$

where $\hat{p}_\varrho = P_\varrho$, $\hat{p}_r = P_r$, $\hat{q}_1 = Q_1, \ldots, \hat{q}_s = Q_s$ are the Laplace transforms of such solutions. It is well known [22] that the corresponding solutions to problems (187) exist and are uniquely determined. Thus, the function $V_0 = V_0(z, x, y)$ is defined and equality (184) due to (185) takes the form

$$\vartheta\, \mathrm{div}_y(A_0 \nabla_y V_0) = \mu\, m_0(zP_r - m_0^{-1} r_0)F + \mu\, m_0(zP_\varrho - \varrho + e^s E^s)\, u_0$$
$$-\,\mu\, m_0(z\, Q_1 - E_1)(zV^1 - e_1 u_0) - \ldots - \mu\, m_0(z\, Q_s - E_s)(zV^s - e_s u_0). \tag{188}$$

Let's multiply this equation by E_i and integrate the resulting equality over Y_0 and parts. Then for the left side of this equality we have

$$-\,\vartheta \int_{\partial Y_0} E_i \Upsilon \cdot A_0 \nabla_y V_0 \, d\tau \;-\; \vartheta \int_{Y_0} (A_0 \nabla_y V_0, \nabla_y E_i)\, d\tau = -\,\vartheta \int_{\partial Y_i} \Upsilon \cdot A_0 \nabla_y V_0 \, d\tau$$

for $i = 1, \ldots, s$ by (176). Consequently, from (188) we conclude that

$$-\,\vartheta \int_{\partial Y_i} \Upsilon \cdot A_0 \nabla_y V_0 \, d\tau = \mu\, S_r^i F + \mu\, S_\varrho^i\, u_0$$
$$-\,\mu\, S_1^i (z\, V^1 - e_1 u_0) - \ldots - \mu\, S_s^i (z\, V^s - e_s u_0), \tag{189}$$

where

$$S_r^i = \int_{Y_0} m_0(z\, P_r - m_0^{-1} r_0)\, E_i\, dy, \quad S_\varrho^i = \int_{Y_0} m_0(z\, P_\varrho - \varrho + e^s E^s)\, E_i\, dy,$$

$$S_1^i = \int_{Y_0} m_0(z\, Q_1 - E_1)\, E_i\, dy, \;\; \ldots, \;\; S_s^i = \int_{Y_0} m_0(z\, Q_s - E_s)\, E_i\, dy.$$

The inverse Laplace transform of such functions, depending only on $z \in \mathbb{C}_0$ in accordance with (186), is defined by the following equalities

$$R_r^i(t) = \int_{Y_0} m_0(p_r(t,y))'_t\, E_i\, dy, \quad R_\varrho^i(t) = \int_{Y_0} m_0(p_\varrho(t,y))'_t\, E_i\, dy, \tag{190}$$

$$M_1^i(t) = \int_{Y_0} m_0(q_i(t,y))'_t\, E_i\, dy, \;\ldots\;, M_s^i(t) = \int_{Y_0} m_0(q_i(t,y))'_t\, E_i\, dy \tag{191}$$

for $i = 1, \ldots, s$. The functions $R_r^i, \ldots, M_s^i$ are defined and depend only on the variable $t \in [0, \infty)$, since $p_r, \ldots, q_s$ are uniquely defined as solutions to problems (187).

Therefore, equation (183) can be rewritten in the following form

$$\hat{m}_i\,(z\,V^i - e_1 u_0) - \mu\,S_1^i(z\,V^1 - e_1 u_0) - \ldots - \mu\,S_s^i(z\,V^s - e_s u_0)$$
$$- \operatorname{div}_x \hat{A}_i(\nabla_x V^i) + \vartheta\,a_i^1\,V^1 + \ldots + \vartheta\,a_i^s\,V^s$$
$$= \operatorname{div}_x \hat{B}_i G - \mu\,S_\varrho^i\,u_0 + \hat{r}_i F - \mu\,S_r^i F, \quad V^i\big|_{\partial\Omega} = 0 \tag{192}$$

for $i = 1, \ldots, s$. Thus, we have obtained s equations for the s unknown functions $V^1(z,x), \ldots, V^s(z,x)$. Such a system of s equations is uniquely solvable for a fixed $z \in \mathbb{C}_0$. But the behavior of the solutions of this system for different $z \in \mathbb{C}_0$ significantly affects the regularity of such solutions.

Consequently, all terms in equality (183) are defined for $i = 1, \ldots, s$. Subtracting these equalities from (182) we obtain solvable equations for $V_2^1(x,y), \ldots, V_2^s(x,y)$ and the construction may be continue. Moreover, one can represent $V_0(z,x,y) = V_0^0(z,x,y) + \varepsilon\,V_0^1(z,x,y) + \ldots$ and determine $V_0^1(z,x,y)$ as in (184). Such constructions will be quite cumbersome, but the interested reader has every opportunity to write the corresponding terms of the expansions.

The system of equations (192), taking into account (190) and (191), defines a homogenized problem for problem (152) in the case under consideration. Indeed, applying the inverse Laplace transform to this equality, we have for the functions $v_1(t,x), \ldots, v_s(t,x)$ the following homogenized initial boundary value problem

$$\hat{m}_1(v_1)'_t - \mu M_1^1 * (v_1)'_t - \ldots - \mu M_s^1 * (v_s)'_t + \vartheta\,a_1^1\,v_1 + \ldots + \vartheta\,a_s^1\,v_s$$
$$- \operatorname{div}_x \hat{A}_1(\nabla_x v_1) = \operatorname{div}_x \hat{B}_1 g - \mu R_\varrho^1\,u_0 + \hat{r}_1 f - \mu R_r^1 * f, \ldots,$$
$$\hat{m}_s(v_s)'_t - \mu M_1^s * (v_1)'_t - \ldots - \mu M_s^s * (v_s)'_t + \vartheta\,a_1^1\,v_1 + \ldots + \vartheta\,a_s^1\,v_s$$
$$- \operatorname{div}_x \hat{A}_s(\nabla_x v_s) = \operatorname{div}_x \hat{B}_s g - \mu R_\varrho^s\,u_0 + \hat{r}_s f - \mu R_r^s * f, \tag{193}$$
$$v_i\big|_{t=0} = e_i u_0 \text{ in } \Omega, \quad v_i = 0 \text{ on } \partial\Omega \times (0,T) \text{ for } i = 1, \ldots, s,$$

where $*$ denotes the convolution operator relative to t, for example, by definition

$$M_i^j * (v_i)'_t = \int_0^t M_i^j(t-\tau)(v_i)'_\tau(\tau,x)\,d\tau \text{ for } i,j = 1, \ldots, s.$$

Let us recall that $e_i = \hat{m}_i^{-1}\hat{m}_\varrho^i$ is the same as in (162) but with the calculation of integrals only over Y_i. Moreover, if $\varrho = 1$, then $e_i = 1$,

$e^s E^s = e_1 E_1 + \ldots + e_s E_s = 1$ and $R^i_\varrho = 0$ according to (187) for $i = 1, \ldots, s$. Thus, the function R^i_ϱ measures the average deviation of $\varrho(y)$ from 1 with weight e_i and introduces non-trivial external forces into the homogenized problem even if $f = 0$ and $g = 0$.

Indeed, even for $s = 1$ the resulting homogenized problem has the form

$$\begin{aligned} &\hat{m}_1 (v_1)'_t - \mu M^1_1 * (v_1)'_t - \operatorname{div}_x \hat{A}_1 (\nabla_x v_1) \\ &= \operatorname{div}_x \hat{B}_1 g - \mu R^1_\varrho u_0 + \hat{r}_1 f - \mu R^1_r * f, \end{aligned} \tag{194}$$

$$v_1\big|_{t=0} = e_1 u_0 \text{ in } \Omega, \quad v_1 = 0 \text{ on } \partial\Omega \times (0, T).$$

Hence, even if $f = 0$ and $g = 0$, this problem will involve the external force $-\mu R^1_\varrho u_0$. From the point of view of elasticity, the latter means that the presence of microstresses in the composite can lead to the action of external forces on it, the so-called effect of residual stresses. In accordance with (162), the effect can be observed only for composites with contrast coefficients for the models under consideration.

Note that the function $R^1_\varrho = R^1_\varrho(t)$ is not very regular, which may prevent the regularity of the solution to the homogenized problem. In fact, it is known [80], [83], that $R^1_\varrho \in L^1(0, \infty)$. Similarly, the functions M^j_i, R^i_r, R^i_ϱ for $i, j = 1, \ldots, s$ are elements of the space $L^1(0, \infty)$ in accordance with [80], [83].

As an optimal result on the regularity of $R^1_\varrho = R^1_\varrho(t)$, one can state that

$$p_\varrho(t, y) \in L^p(0, \infty; W^{1,p}_0(Y_0)) \cap W^{1,p}(0, \infty; L^p(Y_0)) \quad \text{for} \quad 1 < p < 3/2, \tag{195}$$

for example, according to [88], where further details can be found. Such inclusions are accurate and cannot be improved even if $p_\varrho\big|_{t=0} = 1$. This is due to the fact that assuming sufficient regularity of $p_\varrho(t, y)$ and passing to the limit for $t \to 0$, one should obtain $p_\varrho\big|_{t=0} = 0$ on $\bar{Y}_{1s}$, which is not true for $p_\varrho\big|_{t=0} = 1$.

Thus, $R^1_\varrho \in L^p(0, \infty)$ for $1 \leqslant p < 3/2$ by definition (190). For the same reason, similar inclusions are valid for functions M^j_i, R^i_r, R^i_ϱ for $i, j = 1, \ldots, s$. Such inclusions can be significantly improved only by assuming that these functions are equal to zero on $\bar{Y}_{1s}$, which is incompatible, for example, with (176). However, inclusions in $L^1(0, \infty)$ for these functions are sufficient for the correct definition of convolutions in the homogenized problem, as proven [80], [83].

Another obstacle to the regularity of solutions to problem (193) may be the structure of the matrix $\hat{A}_i$ here and below $i = 1, \ldots, s$. It is known [89], that the matrix $\hat{A}_i$ is symmetric, non-negative definite and can have rank

$0, 1, 2, \ldots n$. Moreover, this rank is a topological invariant of Y_i as a submanifold of the torus according to [89]. Namely, the rank is equal to the number of basic cycles of the torus that belong to an Y_i with some open neighborhood.

For example, if F_i is connected, then all basis cycles belong to Y_i and the rank is equal to n, as already noted for the Neumann problem in a perforated domain. Another extreme case, Y_i is compactly embedded in Y (such as Y_2 in **Fig. 3**). Then, the solution to the first problem from (180) is $N_a^i = (-y_1, \ldots, -y_n)$ and therefore $\hat{A}_i = 0$ by definition for the index $i = 1, \ldots, s$ under consideration. The function N_a^i is determined up to a constant. Remark that the choice of the constant, as before, sets the center of gravity of Y_i as a homogeneous body. It is interesting that in a similar elasticity problem with contrast coefficients, considered in [80], the homogenized problem describes the rotations of the body around the center, which indicates the absorption of energy at the microlevel for a large number of the bodies.

It is known [78] and can be directly verified that the matrices $\{-S_i^j\}_{i,j=1}^s$ and $\{a_i^j\}_{i,j=1}^s$ in system (192) are symmetric and non-negative definite and therefore may not be taken into account in a priori estimates. Consequently, such a system of equations has a unique solution for a fixed $z \in \mathbb{C}_0$, since the corresponding differential operator can be inverted on functions with compact support. However, sufficient regularity of solutions may not be satisfied, which is due to the need to consider such equations in anisotropic Sobolev spaces, in which it is very difficult to work with regularity. Whereas in the case when the rank of the matrices $\hat{A}_i$ for $i = 1, \ldots, s$ are equal to n or 0, such problems do not arise and one can use elliptic regularity for second or zero order equations to prove sufficient regularity.

Thus, the asymptotic solution expansion for problem (175) has the form

$$U_a = E_1(x/\varepsilon)V^1 + \ldots + E_s(x/\varepsilon)V^s + P_r(z, x/\varepsilon)F + P_\varrho(z, x/\varepsilon)u_0$$

$$- Q_1(z, x/\varepsilon)(zV^1(z, x) - e_1 u_0) - \ldots - Q_s(z, x/\varepsilon)(zV^s(z, x) - e_s u_0)$$

$$+ \varepsilon N_a^1(x/\varepsilon)\nabla_x V^1 + \varepsilon N_b^1(x/\varepsilon)G + \ldots + \varepsilon N_a^s(x/\varepsilon)\nabla_x V^s + \varepsilon N_b^s(x/\varepsilon)G \tag{196}$$

where solutions to problems (176), (180), (181), (186), and (192) are used. Here, $V_2^1(x, y), \ldots, V_2^s(x, y)$ are chosen to be zero for simplicity, since these functions will not contribute to improving the relevant accuracy estimate.

It is known [26] that the product of two functions depending on $z \in \mathbb{C}_0$ (and other variables as parameters) under the inverse Laplace transform transforms into the convolution of the inverse Laplace transforms of these functions, which in fact has already been used in the transition from problem (192) to problem (193). Consequently, applying this transformation to (196) we obtain the initial terms of the asymptotic solution expansion for problem (152) in the following form

$$u_a = E_1^\varepsilon v_1 + \ldots + E_s^\varepsilon v_s + p_r^\varepsilon * f - q_1^\varepsilon * (v_1)'_t - \ldots - q_s^\varepsilon * (v_s)'_t$$

$$+ p_\varrho^\varepsilon u_0 + \varepsilon N_a^{1\varepsilon} \nabla_x v_1 + \varepsilon N_b^{1\varepsilon} g + \ldots + \varepsilon N_a^{s\varepsilon} \nabla_x v_s + \varepsilon N_b^{s\varepsilon} g, \tag{197}$$

where the notations $E_1^\varepsilon = E_1(x/\varepsilon)$, ..., $q_1^\varepsilon = q_1(t, x/\varepsilon)$, ..., $N_b^{s\varepsilon} = N_b^{s\varepsilon}(x/\varepsilon)$ are introduced and convolution is denoted as for (193).

The fact that this is indeed an approximation for solutions to problem (152) is proven in [78] and [80] under the assumption that the solution to problem (193) is sufficiently regular (the exact assumption is in [80]). Note that such regularity was proven in [90] (for $s = 1$) and in [78], [80] for the case when the $u_0 = 0$ and rank of the matrices $\hat{A}_i$ for $i = 1, \ldots, s$ in (193) is equal to n or 0. In addition, in the case of small μ, it is only assumed that the order of smallness of the value μ does not exceed a power of ε, namely, the positive $\mu = \mu(\varepsilon)$ is a function of the parameter $\varepsilon \in (0, \varepsilon_0]$ and there exists a positive κ such that

$$\varepsilon^\kappa / \mu \to 0 \qquad \text{for} \quad \varepsilon \to 0. \tag{198}$$

This condition can also be written in the form $\mu \in [\varepsilon^{\kappa - \nu}, 1]$ for positive ν, which prevents very rapid degeneration at small μ, for example, exponential decay.

Theorem 7. *Let u be a solution to problem (152) and u_a be defined by equality (197) for regular $v_1, \ldots, v_s$, $u_0 = 0$ and (198) is valid. Then*

$$\mathcal{E}(u - u_a) \leqslant C(\varepsilon + \sigma), \tag{199}$$

where the constant C is independent of μ, ε and σ for $\varepsilon^{\kappa-\nu} \leqslant \mu \leqslant 1$, $0 < \varepsilon \leqslant \varepsilon_0$ $0 < \sigma \leqslant \sigma_0$ with the appropriate ε_0 and σ_0.

This theorem was proven in [78]. Here $\mathcal{E}(u)$ is an energy norm of this form

$$\mathcal{E}(u) = \| u \|^2_{C^0([0,T]; L^2(\Omega \setminus \Omega_0^\varepsilon))} + \mu \| u \|^2_{C^0([0,T]; L^2(\Omega_0^\varepsilon))}$$

$$+ \| \nabla u \|^2_{L^2(0,T; L^2(\Omega \setminus \Omega_0^\varepsilon)^n)} + \sigma \| \nabla u \|^2_{L^2(0,T; L^2(\Omega_0^\varepsilon)^n)}.$$

It should probably be emphasized that problems (174) and (175) are not used in the proof of this theorem in [78], and the solution to problem (152) is defined in the sense of distributions. This definition uses integrate by parts and therefore the equations of problem (174) arise implicitly, since Ω is the union of the sets Ω_0^ε, ..., Ω_s^ε. Further details can be found in [78], where the Lipschitz property of the boundaries is essential for integration by parts, since without this the relevant integrals may simply not be defined. The same property of the boundaries is used in Theorem 6, without which this theorem may not be correct.

The conditions for the regularity of solutions to the homogenized system of equations (193) are essential for obtaining an accuracy estimate as for all the problems considered. This regularity is based on the regularity of the data. If the data is not regular, for example, $f \in L^2(\Omega)$. Then, from Theorem 7 it is possible to deduce that $\mathcal{E}(u - u_a) \to 0$ for $\varepsilon \to 0$. For this, f is approximated by smooth and compactly supported functions for which the accuracy estimate of Theorem 7 can be established and then it is possible to go to the limit. Details can be found in [78] and [90]. By inclusions (195), the condition $u_0 = 0$ in Theorem 7 is also essential. Without this condition, we can only prove convergence in weak norms (defined by (195)), again using approximations by smooth functions and the estimate of Theorem 7. However, it should be noted that the inclusions are relevant only for norms on Ω_0^ε.

Problem (175) was considered in [80] and [90] as an equivalent isomorphic image of problem (152). It was for problem (175) that asymptotic solution expansions were constructed in the works, as well as here. The asymptotic expansions and the accuracy theorems from [80] and [90] are given in dissertation [19]. Problem (175) was also considered in [91], where statements about the two-scale convergence of solutions are given for $\mu = 1$, $\sigma = \varepsilon^2$, $m_\mu^\varepsilon = r_\mu^\varepsilon = 1$ and $b^\varepsilon = 0$. The author of [91] was a referee of dissertation [19], from which the two-scale convergence results were rewritten. The two-scale convergence statements directly follow from the definition and norm convergence proven in [90] for regular boundaries.

But the methods of papers [91] are interesting. For example, formula (3.14) in [91] has the form $\int_F \rho(y)\, b(y) \cdot \nabla\varphi(y) dy = -\int_F \rho(y)\, \varphi(y)\, \mathrm{div}\, b(y)\, dy$ for smooth functions $\rho(y)$, $b(y)$, $\varphi(y)$ and the perforated domains ([91], p.978), where F is an arbitrary open subdomain in $(0,1)^n$. This formula is remarkable already for $n = 1$ and $\rho = b = 1$. Indeed, the Newton-Leibniz formula is $\int_a^c f_y'(y)\, dy = f(c) - f(a)$, but from formula (3.14) in [91] it follows that $\int_a^c f_y'(y)\, dy = 0$, which, together with Fubini theorem, implies that the integral of any smooth function is zero for every n. The interested reader can also deduce from formula (3.14) in [91] that the gradient of the smooth function $\rho(y)$ is zero, and by duality this is also true for distributions.

In addition, the boundary integral over ∂F is ignored for an arbitrary domain F in formula (3.14) of [91]. Such an integral may be undefined, and there are domains whose measure of the boundary is infinite and therefore the boundary integral may be infinite. Thus, in [91], uncertainty and infinity are reduced to zero. However, paper [91] has hundreds of citations and (3.14) from there has been rewritten in dozens of papers. For example, this can be found in [92] and [93], one can find a random analogue of this formula in [94]. Moreover, not only this formula is rewritten in [95], but also the results on two-scale convergence as applied to the elasticity system, which strangely coincide with the results from [80]. In book [96] this formula is reproduced several times and is used in [97], where an expression that is not a norm is called a norm, as in other masterpieces. It is strange that the rewriters of the formula do not respect the Newton-Leibniz and Stokes formulas. The interested reader can find other gems in [91] and from fans of this paper.

For $s = 1$, Theorem 7 was proven in [78] without additional assumptions on μ, that is, uniformly for $0 < \mu \leqslant 1$. Note also that the accuracy estimate (199) of the theorem is uniform in the parameter $\vartheta = \sigma/\varepsilon^2$, on which problems (187) and (193) depend. Nevertheless, in [80] and [90] accuracy estimates for very small and very large ϑ were separately proved. Asymptotic expansions of solutions for such cases are also constructed there without the assumption that $u_0 = 0$. Similar accuracy estimates were proved in [78], as a consequence of Theorem 7.

The case of large ϑ is interesting, for example. Indeed, the matrix $\{a_i^j\}_{i,j=1}^s$ is symmetric and non-negatively definite, as follows from the definitions for (183). It is known [78] and directly verified that the rank of this matrix is equal to $s-1$ and the one-dimensional kernel of this matrix consists of s-dimensional vectors that are multiples of an vector whose components are all equal to 1. Consequently, the terms with this matrix disappear from (193) if and only if $v_1 = v_2 = \ldots = v_s$. Therefore, for very large ϑ, the last equalities are necessarily satisfied and the system of equations from (193) is reduced to one equation as in (162). An exact statement with accuracy estimates can be found in [78], Corollary 2.5.

For $s = 2$, the matrix is defined by one number $a = a_1^1 = -a_1^2 = -a_2^1 = a_2^2 > 0$ and problem (193) takes the following form

$$
\begin{aligned}
&\hat{m}_1(v_1)'_t - \mu M_1^1 * (v_1)'_t - \mu M_2^1 * (v_2)'_t - \operatorname{div}_x \hat{A}_1(\nabla_x v_1) + \vartheta\, a\,(v_1 - v_2)\\
&\qquad = \operatorname{div}_x \hat{B}_1 g - \mu R_\varrho^1 u_0 + \hat{r}_1 f - \mu R_r^1 * f,\\
&\hat{m}_2(v_2)'_t - \mu M_1^2 * (v_1)'_t - \mu M_2^2 * (v_2)'_t - \operatorname{div}_x \hat{A}_2(\nabla_x v_2) - \vartheta\, a\,(v_1 - v_2) \qquad (200)\\
&\qquad = \operatorname{div}_x \hat{B}_2 g - \mu R_\varrho^2 u_0 + \hat{r}_s f - \mu R_r^2 * f,\\
&v_1\big|_{t=0} = e_1 u_0,\ v_2\big|_{t=0} = e_2 u_0 \ \text{ in } \Omega, \quad v_1 = v_2 = 0 \ \text{ on } \partial\Omega \times (0,T).
\end{aligned}
$$

The equations from the system for $\mu = 0$ are called *the reaction-diffusion system*. Note also that the presence of convolutions in such systems is far from trivial, since the spectrum of the operator of problem (192) isometrically equivalent to problem (193) already for $s = 1$ contains gaps, as for the Mathieu problem (121). Accordingly, in the absence of convolutions, for example, when $\mu = 0$, such gaps do not arise. Further details and exact definitions can be found in [78].

It is also useful to note that expansions (196) and (197) contain rapidly oscillating terms. This means that a significant part of the energy is absorbed on the weakly conducting material, which corresponds to Ω_0^ε. Exact expressions for such energy and accuracy estimates are given in [78] and [98]. Perhaps it is precisely this energy absorption that determines the good thermal insulation properties of wood. Of course, wood does not have a periodic structure. But, it is possible to create artificial composites with good thermal insulation properties or wave-absorbing properties, since similar terms were obtained for the wave equation in [80].

Metamaterials are usually called composite materials whose properties are determined by the resonant properties of their constituent elements, the so-called metaatoms. Metamaterials can be synthesized by introducing various periodic structures with a variety of geometric forms into a certain material, which leads to various unusual properties of such materials. It is possible that it is the oscillating terms in the presented asymptotic expansions, in interaction with the convolutions in them and in the equations, that describe such resonant properties. Let us remark in conclusion that the cell problems (180) and (187) obviously depend on the geometric forms of $Y_1, \ldots, Y_s$, as well as on their topology, as already noted.

Acknowledgment

This work has been supported partially by grant of the Ministry of Education and Science of Ukraine for perspective development of a scientific direction “Mathematical Sciences and Natural Sciences" at Taras Shechenko National University of Kyiv.

References

[1] Bahvalov N. S. Averaged characteristics of bodies with periodic structure. *Dokl. Akad. Nauk SSSR* 218 (1974) 1046-1048. MR0395435

[2] Bahvalov N. S. Averaging of partial differential equations with rapidly oscillating coefficients. *Dokl. Akad. Nauk SSSR* 221:3 (1975) 516-519. MR0380075

[3] Bensoussan A., Lions J.-L., Papanicolaou G. Sur quelques phenomenes asymptotiques stationares. *Comp. Rend. Acad. Sci. Paris. Ser. A.* 281 (1975) 89-94.

[4] Bensoussan A., Lions J.-L., Papanicolaou G. Sur quelques phenomenes asymptotiques d’evolution. *Comp. Rend. Acad. Sci. Paris. Ser. A.* 281 (1975) 317-322.

[5] Babushka I. Homogenization approach in engineering. In: Glowinski, R., Lions, J.L. (eds) Computing methods in applied sciences and engineering. *Lecture Notes in Economics and Mathematical Systems*, vol 134. Springer, Berlin, Heidelberg (1976), 137-153. doi: 10.1007/978-3-642-85972-4_8

[6] Bensoussan A., Lions J.-L., Papanicolau G. *Asymptotic analysis for periodic structures*. Amsterdam, North-Holland, 1978. ISBN: 9780080875262

[7] Bakhvalov N. S., Panasenko G. P. *Homogenization: averaging processes in periodic media*. Kluwer, Dordrecht, 1989. doi: 10.1007/978-94-009-2247-1

[8] Allaire G. Homogenization and two-scale convergence. *SIAM J. Math. Anal.* 23 (1992) 1482-1518. doi: 10.1137/0523084

[9] Cioranescu D., Damlamian A., Griso G. *The periodic unfolding method.* Springer Nature, Singapore, 2018. doi: 10.1007/978-981-13-3032-2

[10] Jager W., L. Woukeng L. Homogenization of Richards' equations in multiscale porous media with soft inclusions. *J. Differential Equations.* 281 (2021) 503-549. doi: 10.1016/j.jde.2021.02.012

[11] Gahn M., Neuss-Radu M., Pop I. Homogenization of a reaction-diffusion-advection problem in an evolving micro-domain and including nonlinear boundary conditions. *J. Differential Equations.* 289 (2021) 95-127. doi: 10.1016/j.jde.2021.04.013

[12] Garttner S., Frolkovic P., Knabner P., Ray N. Efficiency and accuracy of micro-macro models for mineral dissolution. *Water Resources Research.* 56:8 (2020). doi: 10.1029/2020WR027585.

[13] Garttner S., Frolkovic P., Knabner P., Ray N. Efficiency of micro-macro models for reactive two-mineral systems. *Multiscale Modeling and Simulation.* 20:1 (2022) 433-461. doi: 10.1137/20M1380648

[14] Cioranescu D., Damlamian A., Griso G. The periodic unfolding method for countably many scales. *Advances in Math. Sciences and Applications.* 32:1 (2023) 17-58.

[15] Sandrakov G. V. Homogenization of variational inequalities for obstacle problems. *Sbornik Mathematics.* 196:(3-4) (2005) 541-560. doi: 10.1070/sm2005v196n04abeh000891

[16] Sandrakov G. V. Homogenization of nonlinear equations and variational inequalities with obstacles. *Doklady Mathematics.* 73:2 (2006) 178-181. doi: 10.1134/S1064562406020062

[17] Sandrakov G. V. Homogenization of variational inequalities and equations defined by pseudomonotone operators. *Sbornik Mathematics.* 199:(1-2) (2008) 67-98. doi: 10.1070/SM2008v199n01ABEH003911

[18] Sandrakov G. V. Averaging principles for equations with rapidly oscillating coefficients. Mat. Sbornik 180:12 (1989), 1643-1679. (Russian); English transl., *Mathematics of the USSR - Sbornik.* 68:2 (1991) 503-553. doi: 10.1070/SM1991v068n02ABEH002111

[19] Sandrakov G. V. *The homogenization of nonstationary equations with strongly varying coefficients*, Doctoral dissertation in Mathematics and Physics, Moscow State Univ., Moscow 1998. (Russian).

[20] Bogoliubov N. N., Mitropolsky Yu. *A. Asymptotic methods in the theory of non-linear oscillations.* Gordon and Breach, New York, 1961. Zbl 0151.12201

[21] Lions J.-L., Magenes E. *Problemes aux limites non homog'enes et applications*, vol. I. Dunod, Paris, 1968.

[22] Duvaut G., Lions J.-L. *Les inequations en mecanique et en physique.* Dunod, Paris, 1972.

[23] Shubin M. A. *Pseudo differential operators and spectral theory.* Springer Berlin, Heidelberg, 2001. doi: 10.1007/978-3-642-56579-3

[24] Rempel S., Schulze B.-W. *Index theory of elliptic boundary problems.* Akademie-Verlag, Berlin, 1982.

[25] Antonevich A. B. *Linear functional equations. Operator approach.* Birkhauser, Basel, 1996. Zbl 0841.47001

[26] Agranovich M. S., Vishik M. I. Elliptic problems with a parameter and parabolic problems of general type. *Russian Math. Surveys.* 19:3 (1964) 53-157. doi: 10.1070/RM1964v019n03ABEH001149

[27] Sandrakov G. V. Diffusion in a strongly inhomogeneous periodic medium. *Russian Math. Surveys.* 51:5 (1996), 977-978. doi: 10.1070/RM1996v051n05ABEH002995

[28] Sandrakov G. V. The homogenization of nonstationary equations with contrast coefficients. *Doklady Mathematics.* 56:1 (1997) 586-589.

[29] Sandrakov G. V. Homogenization of nonstationary problems of the theory of strong nonhomogeneous elastic media. *Doklady Akademii Nauk.* 358:3 (1998) 308-311.

[30] Dolzmann G., Muller S. Estimates for Green's matrices of elliptic systems by Lp theory. *Manuscripta Math.* 88 (1995) 261-273. doi: 10.1007/BF02567822

[31] Sakellaris G. On scale-invariant bounds for the Green's function for second-order elliptic equations with lower-order coefficients and applications. *Analysis & PDE.* 14:1 (2021) 251-299. doi: 10.2140/apde.2021.14.251

[32] Kim S., Sakellaris G. The Neumann Green function and scale invariant regularity estimates for elliptic equations with Neumann data in Lipschitz domains. *ArXiv-Analysis of PDEs* (2023). doi: 10.48550/arXiv.2302.00132

[33] Luo X., Yang L., Cui Y. Microneedles: materials, fabrication, and biomedical applications. *Biomedical Microdevices* 25:20 (2023). doi: 10.1007/s10544-023-00658-y

[34] Peng, T., Chen, Y., Hu, W. et al. Microneedles for enhanced topical treatment of skin disorders: applications, challenges, and prospects. *Engineering* (2023). doi: 10.1016/j.eng.2023.05.009

[35] Sandrakov G. V., Lyashko S. I., Bondar E. S., Lyashko N. I. Modeling and optimization of microneedle systems. *J. Automation and Information Sciences.* 51:6 (2019) 1-11. doi: 10.1615/jautomatinfscien.v51.i6.10

[36] Sandrakov G. V., Lyashko S. I., Bondar E. S., Lyashko N. I., Semenov V. V. Modeling of configurations formed when using microneedle systems. *J. Automation and Information Sciences.* 52:12 (2020) 1-11. doi: 10.1615/JAutomatInfScien.v52.i12.10

[37] Vladimirov V. S. *Equations of mathematical physics.* Dekker, New York, 1971.

[38] Tymoshenko A., Klyushin D., Lyashko S. Optimal control of point sources in Richards-Klute equation. *Advances in Intelligent Systems and Computing.* 754 (2019) 194-203. doi: 10.1007/978-3-319-91008-6_20

[39] Lyashko S., Klyushin D., Timoshenko A. Optimal control of buried point sources in a two-dimensional Richards-Klute equation. *Lecture Notes in Networks and Systems.* 344 (2022) 59-71. doi: 10.1007/978-3-030-89902-8_5

[40] Chemeris A., Sushko S., Reznikova S. Smart tiling for program optimization and parallelization. *Communications in Computer and Information Science.* 1562 (2021) 12-32. doi: 10.1007/978-3-030-98883-8_2

[41] Mokhor V., Korchenko O., Honchar S., Komarov M., Onyskova A. Research of the impact on the ecology of the state of cybersecurity of the critical infrastructure objects. *E3S Web of Conferences.* 280 (2021) 09009. doi: 10.1051/e3sconf/202128009009

[42] Levinson N. The first boundary value problem for $\varepsilon\Delta u + Au_x + Bu_y + Cu = D$ for small ε. *Annals of Mathematics.* 51 (1950) 428-445. doi: 10.2307/1969333

[43] Oleinik O. A. On equations of elliptic type with a small parameter in the highest derivatives. *Mat. Sbornik.* 31(73):1 (1952) 104-117. (Russian) Zbl 0049.07601

[44] Kamenomostskaya S. L. The first boundary problem for equations of elliptic type with a small parameter with the highest derivatives. *Izvestiya. Akad. Nauk SSSR, Ser. Mat.* 19:5 (1955) 345-360. (Russian) Zbl 0049.07601

[45] Kopacek Ju. On the Dirichlet problem for elliptic equations with a small parameter in the highest derivatives. *Uspehi Mat. Nauk.* 12:5(77) (1957) 211-220. (Russian) Zbl 0083.32405

[46] Kellogg R. B., Stynes M. Corner singularities and boundary layers in a simple convection-diffusion problem. *J. Differential Equations.* 213 (2005) 81-120. doi: 10.1016/j.jde.2005.02.011

[47] Hamouda M., Han D., Jung C.-Y., Tawri K., Temam R. Boundary layers for the subcritical modes of the 3D primitive equations in a cube. *J. Differential Equations.* 267:1 (2019) 61-96. doi: 10.1016/j.jde.2019.01.005

[48] Poulton C. G., Botten L. C., McPhedran R. C., Movchan A. B. Source-neutral Green'functions for periodic problems in electrostatics, and their equivalents in electromagnetism. *Proc. R. Soc. Lond. A.* 455 (1999) 1107-1123. doi: 10.1098/rspa.1999.0351

[49] Alexandrov A.D. Conditions for uniqueness and estimates for the solution of the Dirichlet problem. *Vestn. Leningrad. Univ. Ser. mathematics, mechanics, astronomy.* 3 (1963) 5-29. (Russian)

[50] Dong R., Li D. Uniform $C^{1,1}$ estimates of homogenization of elliptic equations in non-divergence form with Dini conditions. *Discrete and Continuous Dynamical Systems.* 44:1 (2024) 318-341. doi: 10.3934/dcds.2023105

[51] Egorov Yu. V. *Lectures on partial differential equations.* Additional chapters. M. Publishing house of Moscow State University. 1985.

[52] Bakhvalov N. S., Eglit M. E. Effective moduli of composites reinforced by systems of plates and bars. *Comput. Math. Math. Phys.* 38:5 (1998) 783-804. Zbl 1003.74061

[53] Egorov Yu. V. *Lectures on partial differential equations.* Additional chapters. Publishing house of Moscow State University, Moscow. 1985.

[54] Grebenikov E. A., Mitropolsky Yu. A., Ryabov, Yu. A. *Asymptotic methods in resonance analytical dynamics. Stability and control: theory, methods and applications.* Chapman and Hall, Boca Raton, 2004.

[55] Hartman P. *Ordinary Differential Equations.* Birkhauser, New Jersey, 1982.

[56] Sandrakov G. V. On a problem of H. Poincare. *Sov. Math., Doklady.* 43:2 (1991) 402-405. Zbl 0778.34025

[57] Shubin M. A. Differential and pseudodifferential operators in spaces of almost periodic functions. *Mathematics of the USSR - Sbornik.* 24:4 (1974) 547-573. doi: 10.1070/SM1974v024n04ABEH001923

[58] Sandrakov G. V. Evolution of resonances under perturbation of a system with constant frequencies. *Sov. Math., Doklady.* 43:2 (1991) 467-470. Zbl 0782.70021

[59] Poschel J. A lecture on the classical KAM theorem. *Proceedings of Symposia in Pure Mathematics.* 69 (2001) 707-732. doi: 10.1090/pspum/069/1858551

[60] Kozlov S. M. Reducibility of quasiperiodic differential operators and averaging. *Transactions of Moscow Mathematical Society.* 2 (1984) 101-126. Zbl 0566.35036

[61] Sandrakov G. V. Averaging of quasiperiodic functions. *Russ. Acad. Sci., Doklady. Math.* 47:2 (1993) 289-293. Zbl 0835.42010

[62] McLachlan N. W. *Theory and application of Mathieu functions.* University Press, Oxford, 1951.

[63] Abramowitz M., Stegun I. A. *Handbook of Mathematical Functions.* Dover, New York, 1968.

[64] Sandrakov G. V., Basileva M.I. Homogenization of the Mathieu equation with a rapidly oscillating potential. *J. computational and applied mathematics.* 1:110 (2017) 63-70. (Russian)

[65] Sandrakov G. V. Asymptotics of the spectrum of the Mathieu equation with a rapidly oscillating potential. *J. computational and applied mathematics.* 1:127 (2018) 66-77. (Russian)

[66] Sandrakov G. V. Asymptotics of bound states for an oscillating potential on an interval. *J. computational and applied mathematics.* 3:129 (2018) 45-51. (Russian)

[67] Titchmarsh E.C. *Eigenfunction expansions associated with second-order differential equations.* Part II. Oxford University Press, Oxford, 1958.

[68] Reed M., Simon B. *Methods of modern mathematical physics.* Vol. IV. Academic Press, New York, 1975.

[69] Krylova A.S., Sandrakov G.V. Homogenization of spectral problem on small-periodic networks. *J. Mathematical Physics, Analysis, Geometry.* 8:4 (2012) 336-356.

[70] Skriganov M.M. The spectrum band structure of the three-dimensional Schrodinger operator with periodic potential. *Inventiones mathematicae.* 80 (1985) 107-122. doi: 10.1007/BF01388550

[71] Nazarov S. A. General averaging procedure for selfadjoint elliptic systems in many-dimensional domains, including thin ones. St. Petersbg. *Math. J.* 7:5 (1996) 681-748. Zbl 0859.35025

[72] Nazarov S. A. The polynomial property of self-adjoint elliptic boundary-value problems and an algebraic description of their attributes. *Russian Mathematical Surveys.* 54:5 (1999) 947-1014. doi: 10.1070/rm1999v054n05abeh000204

[73] Nazarov S. A. New series of asymptotics of eigenvalues of the Sturm-Liouville problem with rapidly oscillating coefficients. Math Notes 52:5 (1992) 1134-1136. doi: 10.1007/BF01236623

[74] Oleinik O.A., Shamaev A.S., Yosifian G.A. *Mathematical problems in elasticity and homogenization.* Elsevier, Amsterdam, 1992.

[75] Sandrakov G. V. Multiphase diffusion models in homogenization. *Doklady Mathematics.* 69:1 (2004) 50-53.

[76] Sandrakov G. V. Multiphase homogenized models for diffusion in highly nonhomogeneous media. *Doklady Mathematics.* 70:1 (2004) 507-511.

[77] Sandrakov G. V. Multiphase models of nonstationary diffusion arising from homogenization. *Computational Mathematics and Mathematical Physics.* 44:10 (2004) 1741-1756.

[78] Sandrakov G. V. Multiphase homogenized diffusion models for problems with several parameters. *Izvestiya: Mathematics.* 71:6 (2007) 1193-1252. doi: 10.1070/IM2007v071n06ABEH002387

[79] Sandrakov G. V. The averaging of dynamic problems in the theory of elasticity with strongly varying coefficients. *Doklady Mathematics.* 58:2 (1998) 238-241.

[80] Sandrakov G. V. Homogenization of elasticity equations with contrasting coefficients. *Sbornik: Mathematics.* 190:12 (1999) 1749-1806. doi: 10.1070/SM1999v190n12ABEH000443

[81] Sandrakov G. V. Averaging of dynamic processes in composites periodically reinforced by fibers. *Doklady Mathematics.* 61:3 (2000) 346-349.

[82] Sandrakov G. V. Homogenization of the non-stationary stokes flow in a periodic porous medium. *Doklady Akademii Nauk.* 347:3 (1996) 312-315.

[83] Sandrakov G. V. Homogenization of non-stationary Stokes equations with viscosity in a perforated domain. *Izvestiya Mathematics.* 61:1 (1997) 113-141. doi: 10.1070/IM1997v061n01ABEH000107

[84] Sandrakov G. V. Influence of viscosity on acoustic phenomena in porous media. *Russian J. Numerical Analysis and Mathematical Modelling.* 13:3 (1998) 245-264. doi: 10.1515/rnam.1998.13.3.245

[85] Mitrea M., Taylor M. Boundary layer methods for Lipschitz domains in Riemannian manifolds. *J. Functional Analysis.* 163:2 (1999) 181-251. doi: 10.1006/jfan.1998.3383

[86] Agranovich M. S. Spectral problems for second-order strongly elliptic systems in smooth and non-smooth domains. *Russian Math. Surveys.* 57:5 (2002) 847-920. doi: 10.1070/rm2002v057n05abeh000552

[87] Temam R. *Navier-Stokes equations. Theory and numerical analysis.* North-Holland, Amsterdam-New York, 1979.

[88] Amann H. Parabolic equations on uniformly regular Riemannian manifolds and degenerate initial boundary value problems. In: Amann H., Giga Y., Kozono H., Okamoto H., Yamazaki M. (eds) Recent Developments of Mathematical Fluid Mechanics. *Advances in Mathematical Fluid Mechanics.* Birkhauser, Basel, (2016) 43-77. doi: 10.1007/978-3-0348-0939-9_4

[89] Sandrakov G. V. A topological interpretation of averaged coefficients for periodic perforated domains. *Doklady Mathematics.* 61:1 (2000) 37-39.

[90] Sandrakov G. V. Homogenization of parabolic equations with contrasting coefficients. *Izvestiya Mathematics.* 63:5 (1999) 1015-1061. doi: 10.1070/im1999v063n05abeh000264

[91] Zhikov V.V. On an extension of the method of two-scale convergence and its applications. *Sbornik Mathematics.* 191:(7-8) (2000) 973-1014. doi: 10.1070/sm2000v191n07ABEH000491

[92] Chechkin G.A., Jikov V.V., Lukkassen D., Piatnitski A.L. On homogenization of networks and junctions. *Asymptotic Analysis.* 30:1 (2002) 61-80.

[93] Cherednichenko K.D., Smyshlyaev V.P., Zhikov V.V. Non-local homogenized limits for composite media with highly anisotropic periodic fibres. Royal Society of Edinburgh. *Proceedings A.* 136:1 (2006) 87-114. doi: 10.1017/S0308210500004455

[94] Zhikov V.V., Pyatnitskii A.L. Homogenization of random singular structures and random measures. *Izvestiya Mathematics.* 70:1 (2006) 19-67. doi: 10.1070/im2006v070n01abeh002302

[95] Cherednichenko K.D. Two-scale asymptotics for non-local effects in composites with highly anisotropic fibres. *Asymptotic Analysis.* 49:(1-2) (2006) 39-59.

[96] Chechkin G.A., Piatnitski A.L., Shamaev A. S. Homogenization. *Methods and applications.* Providence, AMS, 2007.

[97] Bourgeat A., Chechkin G. A., Piatnitski A. L. Singular double porosity model. *Applicable Analysis.* 82:2 (2003) 103-116. doi: 10.1080/0003681031000063739

[98] Sandrakov G.V., Lyashko S.I., Semenov V.V. Simulation of filtration processes for inhomogeneous media and homogenization. *Cybernetics and Systems Analysis.* 59:2 (2023) 212-230. doi: 10.1007/s10559-023-00556-4

Chapter 3

Ultrasonic Homogenization in the Food Industry

Gökçen Yıldız*
Department of Food Engineering, Bursa Technical University, Bursa, Turkey

Abstract

Ultrasound is sound waves with frequency values ranging from 20 kHz to 10 MHz, above what the human ear can perceive. Ultrasound in the food industry compared to thermal processes such as pasteurization and sterilization, causes less damage to food products, shortens the heat treatment time, accelerates heat transfer in thermal processes, increases the chemical reaction rate, and improves the sensory quality of foods. While it is important to improve the sensory quality of foods, the ease of accessibility and low cost of the equipment are reasons for preference. In this review, the effects of ultrasonic homogenization in the food industry have been presented.

Keywords: ultrasonic homogenization, food, quality

Introduction

Throughout history, people have developed many methods to prevent food from spoiling and to preserve it for a long time. The goal was to control the negative changes caused by physical, chemical, biological, and/or microbiological factors on food products. The concept of long-lasting, high-

* Corresponding Author's Email: gokcen.yildiz@btu.edu.tr.

In: A Closer Look at Homogenization
Editor: Gulcin Yildiz
ISBN: 979-8-89113-656-4

quality, and reliable food was adopted. For this purpose, the control of biological and chemical activities of foods is most commonly provided by thermal processes. However, there has been an increasing interest in non-thermal food preservation techniques due to the negative effects on nutritional value and sensory properties resulting from high temperatures and duration in heat treatment applications. Among these new technologies where heat is not applied, ultrasonic sound waves (ultrasound), pulsed electric field, pulsed light, high pressure applications, microfiltration, X-rays, ultraviolet light, high voltage and oscillating magnetic field methods stand out. These non-thermal processes, on the one hand, ensure the inactivation of spoilage microorganisms and undesirable enzymes, on the other hand, the taste, odor and texture properties of the products are preserved due to the low temperature during the processes.

Ultrasound is sound waves with frequency values ranging from 20 kHz to 10 MHz, above what the human ear can perceive. Ultrasound in the food industry compared to thermal processes such as pasteurization and sterilization, causes less damage to food products, shortens the heat treatment time, accelerates heat transfer in thermal processes, increases the chemical reaction rate, and improves the sensory quality of foods. While it is important to improve the sensory quality of foods, the ease of accessibility and low cost of equipment are also significant factors. Ultrasonic homogenization in the food industry is widely used in many fields such as:

- improving extraction efficiency
- drying fruits and vegetables
- in the filtration of various extracts and beverages
- increasing emulsion stability
- in defoaming process in fermented beverages
- in milk coagulation
- in determining the quality of eggs
- on the surface cracks in cheese
- in the texture of biscuits
- in the control of wine fermentation
- in the characterization of dough
- in determining physicochemical properties such as sugar content and acidity.

Ultrasonic Sound Waves

The emergence of ultrasonic sound waves occurred when the sound waves sent by submarines during the world wars caused mass fish deaths. In the 1960s, low-frequency, high-energy ultrasonic sound waves began to be used in industry [1]. The sound range that the human ear can hear is from 16 Hz to 18 kHz. Ultrasonic sound waves are a form of energy produced by sonic waves of 20 kHz and above. Ultrasonic sound waves are acoustic energy [2]. Although the upper limit of the frequency of ultrasonic sound waves cannot be proven precisely, it is expressed as 5 mHz in gases and 500 MHz in solids.

Ultrasonic sound waves are expressed as:

- sound power (W),
- sound intensity (W/m^2), or
- sound energy density (Ws/m^3) [3].

It is possible to classify ultrasonic sound waves under three categories. They are:

- Power ultrasonic sound waves with a frequency range from 20 kHz to 100 kHz,
- High frequency ultrasonic sound waves with a frequency range of 100 kHz - 1 MHz,
- Between 1 and 10 MHz are recognition ultrasonic sound waves.

Ultrasonic sound waves are generally produced by pressure electricity (piezo electricity) or magnetic compression transducers that create high energy vibrations [4]. When ultrasonic sound wave applications are evaluated in terms of their use in the food industry, they are divided into two main groups:

a. High-intensity and low-frequency ultrasonic sound waves
b. Low-intensity high-frequency ultrasonic sound waves.

Low ultrasonic sound waves have a processing intensity of less than 1 W/m^2 and a frequency of more than 100 kHz. This process is used in many applications including:

- cleaning of surfaces
- inactivation of enzymes
- crystallization
- emulsification
- filtration
- freezing
- tenderization of meats.

On the other hand, high ultrasonic sound waves that have an intensity higher than 1 W/m^2 and a frequency between 18-100 kHz is applied in different processes such as:

- deaeration
- oxidation/reduction
- enzyme & protein extraction
- inactivation of enzymes
- nucleation formation in crystallization [5].

The application of high-power ultrasonic sound waves is very effective in cell disruption, particle size reduction and inactivation of bacterial spores [6]. There are many studies in the literature showing that ultrasonic sound waves are effective in different applications [7-26]. In low-intensity ultrasonic sound waves applications, no physical or chemical change occurs in the material processed by the wave. Low-energy ultrasonic sound waves are most commonly used to determine the physicochemical properties (hardness, ripeness, composition, particle size, acidity, etc.) of food products [27]. High energy ultrasonic sound waves are used for microbial and enzymatic inactivation in foods. While high-energy ultrasonic sound wave applications affect food products physically, chemically, and mechanically, such an effect is not observed in low-energy ultrasonic sound wave applications [28].

Ultrasonic sound waves, as they move through a medium, create many physical, chemical and biochemical effects that allow them to be used in different processes and are related to the frequency and amplitude of the sound waves [27]. The main effect of ultrasonic sound waves is the formation of bubbles, called cavitation, when the distance between molecules is above the normal level in the parts where the pressure of the sound wave decreases while passing through a liquid. It is a phenomenon defined as fading towards the interior when they reach a critical volume [29-30]. Energy accumulation occurs in the region where the internal explosion of cavitation bubbles occurs. This energy heats the environment and causes chemical reactions [31]. As a result, due to this high temperature and pressure, turbulence occurs in the cavitation area with high-energy shear waves [32]. Many factors are effective in the formation of cavitation. At high temperatures, vapor pressure increases and surface tension decreases, resulting in more bubbles. Although more bubbles form at higher temperatures, the intensity of these bubbles when they collapse is lower. At low frequencies, bubble formation occurs in larger sizes and more energy is generated when they are damp. If a high intensity cavitation is desired, the magnitude of ultrasonic sound waves should be increased [33].

Cavitation occurs in two different ways: temporary or permanent. Temporary cavitation occurs in systems called probes at frequencies of 20 kHz and above and at a wave intensity of 10 W/cm^2. The bubbles that occur in temporary cavitation expand in volume with a few acoustic oscillations and burst immediately. The size of the bubbles formed is small [30]. In stable cavitation, the growth of bubbles occurs after an average of 1000 acoustic cycles at frequencies of 200 kHz and higher. The bursting intensity of bubbles is slower than temporary cavitation [34].

Another mechanism associated with microbial inactivation is expressed by free radical formation. During the application of ultrasonic sound waves, free radicals such as hydroxyl, hydrogen or hydroperoxides are formed, and these formed compounds have significant bactericidal effects [35-36].

The basic parts that make up the ultrasonic sound waves system are:

- generator or power provider
- transducer
- conductive part [34, 37].

While the generator, called the power provider, converts alternating current into high-frequency alternating current, the transducer converts high-frequency electrical current into mechanical vibrations. These mechanical vibrations are transmitted to the environment where ultrasonic sound waves will be applied in the transmitter part.

Determinants on the application of ultrasonic sound waves are:

- frequency
- wavelength

- amplitude
- ultrasonic power. In addition, reactor design and the way ultrasound are applied also determine the process efficiency.

Ultrasonic sound waves are applied with two different methods:

- ultrasonic bath, and
- ultrasonic probe.

Ultrasonic baths are generally used to remove air in solvents or to clean small glassware. Ultrasonic baths; It is little used in food processing flows because the repeatability of chemical reactions is very low, and the intensity of the energy transferred is low [38]. Another method is to apply ultrasonic sound waves with ultrasonic probes. In this method, since the probe is immersed in the sample, ultrasonic intensity is transferred to a smaller surface (only the probe tip). This method is more effective than ultrasonic baths and is used in small volume production [39]. In terms of cost, probe systems are more expensive than bath systems.

Uses of Ultrasonic Homogenization in the Food Industry

While ultrasonic sound waves are effective alone in some applications, in others they need to be supplemented with parameters such as heat and pressure to reach the required inactivation level. These applications:

- combination of ultrasonic sound waves and temperature (thermosonication, TS),
- combination of ultrasonic sound waves and pressure (manosonication, MS),
- ultrasonic sound waves, pressure and temperature (manothermosonication, MTS) [31, 40].

Thermosonication is used for pasteurization and sterilization purposes by reducing the process temperature and time with the combination of heat-treatment and ultrasonic sound waves applications. When the ambient temperature approaches the boiling point of the liquid material, the intensity of cavitation decreases as a result of the increase in the vapor pressure of the liquid. To solve this problem, thermosonication is performed under pressure (100-700 kPa) and this process is called manothermosonication. The effect created by the manothermosonication process is synergistic. In other words, the total effect created is much greater than the sum of the effects created by each parameter (ultrasonic sound waves, heat treatment, pressure) that creates this effect if they are used alone. While ultrasonic sound waves sensitize target molecules and structures, their effects may be greater with heat and pressure [31].

Effect of Ultrasonic Homogenization on Microorganisms

Ultrasonic sound waves inactivate microorganisms with the cavitation effect they create, causing less change in foods compared to heat treatment applications. The power of ultrasonic sound waves to kill microorganisms occurs as a result of these sound waves breaking down the cell walls [41]. As a result of sonication, cavitation and gas bubbles occur in this environment. Since the energy of these ultrasonic sound waves cannot effectively retain the vapor in the gas bubbles, the gas bubbles burst, and the medium condenses. The cell walls of the microorganisms in this environment are damaged and become inactive. This area, where the temperature is increased, causes the bacteria to die, but since it contains a limited area, it cannot kill all the bacteria [41]. Therefore, the sonication process must be ensured to reach all parts of the liquid homogeneously.

Research shows that using ultrasonic sound waves together with heat and pressure processes inactivates microorganisms much more effectively [41]. In particular, the MTS process is more effective on heat-resistant microorganisms than traditional methods [9]. Scherba and his colleagues stated that ultrasonic sound wave treatment can destroy *Escherichia coli, Staphylococcus aureus, Bacillus subtilis* and *Pseudomonas aeruginosa* in aqueous suspension [42]. Additionally, it has been stated that the application of ultrasonic sound waves can destroy Trichophyton mentagrophytes molds and Feline herpesvirus type 1 virus. Gram-positive cells are much more resistant than gram-negative cells because their walls are thicker. In a study, the inactivation of Streptococcus faecium and Streptococcus durians was studied with an application of 20 kHz/160 W at different temperatures between 5 and 62°C [43].

Ultrasonic sound waves were applied on *Bacillus subtilis* spores in milk at temperatures between 70 and 95°C [44]. While ultrasonic sound waves alone had no effect, ultrasonic sound waves combined with heat reduced the spore population by around 63-73%. In another study, ultrasonic sound waves were applied to UHT milk with previous norms for Staphylococcus aureus [45]. It has been observed that the heat treatment applied with ultrasonic sound waves reduces the D value by 43% more than the thermal method alone. It has been stated that ultrasonic sound waves applied together with high pressure and thermal processes are effective on Listeria monocytogenes. Since dense liquids and solids prevent the propagation of ultrasonic sound waves, this method is thought to be more effective in the sterilization of milk and fruit juices [42]. In a study investigating the effect of ultrasonic sound waves on the quality characteristics of peach juice, a five-minute ultrasound application was effective on *Eschericha coli* bacteria, and a five-log reduction factor reduction (99.999%) was achieved in the number of microorganisms determined by the American Food and Drug Administration [17]. In a similar study, Yildiz and Aadil [18] stated that the ultrasonic sound waves of different durations (5, 10 and 15 min) they applied to strawberry juice achieved a reduction of 5 log reduction factor in the number of microorganisms determined as the target.

Effect of Ultrasonic Homogenization on Food Enzymes

It is known that ultrasonic sound waves are effective on the inactivation of enzymes. There was disagreement among the researchers: some claimed that the enzyme is unaffected by ultrasonic sound waves alone, while others claimed that the enzyme is affected both chemically and physically by acoustic cavitation. This effect varies depending on the enzyme and process parameters used. But in general, enzyme inactivation with ultrasonic sound waves is explained by the fact that ultrasonic sound waves depolymerize macromolecules. It has been stated that ultrasonic waves are more effective when applied with different parameters such as heat and pressure [9] Ultrasound process, along with thermosonication, monosonication and monothermosonication processes, appears as an effective method against enzymes such as lipoxygenase (LOX), polyphenoloxidase (PPO), peroxidase (POD) found in milk and fruits, and lipase and protoase enzymes that are resistant to heat [35].

In a study, ultrasonic sound waves were applied to tomato peroxidase with 15%-75% ultrasonic power and 20-150 seconds, and it was observed that the enzyme was completely inactive at the end of 50%-75% ultrasonic power and 150 seconds. The study also stated that the loss of vitamin C in tomato juice to which ultrasonic sound waves were applied was less than the loss as a redult of heat treatment [46]. In another study, the LOX enzyme in soybean oil was inactivated by 75 - 85% with ultrasonic sound waves. While pathogenic and spoilage microorganisms in milk can be easily inhibited by ultrasonic sound waves in daily milk, many microorganisms in the environment can reduce the quality and drinkability of milk produced with UHT by producing extracellular lipase and protease [47].

Aadil et al. [48] examined quality criteria such as color, acidity, total soluble matter, antioxidant capacity, total phenol, and ascorbic acid in grape juice, which they treated with ultrasonic sound waves for 30, 60, and 90 minutes with a frequency of 28 kHz. Significant differences were observed in most of the parameters compared to freshly squeezed fruit juice. While a significant increase was achieved in the total antioxidant amount, ascorbic acid and total phenol content of grape juices treated with ultrasound, no significant difference was observed in the acidity and total soluble amounts of the juice. As a result of the study, it was concluded that the applied ultrasound technique could be successful in processing grape juice on an industrial scale.

Abid et al. [49] examined the effects of ultrasonic sound waves on polyphenolic substances, sugar, carotenoids and minerals in apple juice. They applied ultrasonic sound waves to freshly squeezed apple juices with a frequency of 25 kHz and 70% vibration amplitude for 30 and 60 minutes. As a result of the analyses, they achieved a high increase in the amounts of polyphenolic substances and sugars in apple juices to which ultrasonic sound waves were applied for 30 minutes, while a significant improvement was observed in the total carotenoids and sodium, potassium and calcium minerals and viscosity values as a result of the ultrasonic sound waves applied for 60 minutes.

No change was recorded in the electrical conductivity of apple juices subjected to ultrasonic sound waves. Study findings showed that ultrasound application significantly increased the nutritional elements already present in apple juice.

Cruz-Cansino et al. [50] examined the changes in physicochemical, microbiological and antioxidant properties of pear juices using ultrasonic sound waves. They observed the properties of pear juices, which were subjected to ultrasound treatment for 15 and 25 minutes at a frequency of 20 kHz and 80% vibration amplitude, during 14 days of storage. They observed that 25 minutes of ultrasound treatment resulted in a greater increase in the color stability and viscosity of pear juices compared to just 15 minutes. They observed a very small increase in the pectinmethylesterase activity of pear juices treated with ultrasonic sound waves during storage. Additionally, with the ultrasound conditions they applied, they achieved a microbiological reduction equivalent to fruit juice treated with pasteurization. Finally, after 14 days, a significant increase was recorded in the phenolic substance and antioxidant activity of pear juices, especially those subjected to ultrasound treatment for 25 minutes.

Rojas et al. [51] in their study on the physical quality and stability of peach juice using ultrasound technology, they applied ultrasonic sound waves with a frequency of 20 kHz for 3, 6, 10 and 15 min. The findings showed that the ultrasound process positively contributed to both the quality criteria, durability and sensory properties of the product. Color changes in fruit juices treated with ultrasonic sound waves were examined and no significant change was observed during storage. In a study on the quality of fruit and vegetable juices, Khandpur and Gogate [52] studied orange, sweet lemon, carrot, and spinach juices. In this study, fruit and vegetable juices to which ultrasonic sound waves were applied at 20 kHz frequency and 15 minutes were compared with the traditional pasteurization process applied at 10 minutes and 80 °C. Total phenolic substance, antioxidant activity, Vitamin C and carbohydrates were evaluated as quality criteria. Although the samples treated with ultrasonic sound waves preserved most of their nutritional properties, their shelf life was found to be 18 days longer than the samples treated with pasteurization.

In a study conducted with ultrasonic cutting, which is another area of use of ultrasonic sound waves, red delicious and golden delicious apple varieties were subjected to ultrasonic cutting, and it was determined that the products cut by applying this technology were effective against the polyphenol oxidase enzyme for a 2-week storage period [13].

Effect of Ultrasonic Homogenization on Freezing Technology

It has been shown that ultrasonic sound waves give positive results on the freezing process, as in many other areas. The benefit of the sound energy produced by ultrasonic sound waves technology can be explained by the different effects it creates when passing through the environment. The most important effect caused by sound waves is the cavitation effect, as mentioned before. The gas bubbles created by ultrasound are small in diameter and this application continues continuously. Meanwhile, ice nuclei (icing) are formed. These crystals help the freezing process to be much more efficient.

On the other hand, when dense and incompressible materials are subjected to variable acoustic stress, ice crystals break down and cause ice crystal particles of equal sizes to form. This phenomenon is a desired event in the freezing process [53].

Ultrasound process has benefits such as:

- formation of a frozen surface in partially frozen material,
- reducing ice crystals and controlling the size of the crystals,
- preventing crust formation on the frozen surface.

A freezing process involving ultrasound treatment contributes to the formation of both a high-quality food product and a pharmaceutical product [53]. In a study examining the effect of ultrasound on freezing time, it was observed that ultrasound application shortened the freezing time of the product in the freezing of potato chips [54].

Effect of Ultrasonic Homogenization on Emulsification

A good emulsification process occurs when ultrasonic sound waves are applied to two immiscible liquids. When the bubbles that occur due to the cavitation caused by the ultrasound process burst between these two liquids, a shock effect occurs, and these liquids are mixed more homogenized. Compared to the low-energy ultrasound process, high-energy ultrasound application creates an emulsion with a more stable structure [36]. It has been observed that palm oil can make the coagulated droplets very small after the ultrasound treatment applied in a mixing process with ß-lactoglobulin alginate, creating a suitable environment for emulsion [55]. A study investigated the effects of ultrasound processing on the droplet size in the dispersed phase in oil, the volume and physicochemical properties of the dispersed phase, and time and power parameters. Ultimately, it was determined that the droplets produced by applying ultrasonic sound waves had smaller sizes compared to the normally formed droplet sizes. It has also been declared that the application of ultrasonic sound waves in the production of sauces such as ketchup and mayonnaise can increase emulsion stability [56].

Lee et al. [8] significantly reduced the particle size in soy protein emulsion by using ultrasonic sound waves and pH change processes together in a study they conducted with soy protein. The particle size of the soy protein solution, which is normally around 376 nm, was reduced to around 21 nm with the help of ultrasonic sound waves. Similarly, while the particle size of untreated soy protein isolate samples was 192 nm, this size was reduced to around 40 nm with the MTS process, and by going one step further, the MTS process was applied by combining it with the effect of pH, and as a result of the process, the particle size was observed to be 27 nm [9].

Effect of Ultrasonic Homogenization on Meat Technology

The crispness and color parameters of meat are among the most important features sought by the consumer. Since the existing properties of food can be changed with sound waves, ultrasonic sound waves are a method used to add crispness to meat. It has been determined that with this method, the extraction of muscle proteins is increased, and the texture of the cured ham is improved [57].

In meats exposed to ultrasonic sound waves, redder color and fresher muscle appearance were obtained without adding salt, and it was observed that the muscle fibers were easily separated [58].

In another study, it was stated that short-term, low-frequency and high-intensity ultrasound treatment applied to whole meat added to degassed mineral salt solution had less hardness than meat not placed in mineral salt solution for the same period of time (Smith et al., 1991). It has also been reported that ultrasound treatment applied to dry ham (with the addition of 2% salt) can increase intermolecular adhesion [58].

Effect of Ultrasonic Homogenization on Filtration Process

During the filtration process, solid substances form precipitates on the surface of the filter membrane, clogging the filter pores and negatively affecting the filtration process. In the filtration process where ultrasound is applied, the flow is faster by breaking down the polarized materials formed on the surface of the filter membrane. In addition, it has been observed that ultrasonic sound waves applied with a membrane filter give more effective results than the application of ultrasonic sound waves alone [39].

Effect of Ultrasonic Homogenization on the Extraction Process

Extraction has been used in processes such as food, perfume and cosmetics since ancient times. Today, it is used in the food, pharmaceutical, cosmetics, nutraceutical, or bioenergy industries (maceration, solvent extraction, steam and hydro distillation, cold pressing, squeezing, etc.). Extraction of bioactive compounds; It depends on various factors such as extraction technique, raw material and extraction solvent used. Modern or non-traditional extraction techniques are known as green or clean techniques because they reduce energy use and apply organic solvents that are not environmentally harmful [59].

To reduce increasing energy costs, greenhouse gas emissions and meet legal requirements, food and plant-based chemical industries are turning to green technologies. The application of these technologies is aimed at protecting the environment and natural resources. Basically, green technologies include improving extraction efficiency, shortening processing time and reducing environmental damage caused by toxic solvents.

The ultrasound process mechanically breaks down the cell wall and enables material transfer. With the breakdown of the cell wall, the liquid extract inside the cell can easily leave the cell. As the particle diameter decreases as a result of ultrasonic application, the surface area between the solid and liquid parts increases [42]. The mechanical activity of ultrasound accelerates the distribution of the solvent towards the tissues and the intracellular component easily passes into the solvent [60]. Factors that determine the effect of ultrasound: temperature, pressure, frequency and sonication time [61]. Ultrasound is used to obtain polysaccharides, essential oils, proteins, peptides and bioactive substances [62]. The application of ultrasound energy is considered an alternative for the extraction of bioactive compounds as it increases the mass transfer coefficient, rate and yield [63].

Xu et al. [64] investigated the extraction of natural antioxidants from the Eucommia oliver plant by ultrasound-assisted extraction and reported that the use of ultrasound increased the effect of traditional methods and provided high efficiency and selectivity of natural antioxidants. Another research is about the extraction of bioactive compounds from black chokeberry fruit. The extraction time (0-240 min), solvent composition (ethanol-water) and ultrasound energy level (0-100 W) indicate that ultrasound application is a suitable technology for total anthocyanins and polyphenols [65].

Sivakumar et al. [66] examined the effects of ultrasound on the removal of tannins in Avaram bark (Cassia auriculata). They state that there is a 160% improvement when using ultrasound at 100 W compared to magnetic stirring, and the reason is linked to improved mass transfer in the leaching of tannins. Khan et al. [67] investigated the extraction of polyphenols, especially flavonoids, from orange peel (Citrus sinensis L.) and stated that ultrasound provided the best conditions and that there was an increase in the yield of bioactive compounds and antioxidant activity in the obtained extracts and confirmed the suitability of this technique for preparing fruit extracts. There are other studies reporting that the ultrasound-assisted extraction method increases the yield and bioactive content in the extraction of bioactive components [59]. Paula et al. [68] used the ultrasound-assisted extraction technique to obtain bioactive compounds from the dried fruits of the rosary tree (Meliaceae), reporting that this method is more efficient, simpler, faster and cheaper in terms of bioactive components. In another study where the same technique was applied to obtain polysaccharides with antioxidant activity from black mulberry fruit, it was stated that the yield increased by 3.13% compared to other methods [69].

The application of ultrasonic sound waves mechanically disrupts the cell wall and causes material transfer. Due to the cell wall breakdown, the liquid extract inside the cell can easily come out. As a result of the decrease in particle diameter because of the application of ultrasonic sound waves, the surface area between the solid and liquid phases increases [42]. The mechanical effect of ultrasonic sound waves accelerates the movement of the solvent into the cell tissue and the intracellular component easily passes into the solvent [60].

Positive results were observed when ultrasound was applied to pickle brine. Due to salt losses occurring during brining, salt is constantly added to the brine, and this practice causes both time and financial losses. In pickles where ultrasonic sound waves were applied, the penetration of salt into the materials was accelerated and the brining process was carried out in a short time [39].

Significant and positive effects of ultrasonic sound waves on protein solubility have been observed. The solubility in mixtures of pea proteins with pectin was increased from 8.0% to 60% with the application of ultrasonic sound waves [11]. In the same study, pea proteins formed a complex with modified starch, where the solubility was increased from 8% to approximately 73% [11].

Effect of Ultrasounic Homogenization on Defoaming

Foam is a gas distribution in a liquid in which the distance between bubbles is small, approaching its density. Foams are involved in many industrial processes, including food production and cosmetic products. Intense foaming in industry causes product losses, decreased equipment efficiency and environmental pollution. Mechanical breakers or chemical defoamers are used to remove foam in the food industry [70]. In recent years, with the knowledge of the defoaming potential of sound energy, methods based on aerodynamic acoustic sources have been developed, but they have not been used industrially due to the disadvantages they cause. The importance of ultrasound in foam breaking has increased because high-intensity ultrasonic waves prevent chemical contamination and can be applied under sterile conditions. It attracts great attention especially in the food and pharmaceutical industries.

A defoamer system based on a new type of ultrasonic generator has been developed and it has been determined that this system can be successfully used to control excess foam that occurs in applications such as high-speed bottling and canning of carbonated beverages [71]. The focused air source ultrasonic emitter is connected to an electronically controlled rotation system. Most of the bubbles are broken by the movement of the ultrasonic transducer as it rotates at different speeds. Acoustic density appears to be an important criterion for a perfect defoaming effect [72].

Effect of Ultrasounic Homogenization on Degassing (Aeration)

Common methods for degassing are boiling and depressurization. Ultrasonic degassing is done using potable tap water, just like ultrasonic bath application. Acoustic waves and rapid vibrations caused by bubbles occur when they defy gravity and reach a large size until they reach the surface [73].

In the food industry, this technique is used to degas (defoam) carbonated beverages such as beer before bottling [74]. It aims to remove air from the liquid surface to prevent organoleptic damage of the product by bacteria and oxygen in the processing of carbonated drinks. Low-intensity ultrasound application used during the fermentation of sake, beer and wine shortens the time by 35-60%. While ultrasonic degassing shortens the time in aqueous systems, it is a difficult technique to apply in high viscosity liquids such as melted chocolate.

Effect of Ultrasounic Homogenization on Depolymerization

One of the oldest applications of ultrasound is the degradation of polymers [75]. The depolymerization process occurs with the effect of cavitation and includes two possible mechanisms. These are the chemical degradations that occur due to the mechanical disintegration of polymers in the collapsed cavitation bubble and the chemical reaction between them. Hydroxyl radicals formed as a result of cavitation are given as an example [76].

In the food industry, ultrasound method is used to depolymerize starch. Sonication is generally performed by applying simple immersion probes at 20 kHz and provides efficient depolymerization. This method is used in food processing in addition to its positive effects on ultrasound power, processing time and other parameters, it also causes improvements in food quality [77].

Effect of Ultrasounic Homogenization on Extrusion

In the food industry, various difficulties arise in removing the cooked product from the mold due to the product sticking to the mold. To eliminate these difficulties, the surface of the molds is coated with silicone or PTFE (polytetrafluoroethylene), but this process must be constantly repeated or changed. In addition to these expensive traditional methods, ultrasound applications provide advantages such as eliminating surface coatings and cleaning all kinds of materials from the mold.

The use of ultrasound technology in food industries aims to minimize product waste and maintenance costs [7]. Ultrasonic cutting takes place with the help of a blade connected to the ultrasonic source via a shaft. Ultrasonic cutting depends on the food type and condition [13]. This technology is most used in special cases such as fragile and heterogeneous products (cakes, bakery products), oily or sticky products (cheese) [78].

Effect of Ultrasounic Homogenization on Sterilization

Ultrasound technology is an effective auxiliary sterilization method. It has been successfully used in the food industry to sterilize liquid foods and ready-to-eat fresh foods. Ultrasonic treatment applied to freshly harvested strawberries; While it significantly reduced the rate of bacteria, yeast and mold, it preserved fruit firmness and increased total antioxidant activity [79]. In a study, Zhang et al. [80] combined ultrasound with modified atmosphere packaging and observed that this combined process slowed down the losses of ascorbic acid and chlorophyll. Another study conducted; reported that there was a significant increase in vitamin C, total flavonoid and total phenolic contents in loquat fruit where peracetic acid and ultrasound were applied together. It was observed that the application of ultrasound combined with Ultraviolet-C radiation on tomatoes stored for 28 days increased the lycopene, total phenolic content and vitamin C content by 90%, 30% and 60%, respectively [81].

It is reported that approximately 1.3 billion tons of food is lost every year worldwide, and more than 40% of these are fruits and vegetables [82]. Storage methods for fruits and vegetables are basically refrigerated storage, hypobaric storage, controlled atmospheric storage, modified atmosphere, natural/synthetic protective agents, edible coatings and films, and ultraviolet applications. In addition, the use of green technology provides foods with fewer artificial additives and higher nutritional value and quality [83]. Consuming fruits and vegetables containing pesticide residues endangers human health [84]. Dust and microorganisms on the surface of fruits and vegetables cause decay. It is found in literature studies that pesticides in fruits and vegetables are destroyed by applying ultrasound technology [85].

Ultrasonication is used to preserve pre- and post-harvest quality characteristics of fresh fruits and vegetables [86]. They also play a role in washing fruits and vegetables in the food industry [87]. Ultrasonication technology is effective to improve the nutritional value of fruit juices in meeting consumer needs, and it is stated that it protects the quality, nutritional value and microbiological safety of guava juice [88]. At the same time, ultrasound can be used to compensate for nutritional losses occurring during bleaching [89]. It has been stated that the application of ultrasonication cleaners (20-400 kHz) is effective for fruits and vegetables free from contamination, and at 40 kHz, rot and infection in strawberries are significantly reduced [90-91]. Ultrasound is a positively developing technology that reduces nutritional losses of fruits and vegetables both before and after harvest. Research shows that the firmness of fresh fruits and vegetables can be affected by ultrasound applications, which vary with intensity. They are also responsible for color changes and browning reactions as a result of the inactivation of some phenol peroxidase and polyphenol oxidase enzymes. Sonication prevents free radical formation and provides a cavitation-induced antimicrobial effect [92].

Effect of Ultrasonic Homogenization on Drying Food

The process of removing water from a material by means of a high-intensity sound field is called "Acoustic Drying." Ultrasound energy is very suitable for drying heat-sensitive materials because it provides a significant effect at low temperatures [29]. When high intensity ultrasonic sound waves are applied directly to the product to be dried, it creates a series of compression and expansion movements within the product, similar to the compression and release of a sponge. The effect that occurs with this mechanism easily removes water from the material through microscopic channels that arise as a result of creating a force greater than the surface tension that holds the water in the product [93].

In addition, cavitation, which occurs with ultrasound energy, is thought to be an effective factor in the removal of water [94]. In addition to the use of ultrasound energy alone is also possible to apply it together with processes such as:

- hot weather,
- irradiation,
- freeze drying,
- high pressure, and
- osmotic drying [95].

In drying food, ultrasound energy is often combined with different drying techniques. These techniques are used either simultaneously with ultrasound or as pre-treatment. Ultrasonic sound waves applied simultaneously in osmotic drying have been applied as a pre-treatment in other drying techniques [14].

In ultrasound and osmotic drying applications, an ultrasonic bath device is more often used, and this process is applied simultaneously. In a study conducted with osmotically dried apple particles, it was stated that as a result of the osmotic drying process applied in an ultrasonic bath with a frequency of 50 kHz and 150 W energy, ultrasonic sound waves were effective in mass transfer and increased water loss and sugar gain [96]. In hot air drying, ultrasonic sound waves can be applied as a pre-treatment before drying with an ultrasonic bath or can be performed at the time of drying with a probe placed in the dryer. As a result of their research, Fernandes and Rodrigues [97] applied ultrasonic sound waves as a pre-treatment in drying banana fruit with hot air and investigated the effects of this application. It has been reported that bananas that were pre-treated with ultrasound at 30 oC for 10, 20 and 30 minutes with an ultrasonic bath with a frequency of 25 kHz increased the diffusion of water after 20 minutes of application and shortened the total drying time by around 10.3% [97]. In addition, as a result of the ultrasound process, there was a decrease of around 21.3% in the sugar content of banana fruit. It has been stated that ultrasound processing can enable the production of dried products with reduced sugar content [98]. In another study, during the drying of strawberry fruit with the hot air method at different degrees, an increase in the water diffusion coefficient and mass transfer coefficients occurred, this increase decreased with increasing drying degrees, and in addition, with the application of ultrasonic sound waves, the total drying time at various drying temperatures was between 13% and 44%. Shortening at varying rates has been reported [99].

In another study investigating the effect of ultrasonic sound waves on the drying kinetics of potato cubes by hot-air drying, it was stated that the drying kinetics of potatoes showed a statistically significant difference with ultrasonic sound waves. It has been determined that ultrasonic sound waves with a frequency of 21.8 kHz and a power of 37 kW/m^2 reduce the drying time by around 40%, increase the water diffusion coefficient by 64% and the mass transfer coefficient by 58% compared to samples dried without using ultrasound [100].

In the drying of starch isolated from raw banana fruit, in which ultrasonic sound waves were used as pre-treatment, a vibrating probe at one-minute intervals and ultrasound at 24 W energy with a frequency of 20 kHz were applied for 60 minutes. Immediately afterwards, drying was carried out in a mini spray dryer. At the end of the study, it was determined that deterioration occurred in the structure of starch molecules due to cavitation caused by ultrasound application, and the diffusion of water increased the water absorption capacity of starch with its swelling feature. In addition, in the same study, it was reported that the vibrations caused by ultrasonic sound waves reduced the threshold shear stress and consistency coefficient of starch gels [99].

Jambrak et al. [77] carried out a study in which ultrasonic sound waves were applied to mushrooms, brussels sprouts and cauliflower products with a 20 kHz frequency probe and a 40 kHz frequency ultrasonic bath for 3 and 10 minutes. In the products dried by hot air (60°C) and freeze drying (-45°C) methods, ultrasonic sound waves increased the drying speed in all products except the pre-treatment applied in an ultrasonic bath for 10 minutes with a frequency of 40 kHz, and it was observed that the freeze-dried products were dried with a low frequency (20 kHz). It was determined that the rehydration properties of the samples were much higher in the process [77].

García et al. [101] examined the effect of ultrasonic sound waves pre-treatment applied before freeze-drying on the color characteristics of strawberry fruit, and as a result, it was reported that strawberries subjected to ultrasound had a more reddish and brighter color than samples without ultrasound pre-treatment [101].

In a study, Yildiz and Izli [14] examined the effect of ultrasound pre-treatment applied before freeze drying on the color and bioactive components of quince, and as a result, they stated that quinces subjected to ultrasonic sound waves had better color properties, lower water activity and higher antioxidant capacity.

In a study in which the effect of ultrasonic sound waves on drying and structural properties in infrared drying of pear fruit was studied, pears were exposed to ultrasonic sound waves at 25%, 50%, 75% and 100% amplitude for 5 minutes with an ultrasonic probe with a frequency of 24 kHz and immediately afterwards at 70 °C. Infrared drying process was applied. As a result, it was observed that ultrasound pretreatment decreased the drying time at increasing amplitudes due to the changes resulting from cavitation. In this study, thanks to texture analysis, it was observed that hardness and elasticity decreased with increasing amplitude, and it was reported that the applied ultrasound pre-treatment gave the most efficient results in terms of texture properties [4].

Conclusion

Thermal processes used in the food industry, especially pasteurization and sterilization, negatively affect the general quality and nutritional values of foods. New non-thermal technologies have been developed to minimize these negativities. Ultrasonic sound wave is of great importance due to:

- Less damage on the product
- ease of applicability, and
- its environmental friendliness.

Ultrasonic sound waves may not be sufficient for food preservation when used alone. That's why ultrasound technology when used together with

- ultrasound-heat treatment (TS),
- ultrasound-pressure (MS),
- ultrasound-heat treatment-pressure (MTS), is effective in obtaining quality products in terms of sensory and nutritional composition.

With these applications, the negative effects such as loss of volatile taste and odor substances, vitamins, and other nutrients because of high temperature on the food can be eliminated.

References

[1] Abramov, O. V. (1998). *High-Intensity Ultrasound: Theory and Industrial Applications*. London: Gordon and Breach.

[2] Earnshaw, R. G., Appleyard, J., & Hurst, R. M. (1995). Understanding Physical Inactivation Proces: Combined Preservation Opportunites Using Heat, Ultrasound and Pressure. *International Journal of Food Microbiology* 28(2), 197-219.

[3] Dolatowski, Z. J., Stadnik, J., & Stasiak, D. (2007). Applications of Ultrasound in Food Technology. *ACTA Scientiarum Polonorum – Technologia Alimentaria*, 6(3), 89-99.

[4] Jambrak, A. R, Herceg, Z., Subaric, D., Babic, J., Brncic, M., Brncic, S. R., Bosiljkov, T., Cvek, D., Tripalo, B., & Gelo, J. (2010). Ultrasound Effect on Physical Properties of Corn Starch. *Carbohydrate Polymers*, 79, 91-100.

[5] Thakur, B. R., & Nelson, P. E. (1997). Inactivation of Lipoxygenase in Whole Soy Flour Suspension by Ultrasonic Cavitation. *Moleculer Nutrition*, 41(5), 299-301.

[6] Murphy, R., Beard, B., Macry, J., & Berrang, M. E. (2009). *Application of Ultrasonic Technology for Killing Salmonella and Listeria monocytogenes in Fluid System. Institute of Food Technology*. June 6-9, 2009. Anaheim, CA. 236-01.

[7] Yildiz, G., Rababah, T., & Feng, H. (2016). Ultrasound-Assisted Cutting of Cheddar, Mozzarella and Swiss Cheeses – Effects on Quality Attributes during Storage. *Innovative Food Science and Emerging Technologies*, 37, 1-9.

[8] Lee, H., Yildiz, G., Dos Santos, L. C., Jiang, S., Andrade, J., Engeseth, N. C., & Feng, H. (2016). Soy protein nanoaggregates with improved functional properties prepared by sequential pH treatment and ultrasonication. *Food Hydrocolloids*, 55, 200–209.

[9] Yildiz, G., Andrade, J., Engeseth, N. C., & Feng, H. (2017). Functionalizing soy protein nano-aggregates with pHshifting and mano-thermo-sonication. *Journal of Colloid and Interface Science,* 505, 836-846.

[10] Yıldız, G. (2018). Physicochemical properties of soy protein concentrate treated with ultrasound at various amplitudes. *Journal of the Institute of Science and Technology,* 8(4), 133-139.

[11] Yildiz, G., Ding, J., Andrade, J., Engeseth, N. J., & Feng, H. (2018). Effect of plant protein-polysaccharide complexes produced by mano-thermo-sonication and pH-

shifting on the structure and stability of oil-in-water emulsions. *Innovative Food Science and Emerging Technologies*, 47, 317-325.

[12] Jiang, S., Yildiz, G., Ding, J., Andrade, J., Rababah, T. M., Almajwalc, A., Abulmeatyc, M. M., & Feng, H. (2019). Pea Protein Nanoemulsion and Nanocomplex as Carriers for Protection of Cholecalciferol (vitamin D_3). *Food and Bioprocess Technology*, 12(6), 1031-1040.

[13] Yildiz, G., Palma, S., & Feng, H. (2019). Ultrasonic cutting as a new method to produce fresh-cut red delicious and golden delicious apples. *Journal of Food Science*, 84(12), 3391-3398.

[14] Yildiz, G., & Izli, G. (2019a). The effect of ultrasound pretreatment on quality attributes of freeze-dried quince slices: Physical properties and bioactive compounds. *Journal of Food Process Engineering*, 42(5), e13223.

[15] Yildiz, G., & Izli, G. (2019b). Influence of microwave and microwave-convective drying on the drying kinetics and quality characteristics of pomelo. *Journal of Food Processing and Preservation*, 43(6), e13812.

[16] Yildiz, G., & Feng, H. (2019). Sonication of Cherry Juice: Comparison of Different Sonication Times on Color, Antioxidant Activity, Total Phenolic and Ascorbic Acid Content. *Latin American Applied Research Journal*, 49(4), 255-260.

[17] Yildiz, G. (2019). Application of ultrasound and high-pressure homogenization against high temperature-short time in peach juice. *Journal of Food Process Engineering,* 42(3), e12997.

[18] Yildiz, G., & Aadil, R. M. (2020). Comparison of high temperature-short time and sonication on selected parameters of strawberry juice during room temperature storage. *Journal of Food Science and Technology*, 57(4), 1462-1468.

[19] Yıldız, G., İzli, G., Çavuş, M., & Ceylan, M. M. (2021). The Effect of Ultrasound Pre-treatment on the Quality Charcteristics of Dried Iğdır Apricot. *Journal of the Institute of Science and Technology*, 11(1), 303-313.

[20] Yildiz, G. (2021). The Effect of High Intensity Ultrasound Pre-treatment on the Functional Properties of Microwave-dried Pears (Pyrus communis). *Latin American Applied Research Journal,* 51(2), 133-137.

[21] İzli, G., & Yildiz, G. (2021). Evaluation of the high intensity ultrasound pre-treatment effects on the physical properties and bioactive compounds of convective dried quince samples. *International Journal of Fruit Science*, 21(1), 645-656.

[22] Yıldız, G. (2022). Changes in Emulsifying Properties, Droplet Size, Turbidity and Lipid Oxidation of Pea Protein Nanoemulsions Exposed to High-Intensity Ultrasound and High-Pressure Homogenization during Storage. *Academic Food Journal*, 20(4), 336-342.

[23] Yıldız, G., Yildiz, G., Khan, M. R., Aadil, R. M. (2022). High-intensity ultrasound treatment to produce and preserve the quality of fresh-cut kiwifruit. *Journal of Food Processing and Preservation*, 46(5), e16542.

[24] Yildiz, G., Izli, G., & Aadil, R. M. (2020). Comparison of chemical, physical, and ultrasound treatments on the shelf life of fresh-cut quince fruit (Cydonia oblonga Mill.). *Journal of Food Processing and Preservation*, 44(3), e14366.

[25] Yildiz, G., & Aadil, R. M. (2022). Comparative Analysis of Antibrowning Agents, Hot Water and High-intensity Ultrasound Treatments to Maintain the Quality of Fresh-cut Mangoes. *Journal of Food Science and Technology*, 59(1), 202-211.

[26] Yildiz, G., & Yıldız, G. (2023). A new approach to enhance quinoa protein nano-aggregates: Combined pH shifting – high pressure homogenization. *Food Chemistry*, 415, 135800.

[27] Knorr, D., Zenker, M., Heinz, V., & Lee, D. (2004). Applications and Potential of Ultrasonics in Food Processing. *Trends in Food Sciense and Technology*, 15(5), 261-266.

[28] Demirdöven, A., & Baysal, T. (2009). The Use of Ultrasound and Combined Technologies in Food Preservation. *Food Reviews İnternational*, 25, 1-11.

[29] Kantaş, Y. (2007). Effect of Ultrasound on Drying Rate of Selected Produce. *Middle East Technical University*, *Natural and Applied Sciences*, Ankara.

[30] Uzunoğlu, T. P. (2012). Effect of Ultrasound Pretreatment on Drying Performance of Okra and Apple. *ITU Institute of Science and Technology*, İstanbul, Turkey.

[31] Rahman, S. M. (2007). *Handbook of Food Preservation*. (2nt Ed.). CRC Press. London. 713-739.

[32] Patist, A., & Bates, D. (2008). Ultrasonic innovations in the food industry: From the laboratory to commercial production. *Innovative Food Science and Emerging Technology*, 9, 147-154.

[33] Tüfekçi, S., & Özkal, S. G. (2015). Application of Ultrasound in Food Drying. *Pamukkale University Journal of Engineering Sciences*, 21(9), 408-413.

[34] Türkmen, F. (2012). The effect of high power ultrasound treatment on physicochemical and gelation properties of milk. *ITU Institute of Science and Technology*, İstanbul, Turkey.

[35] O'Donnell, C. P., Tiwari, B. K, Bourke, P., & Cullen, P. J. (2010). Effect of ultrasonic processing on food enzymes of industrial importance. *Trends in Food Science and Technology*, 21, 358-367.

[36] Soria, A. C., & Willamiel, M. (2010). Effect of ultrasound on the technological properties ans bioactivity of food: A review. *Trends in Food Science & Technology,* 21, 323-331.

[37] Mason, T.J. (1998). *Power Ultrasound in Food Processing-The Way Forward: In Ultrasound in Food Processing*. Blackie Academic & Professional, London, 105-126.

[38] Wen, C., Zhang, J., Zhang, H., Dzah, C. S., & Zandile, M., Duan, Y., & Luo, X. (2018). Advances in ultrasound assisted extraction of bioactive compounds from cash crops-A review. *Ultrasonic Sonochemistry*, 48, 538-549.

[39] Chemat, F., Zill-e-Huma, & Khan, M. K. (2011). Applications of Ultrasound in Food Technology: Processing, Preservation and Extraction. *Ultrasonics Sonochemistry,* 18, 813-835.

[40] McClements, D. J. (1995). Advances in the Application Ultrasound in Food Analysis and Processing. *Trends in Food Science and Technology*, 6, 293-299.

[41] Piyasena, P., Mohareb, E., & McKellar, R. C. (2003). Inactivation of microbes using ultrasound. *International Journal of Food Microbiology*, 87, 207-216.

[42] Bayraktaroğlu, G., & Obuz, E. (2006). *Principles of the ultrasound method and its use in foodstuffs.* 9th Food Congress, Bolu, Turkey.

[43] Ordonez, J. A., Sanz, B., Hernandez, P. E., & Lopez-Lorenzo, P. (1984). A Note on the Effect of Combined Ultrasonic and Heat Treatments on the Survival of Thermoduric Streptococci. *Journal of Applied Bacteriology*, 56(1), 175-177.

[44] Garcia, M. L., Burgos, J., Sanz, B., & Ordonez, J. A. (1989). Effect of heat and ultrasonic waves on the survival of two strains of Bacillus subtilis. *Journal of Applied Bacteriology*, 67, 619-628.

[45] Ordonez, J. A., Aguilera, M. A., Garcia, M. L., & Sanz, B. (1987). Effect of combined ultrasonic and heat treatment (thermoultrasonication) on the survival of a strain of Staphylococcus aereus. *Journal of Dairy Research*, 54(1), 61-67.

[46] Ercan, S. Ş., & Soysal, Ç. (2010). Effect of ultrasound and temprature on tomato peroxidase. *Ultrasonics Sonochemistry*, 18, 689-695.

[47] Lopez, P., & Burgos, J. (1995). Lipoxygenase inactivation by monothermo-sonication: Effect of sonication physical parameters, pH, KCI, sugars, glycerol and enzyme concentration. *Journal of Agricultural and Food Chemistry*, 43, 620-625.

[48] Aadil, R. M., Zeng, X. A., Han, Z., & Sun, D. W. (2013). Effects of ultrasound treatments on quality of grapefruit juice. *Food Chemistry*, 141, 3201–3206.

[49] Abid, M., Jabbar, S., Wu, T., Hashim, M. M., Hu, B., Lei, S., & Zeng, X. (2014). Sonication enhances polyphenolic compounds, sugars, carotenoids and mineral elements of apple juice. *Ultrasonics Sonochemistry*, 21, 93–97.

[50] Cruz-Cansino, N. S., Ramírez-Moreno, E., León-Rivera, J. E., Delgado-Olivares, L., Alanís-García, E., & Ariza-Ortega, J. A. (2015). Shelf life, physicochemical, microbiological and antioxidant properties of purple cactus pear (Opuntiaficus indica) juice after thermoultrasound treatment. *Ultrasonics Sonochemistry*, 27, 277–286.

[51] Rojas, M. L., Leite, T. S., Cristianini, M., Alvim, I. D., & Augusto, P. E. D. (2016). Peach juice processed by the ultrasound technology: changes in its microstructure improve its physical properties and stability. *Food Research International*, 82, 22–33.

[52] Khandpur, P., & Gogate, P. R. (2016). Evaluation of ultrasound-based sterilization approaches in terms of shelf life and quality parameters of fruit and vegetable juices. *Ultrasonics Sonochemistry*, 29, 337–353.

[53] Zheng, L., & Sun, D. W. (2006). Innovative Applications of Power Ultrasound During Food Freezing Processing-a Review. *Trends in Food Science & Technology*, 17, 16-23.

[54] Li, B., & Sun, D. W. (2002). Effects of power ultrasoun freezing rate during immersionfreezing. *Journal of Food Engineering*, 55(3), 277-282.

[55] Pongsawatmanit, R., Harnsilawat, T., & McClements, D. J. (2006). Influence of Alginate, pH and Ultrasound Treatment on Palm Oil in water Emmulsions Stabilized by B-Lactoglobulin. *Colloids and Surfaces A: Physicochemical and Engineering Aspects*, 287, 59-67.

[56] Povey, M. J. W. (1998). Ultrasonic of Food. *Contemporary Physics*, 39(6), 467-478.

[57] Pohlman, W. F. (1994). *Ultrasound uses for cookery and to improve cooking, textural, sensory and shelf-life stability properties of beef muscle*. Kansas State University. Manhattan, Kansas.

[58] Reynolds, J. B., Anderson, D. B., Schmidt, G. R., Theno, D. M., & Siegal, D. G. (1978). Effects of Ultrasonic Treatment on Binding Strenght in Curet Ham Rolls. *Journal of Food Science*, 43(3), 866-869.

[59] Rodriguez-Perez, C., Quirantes-Piné, R., Fernandez-Gutierrez, A., & Segura-Carretero, A. (2015). Optimization of extraction method to obtain a phenolic compounds-rich extract from Moringa oleifera Lam leaves. *Industrial Crops and Products,* 66, 246-254.

[60] Kim, S. M., & Zayas, J. F. (1989). Processing parameter of chymosin extraction by ultrasound. *Journal of Food Science*, 54, 700.

[61] Rajha, H. N., Boussetta, N., Louka, N., Maroun, R. G., & Vorobiev, E. (2015). Effect of alternative physical pretreatments (pulsed electric field, high voltage electrical discharges and ultrasound) on the dead-end ultrafiltration of vine-shoot extracts. *Separation and Purification Technology*, 146, 243–251.

[62] Tiwari, B. K. (2015). Ultrasound: A clean, green extraction technology. *Trends in Analytical Chemistry*, 71, 100–109.

[63] Riera, E., Golas, Y., Blanco, A., Gallego, J. A., Blasco, M., & Mulet, A. (2004). Mass transfer enhancement in supercritical fluids extraction by means of power ultrasound. *Ultrasonics Sonochemistry*, 11(3-4), 241-244.

[64] Xu, J. K., Li, M. F., & Sun, R. C. (2015). Identifying the impact of ultrasound-assisted extraction on polysaccharides and natural antioxidants from Eucommia ulmoides Oliver. *Process Biochemistry*, 50(3), 473–481.

[65] D'Alessandro, L.G., Dimitrov, K., Vauchel, P., & Nikov, I. (2014). Kinetics of ultrasound assisted extraction of anthocyanins from Aronia melanocarpa (black chokeberry) wastes. *Chemical Engineering Research and Design*, 92(10),1818 – 1826.

[66] Sivakumar, V., Ilanhtiraiyan, S., Ilayaraja, K., Ashly, A., & Hariharan, S. (2014). Influence of ultrasound on Avaram bark (Cassia auriculata) tannin extraction and tanning. *Chemical Engineering Research and Design*, 92(10), 1827 - 1833.

[67] Khan, M. K., Abert-Vian, M., Fabiano Tixier, A. S., Dangles, O., & Chemat, F. (2010). Ultrasound-assisted extraction of polyphenols (flavanone glycosides) from orange (Citrus sinensis L.) peel. *Food Chemistry*, 119(2), 851 – 858.

[68] Paula, J. A. M., Brito, L. F., Caetano, K. L. F. N., de Morais Rodrigues, M. C., Borges, L. L., & da Conceição, E. C. (2016). *Ultrasound-assisted extraction of azadirachtin from dried entire fruits of Azadirachta indica A. Juss. (Meliaceae) and its determination by a validated HPLC-PDA method.* Talanta, 149, 77-84.

[69] Chen, C., You, L. J., Abbasi, A. M., Fu, X., & Liu, R. H. (2015). Optimization for ultrasound extraction of polysaccharides from mulberry fruits with antioxidant and hyperglycemic activity *in vitro*. *Carbohydrate Polymers*, 130, 122-132.

[70] Morey, M. D., Deshpande, N. S., & Barigou, M. (1999). Foam Destabilization by Mechanical and Ultrasonic Vibrations. *Journal of Colloid and Interface Science*, 219, 90-98.

[71] De-Sarabia, E. R. F., Gallego-Juarez, J. A., & Mason, T. J. (2006). Airborne ultrasound for the precipitation of smokes and powders and the destruction of foams. *Ultrasonics Sonochemistry*, 13, 107–116.

[72] Laborde, J. L., Bouyer, C., Caltagirone, J. P., & Gerard, A. (1998). Acoustic bubble cavitation at low frequencies. *Ultrasonics Sonochemistry*, 36, 589–594.

[73] Tervo, J. T., Mettin, R., & Lauterborn, W. (2006). Bubble cluster dynamics in acoustic cavitation. *Acta Acustica united with Acustica*, 92(1), 178-180.

[74] Brown, B., & Goodman, J. E. (1965). *High-Intensity Ultrasonics, Industrial Applications*. Published Iliffe Books Ltd., London: First Edition.

[75] Smith, N. B., Cannon, J. E., Novakofsky, J. E., MacKeith, F. K., & O'Brien, W. D. (1991). *Tenderization of Semitendinosus Muscle Using High Intensty Ultrasound. Ultrasonic Symposium*, Lake Beuna Vista, Florida. 2, 1371-1373.

[76] Grönroos, A., Pirkonen, P., & Ruppert, O. (2004). Ultrasonic depolymerization of aqueous carboxymethylcellulose. *Ultrasonics Sonochemistry*, 11, 9-12.

[77] Jambrak, A. R., Mason, T. J., Paniwyk, L., & Lelas, V. (2007). Accelerated Drying of Button Mushrooms, Brussels Sprouts and Cauliflower by Applying Power Ultrasound and its Rehydration Properties. *Journal of Food Engineering*, 81(1), 88-97.

[78] Arnold, G., Leiteritz, L., Zahn, S., & Rohm, H. (2009). Ultrasonic cutting of cheese: Composition affects cutting work reduction and energy demand. *International Dairy Journal,* 19, 314–320.

[79] Gani, A., Baba, W. N., Ahmad, M., Shah, U., Khan, A. A., Wani, I. A., Masoodi, F. A., & Gani, A. (2016). Effect of ultrasound treatment on physico-chemical, nutraceutical and microbial quality of strawberry. *LWT-Food Science and Technology*, 66, 496–502.

[80] Zhang, X., Zhang, M., Devahastin, S., & Guo, Z. (2019). Effect of Combined Ultrasonication and Modified Atmosphere Packaging on Storage Quality of Pakchoi (Brassica chinensis L.). *Food Bioprocess Technology*, 12, 1573 – 1583.

[81] Esua, O. J., Chin, N. L., Yusof, Y. A., & Sukor, R. (2019). Effects of simultaneous UV-C radiation and ultrasonic energy postharvest treatment on bioactive compounds and antioxidant activity of tomatoes during storage. *Food Chemistry*, 270, 113–122.

[82] FAO. (2020). *Turkey's National Strategy Document on Prevention, Reduction and Monitoring of Food Loss and Waste and its Action Plan*. Ankara, Turkey.

[83] Jiang, Q., Zhang, M., & Xu, B. (2020). Application of ultrasonic technology in postharvested fruits and vegetables storage: A review. Ultrasonics Sonochemistry, 69, 105261.

[84] Valcke, M., Bourgault, M. H., Rochette, L., Normandin, L., Samuel, O., Belleville, D., Blanchet, C., & Phaneuf, D. (2017). Human health risk assessment on the consumption of fruits and vegetables containing residual pesticides: A cancer and non-cancer risk/benefit perspective. *Environment International,* 108, 63-74.

[85] Zhu, Y., Zhang, T., Xu, D., Wang, S., Yuan, Y., He, S., & Cao, Y. (2019). The removal of pesticide residues from pakchoi (Brassica rape L. ssp. chinensis) by ultrasonic treatment. *Food Control*, 95, 176 –180.

[86] Gallego-Juárez J, Rodriguez G, Acosta V, Riera E. 2010. Power ultrasonic transducers with extensive radiators for industrial processing. *Ultrasonics Sonochemistry*, 17, 953 – 964.
[87] Alexandre, E. M. C., Brandao, T. R. S., & Silva, C. L. M. (2013). Impact of non-thermal technologies and sanitizer solutions on microbial load reduction and quality factor retention of frozen red bell peppers. *Innovative Food Science and Emerging Technologies,* 17, 199–205.
[88] Cheng, L. H., Soh, C. Y., Liew, S. C., & Teh, F. F. (2007). Effects of sonication and carbonation on guava juice quality. *Food Chemistry*, 104, 1396-1401.
[89] Jabbar, S., Abid, Hu, B., Hashim, M. M., Lei, S., Wu, T., & Zeng, X. (2015). Exploring the potential of thermosonication in carrot juice processing. *Journal of Food Science and Technology,* 52, 7002-7013.
[90] Lin, I., & Erel, D. (1992). Dynamic ultrasonic cleaning and disinfecting device and method. US Pat. No. 5,113,881. Washington, DC: United States and Trademark Office.
[91] Cao, S., Hu, Z., Pang, B., Wang, H., Xie, H., & Wu, F. (2010). Effect of Ultrasound Treatment on Fruit Decay and Quality Maintenance in Strawberry After Harvest. *Food Control*, 21(4), 529-532.
[92] Nicolau-Lapeña, I., Abadias, M., Bobo, G., Aguiló-Aguayo, I., Lafarga, T., & Viñas, I. (2019). Strawberry sanitization by peracetic acid washing and its effect on fruit quality. *Food Microbiology*, 83, 159-166.
[93] Rastogi, N. K. (2011). Opportunities and challenges in application of ultrasound in food processing. *Critical Reviews in Food Science and Nutrition*, 51, 705-722.
[94] De La Fuente-Blanco, S., Riera-Franco de Sarabia, E., Acosta-Aparicio, V. M., & Gallego-Juarez, J. A. (2006). Food Drying Proces by Power Ultrasound. *Ultrasonics Sonochemistry,* 44, 523-527.
[95] Topdaş, E. F., Başlar, M., & Ertugay, M. F. (2011). Effect of Ultrasound Process on Osmotic Drying of Apples. *Academic Food Journal*, 9(5), 6-10.
[96] Simal, S., Benedito, J., & Rosello, C. (1998). Use of ultrasound to Increase Mass Transport Rates during Osmotic Dehydration. *Journal of Food Engineering*, 36(3), 323-336.
[97] Fernandes, F. A. N., & Rodrigues, S. (2007). Ultrasound as pretreatment for drying of fruits: Dehydration of banana. *Journal of Food Engineering*, 82, 261-267.
[98] Fernandes, F. A. N., Gallão, M. I., & Rodrigues, S. (2009). Effect of osmosis and ultrasound on pineapple cell tissue structure during dehydration. *Journal of Food Engineering*, 90, 186-190.
[99] Izidoro, D. R., Sierakowski, M. R., Haminiuk, C. W. I, Souza, C. F., & Scheer, A. G. (2011). Physical and Chemical Properties of Ultrasonically Spray-Dried Green Banana Starch. *Journal of Food Engineering*, 104(4), 639-648.
[100] Ozuna, C., Carcél, J. A., Garcia-Perez, J. V., & Mulet, A. (2011). Improvement of water transport mechanisms during potato drying by applying ultrasound. *Journal of Science Food and Agriculture*, 91, 2511-2517.

[101] Garcia, J., Oliveira, F. I. B., Weller, C. L., Rodrigues, S., Fernandes, F. A. N. (2014). Effect of Ultrasonic and Osmotic Dehydration Pre-treatments on the Colour of Freeze-Dried Strawberries. *Journal of Food Science and Technology*, 51(9), 2222-2227.

Chapter 4

Mathematical Modeling and Artificial Intelligence Applied to the Homogenization of Grain Storage Conditions

Manuel Osório Binelo*
and Marcia de Fatima Brondani Binelo
Department of Exact Sciences and Engineering,
Regional University of the Northwest of Rio Grande do Sul, Ijuí, Brazil

Abstract

After harvesting, grains such as soybeans, corn, rice, wheat, beans, and oats need to be stored in silos for extended periods, which can exceed one year. During this time, certain properties of the grains are crucial for preserving their nutritional quality and market value, as grain deterioration can render them unsuitable for consumption. Among these properties, the two most important ones are moisture and temperature. High levels of moisture can promote the emergence of pests, such as fungi and insects, in addition to increasing the biological activity of the grains. The temperature of the grains, besides also promoting pest growth and increasing biological activity, serves as an important indicator of the condition of the stored grains, since the aforementioned problems also lead to temperature rise. The grain is a living organism with its own metabolism, and its properties depend on its environment. The primary mechanism for homogenizing grain properties is the aeration process, which involves the forced passage of ambient air through the porous medium formed by the bulk-stored grains. This air flow is generated by electric fans located at the base of storage bins and silos. The

* Corresponding Author's Email: manuel.binelo@unijui.edu.br.

In: A Closer Look at Homogenization
Editor: Gulcin Yildiz
ISBN: 979-8-89113-656-4

optimization of the aeration process depends on factors such as ambient air temperature and humidity, aeration time, air pressure and velocity, and distribution of airflow through the porous medium. Optimizing this process is crucial because poor aeration will fail to maintain the grain properties within desired limits, while excessive aeration will have a high energy cost. This chapter will address the state of the art in mathematical models and methods based on artificial intelligence applied to the optimization of the aeration process, considering the design of warehouses and aeration systems, as well as their operation. Among the mathematical models for the aeration process, we can highlight those based on partial differential equations that simulate the pressure and velocity field of air in the porous medium. Among the artificial intelligence models, evolutionary algorithms stand out, and are applied to optimize factors such as the location of air inlets, air pressure, and the geometry of silos.

Keywords: mathematical modeling, computer simulations, optimization, silos, storage bin

Introduction

Grain storage is essential to ensure food security worldwide, guaranteeing the availability of food throughout the year, regulating prices, and facilitating the logistics of grain transportation between countries. Therefore, optimizing the storage process is crucial to increase food supply and combat hunger. According to FAO [1], approximately 800 million people were affected by hunger in 2021, and all efforts towards eradicating this issue are significant.

Various types of grains, such as soybeans, corn, wheat, rice, and beans, need to be stored after harvest for periods ranging from a few months to over a year. This storage typically occurs in bulk within large storage bins and silos. To ensure the quality of the stored product, several techniques are employed to control its properties, especially temperature and moisture content. Because grains are living organisms, their metabolism is affected by the conditions of their environment. Higher temperatures and moisture content levels increase the metabolism of grains, causing their nutrients to be consumed and reducing their nutritional quality. Elevated levels of moisture content and temperature also promote the growth of bacteria, fungi, and insects that feed on the grains and contaminate them. High levels of biological activity can raise the temperature to the point of charring the grains, potentially resulting in the loss of the entire stored grain. Therefore, hot spots in the mass of grains, while worsening the environment, are also a symptom of increased biological activity and the emergence of pests.

In order to maintain homogeneous properties of grains at appropriate levels, various procedures are adopted in the post-harvest process, whether on the farm or at storage facilities. The first of these is pre-cleaning, where sieve systems separate the grains from impurities. Next, the grains go through the drying process, where their moisture content is reduced to levels considered appropriate, depending on the type of grain. Finally, the grains are transported by elevators and conveyor belts to the storage bin, and during the time they are stored, they undergo aeration, a process in which electric fans blow air from the base of the storage bin. The air passes through the porous spaces between the grains, facilitating the exchange of temperature and moisture content, and exits through the top of the storage bin.

All these processes consume a significant amount of energy and can still harm the quality of the grain if executed improperly. Mathematical and computational modeling have been employed by various researchers to optimize these processes, both for operational improvement and the design of more efficient equipment. A common point in different research studies is homogenization. Proper pre-cleaning is one that leaves the mass of grains with the minimum amount of impurities, making it more homogeneous. Efficient drying is one that ensures all grains have the desired moisture content, without grains being too dry or too moist, in other words, with homogeneous moisture content. Efficient storage is one that maintains the entire mass of grains at suitable and uniform temperature and moisture levels, without hotspots of heat and moisture. This chapter provides a review of the main studies that have applied mathematical and computational modeling to enhance the homogenization of stored grains, considering processes from pre-cleaning to grain aeration in the storage bin.

Pre-Cleaning

Before the drying process, grains go through a preliminary cleaning process where lightweight residues like straw and leaves, as well as heavy residues like small stones, are removed using a system of sieves and airflow. Mathematical and computational modeling of this process is carried out using models based on Computational Fluid Dynamics (CFD) [2] and the Discrete Element Method (DEM) [3]. These two methods are also the primary ones employed in the other phases of the process, which include drying and storage.

This is due to the granular nature of the grain mass, making it convenient to model each grain as a discrete element subject to the laws of motion, and because airflow is present in all these processes. However, as emphasized by Badretdinov et al. [4], solving the airflow coupled with the particle flow is quite complex, leading some researchers to develop specific analytical models for these problems.

Badretdinov et al. [4] presented the mathematical modeling of grain material separation in the pneumatic system of a grain-cleaning machine, where a model based on differential equations using the Lagrange equations of particle motion is adopted. One advantage of this approach is the ability to account for the flow of particles in a non-equilibrium state, considering particles with different shapes, masses, the mutual influence of particles, and the turbulence of the fluid medium. These aspects are crucial for describing the mechanics of the grain and impurity separation process. This separation primarily occurs when the drag force of air overcomes gravity for the case of lighter particles (impurities). The authors modeled this effect by applying the Reynolds number of particles to the Lagrange equations. The dynamic inertia of the particles is modeled as the exposure time of the particles to the airflow, defined as the relaxation time. This relaxation time is corrected by a function that takes into account the effects of inertial forces, such as when the particle moves, therefore, its inertia depends on its Reynolds number.

The equipment modeled by Badretdinov et al. [4] features a closed spiral flow, and the airflow from the fan was modeled using a rotating coordinate system, with an approximation of the one-component, incompressible Navier-Stokes model, considering the effects of turbulence. In this work, the VisionFlow software was used to perform the calculations and visualize the results. The simulation results demonstrated good agreement with experimental data, and the method developed by the authors allowed for recommendations of equipment improvements to make the airflow more homogeneous, thus enhancing its performance.

In Adamchuk et al. [5], a theoretical study of the pneumatic separation of granular mixtures in vortex flow is conducted. This work presents a system of differential equations for the distribution and movement of particles based on their physical and mechanical properties. The developed model considers a cylindrical coordinate system, as the equipment system consists of a rotor. The system of differential equations takes into account the drag force on the particles and the equations of motion considering inertial forces, but it does not consider more complex aspects such as particle interactions or turbulence. Nevertheless, even with various simplifications of the phenomenon, the researchers were able to determine the trajectories of the particles and, based on that, make recommendations for the equipment dimensions according to the desired grain and impurity separation rate. A similar approach is adopted in Priporov et al. [6], where an analytical model is used to estimate the separation between sunflower seeds and impurities in air-sieve equipment. However, in addition to not considering interactions between the particles, individual trajectories are not calculated. Instead, proportional coefficients of airflow for whole seeds, broken seeds, and impurities are obtained.

When more detailed aspects of the process need to be considered, more comprehensive models that take into account factors like particle interactions are required, such as the Discrete Element Method (DEM), initially introduced for soil mechanics applications and widely used in agriculture [7]. Its basic principle involves treating each particle as a basic geometric abstraction, typically a sphere, and the domain boundaries, such as equipment walls, as planes or triangles in three-dimensional space. Interactions between particles and between particles and walls are modeled using a contact model, which defines the normal and shear forces acting on the particles. These forces are integrated to calculate particle movement over a small timestep, determining their new positions.

A commonly used contact model is the linear Cundall model [8], which determines contact force based on the intersection length between spheres and calculates the respective forces considering a mass-spring model. An interesting aspect of DEM is that its contact behavior parameters are based on physical parameters, such as the Poisson's ratio and Young's modulus, which have already been extensively researched. DEM is an iterative and transient mathematical and computational model that performs the following steps for each time step:

- determining contacts by checking for intersections between elements;
- calculating forces according to the contact model;
- integrating the forces to determine the new positions of the elements.

In Xu et al. [9], DEM is applied to model a seed cleaning and classification machine with vibrating screens equipped with a rubber ball rack that constantly hits the screen to minimize clogging. The theoretical analysis of ball motion and its effect on the screen would be very complex due to the chaotic nature of interactions. Several simulations were conducted, considering different configurations of the rack's angulation and using wheat, rice, and corn seeds. The main objective of the study is to propose a method for evaluating the impact of the balls on the screen and their effect on the classification performance, addressing the following questions:

- expressing the impact of the balls on the screen;
- determining the correlation between the impact force of the balls on the screen and its classification performance;
- improving the impact of the balls on the screens, thereby enhancing the classification performance.

To achieve these objectives, three DEM simulation models were created, considering different system characteristics. From the results of these models, relationships between equipment features and screen performance were established. Finally, an optimal configuration was proposed, defining the number of ball grooves, the distance between grooves, and the optimal surface inclination angle. The results were validated through experiments, including the proposed optimal configuration. The simulation results also demonstrated that wheat and rice seeds are more affected by the impact of the balls. This work serves as an interesting example of how more sophisticated simulation models allow for a broader generalization by simulating various aspects of the equipment, leading to better prospects for optimization.

In Wang et al. [10], DEM was coupled with a multibody kinematics (MBK) model for computer simulation of a swing-bar sieve machine. This type of equipment is commonly used for grain classification and cleaning, featuring complex planar motion. In addition to the complexity of the motion, the finite wall contour model typically used in DEM struggles to model circular sieve openings. These openings often need to be approximated by polygons, which can affect the results for particles close in size to the openings. To address this, the study proposes the use of a real-virtual boundary model to simulate the screen deck. The real-boundary represents the surface of the screen deck but without considering the circular openings, while the virtual-boundaries represent the circles that represent the openings on the screen deck. The grain studied was soybeans, which have a shape close to spherical and are usually represented in DEM as an equivalent sphere. When this sphere comes into contact with the real-boundary, contact forces are generated. However, if it is simultaneously in contact with a virtual-boundary, these forces are nullified, allowing the element to pass through the screen.

This approach is highly efficient because modeling the screen with polygons to approximate circular holes would require a large number of triangular elements to achieve a reasonable approximation, significantly increasing the computational cost of the simulation. The authors concluded that the coupled DEM and MBK model showed good agreement with experimental data and that the use of the real-virtual boundary method was valid for the problem. This approach can be employed for optimizing the design of this type of equipment.

The works presented in this section demonstrate that various modeling approaches, ranging from those with a higher level of abstraction to those simulating intricate process details, can be valuable for optimizing pre-cleaning equipment.

Drying

The drying of grains aims to homogenize the moisture content of the grains to levels lower than those at harvest. The most common types of dryers are continuous dryers with crossflow, concurrent, countercurrent, or mixed airflow. In these dryers, grains are loaded at the top of the equipment and flow by gravity to the bottom, where a system controls the discharge rate. During this path, the grains are exposed to a heated airflow that passes through the porous medium, removing moisture content from the grains [11]. The modeling of grain dryers is complex, and three main aspects can be identified:

- the flow of grains (and impurities) throughout the equipment;
- the airflow through the grain mass;
- and heat and mass transfer during drying.

While these three factors are interconnected, it is common for scientific studies to address these issues separately, simplifying other factors due to the complexity of the problem. All three factors influence the homogeneity of drying. Since dryers have complex internal geometry, with different points of air inlet and various chute geometries through which grains flow, grains may traverse the equipment at different velocities. Similarly, the airflow can take various paths, as will be demonstrated in the works to be discussed. The air used in drying can also be uneven since the use of wood-fired furnaces to heat the air is very common, especially in developing countries. All these factors can result in non-uniform grain drying.

Iroba et al. [12] and Mellmann et al. [13] emphasized that the design characteristics of the dryer can lead to unequal residence times for grains. Residence time is the duration that grains take to traverse the entire equipment while being exposed to the drying airflow. Non-uniform residence times will lead to uneven drying, compromising the quality of the grains. In both studies, experiments were conducted using the flow of wheat in a scaled-down dryer model, with a transparent acrylic front wall to visualize grain flow. The modeling of the grain discharge process was done using a two-dimensional DEM. The results obtained indicate that the studied dryer model has regions near the walls and between the air ducts where grains have lower velocity, resulting in significantly longer residence times and leading to over-drying. Overly dried grains decrease in size and weight, resulting in reduced commercial value and potential damage from excessive heat. Additionally, non-uniform drying leads to increased energy consumption.

Three-dimensional simulations of soybean grain flow in mixed-flow dryers were carried out in [14]. An important challenge in applying DEM is its computational cost, especially in three-dimensional simulations. In this work, the authors simulated a small section of the dryer, whose geometry repeats throughout the equipment, and used periodic boundary conditions at the top and bottom of the domain, thus simulating a continuous flow. The model was implemented in the open-source software Yade [3], which offers two other interesting advantages. Firstly, simulations are created using Python scripts, making the configuration of the simulation and interactions highly flexible. Secondly, being open-source allows for continuous study and improvement by the scientific community. Iroba et al. [12], Mellmann et al. [13] (with wheat), and Khatchatourian et al. [14] also found regions of low flow velocity that could result in non-uniform residence times.

Yade was also used by Lorenzoni et al. [15] to simulate soybean flow in dryers. In this work, to address the issue of computational cost in 3D simulations, a quasi-2D approach was proposed. While the simulation remains three-dimensional, one of the dimensions of the simulation domain is reduced to the size of a few particles, and periodic boundaries are employed to eliminate the wall effect. The results indicate that a simulation domain thickness equivalent to five times the particle diameter is sufficient to approximate the results of the full-dimensional simulation. This study also analyzed the porosity of grains across the domain during flow, demonstrating that the dryer's geometry significantly influences porosity variation. Non-uniform porosity leads to non-uniform airflow and, consequently, non-uniform drying. The relative residence times observed in the simulations ranged from 0.8 to 1.7 times the average residence time. It was also observed that the natural variation in soybean grain size does not significantly affect residence time, whereas moisture variation has a substantial impact, with longer residence times for moister grains.

The airflow within the porous medium of the grain mass passing through the dryer has a significant impact on the homogenization of drying. Cao et al. [16] developed a two-dimensional model combining the Finite Difference Method (FDM) and the Finite Element Method (FEM) to solve 2D airflow in a mixed-flow dryer. This work also considered grain flow and heat and mass transfer. From the simulations and regression analysis, three equations were obtained for predicting drying capacity, specific heat consumption, and maximum grain temperature under different drying conditions. These equations were used to propose an optimized mixed-flow dryer model. The proposed dryer model achieved a 30% energy savings compared to conventional models used in China, providing more uniform grain drying.

Scaar et al. [17] conducted an experimental and numerical study of airflow in mixed-flow dryers to investigate the effect of different air inlet and outlet arrangements on the drying of wheat and rapeseed. Since this study focused on airflow, isothermal flow conditions were considered. According to Scaar et al. [17], the complexity of the grain drying process necessitates a compartmentalized analysis of the phenomenon, examining each of the aspects in greater detail: grain flow, airflow, and heat and mass transfer, as observed in other works. Airflow simulations were carried out using Computational Fluid Dynamics (CFD) with ANSYS CFX software. The experimental airflow data demonstrated that irregularities in grain accommodation and impurity distribution significantly affect the airflow, as indicated by the isobaric lines. It was also observed that a reduction in porosity, as seen in rapeseed, increases pressure drop, as expected. The simulation results showed good agreement with the experimental data. The study also concluded that a horizontal arrangement of air inlets provides a more uniform airflow than a diagonal arrangement.

Khatchatourian et al. [18] conducted a mathematical modeling and computational simulation of a grain dryer. According to the authors, various models can be employed to describe the grain drying process, considering heat and mass transfer between grain and air, and within the grain. Most of these models consist of systems of differential equations that describe the energy and mass transfer between air and a grain at the surface layer, as well as energy and mass conservation at the boundary between the grain and air. Since these are nonlinear systems, analytical solutions are infeasible, and numerical methods like Finite Element Method (FEM) and Finite Difference Method (FDM) are commonly used. The study analyzed different dryer configurations, always considering a cooling stage and one, two, three, or four drying stages with various air recirculation schemes. The simulations using the differential equation-based model, solved by the two-dimensional MacCormack's method, were validated against laboratory experimental data, demonstrating good agreement. The non-uniformity of temperature and moisture distribution in the dryer and energy savings were analyzed for the different simulated schemes. The study concluded that a three-stage dryer with upward air re-circulation between the stages offers greater drying uniformity and energy savings.

The development of research for procedural optimization and process development in agricultural drying was presented in Mellmann et al. [19]. According to them, continuous flow dryers are the main type of dryer used in Europe and Latin America, particularly mixed-flow dryers, which have the capacity to process large volumes of grain with high moisture content rates. The work shows the progress of research on grain flow using DEM and air flow using CFD. Based on the models used, new optimized dryer models are proposed, with the main features being a change in the angle of the air inlet ducts near the walls and the addition of a line of closed air ducts. When all the air ducts have the same roof angle, the grains near the side walls have a lower flow velocity. By using narrower angles in the ducts near the walls, the researchers achieved a more uniform flow of grains, reducing the variation in residence time. The closure of air duct lines was proposed to create an equalization zone for the airflow after each section of four horizontal lines of air inlet and outlet ducts, and to retain a more homogeneous airflow. This research demonstrates that mathematical and computational modeling of the grain drying process can lead to substantial improvements in equipment design that may not be intuitive at first glance or that are too complex and expensive to obtain through prototype building and testing.

Models, like the ones discussed so far, have the ability to predict the operation of a dryer, and if all three aspects are considered, grain flow, airflow, and heat and mass transfer, they can predict the moisture content of the grains leaving the dryer given the equipment design and operation parameters. Therefore, it seems like an interesting idea to use a predictive model like this in the automatic control system of a dryer, acting on the air flow and grain discharge rate to achieve the desired drying. Unfortunately, the models discussed are complex and have high computational costs, and control systems need to operate in real time. An important trend in this field is the use of Artificial Intelligence (AI) to create computationally efficient models for real-time control.

In Jin et al. [20], an Artificial Neural Network (ANN) is proposed to predict the moisture content of the grains based on the dryer's operating parameters. A multi-layer feed forward ANN was used to learn the behavior of a rice dryer from a series of data collected during drying experiments. The inputs to the ANN include ambient air temperature and humidity, initial grain moisture content and temperature, air humidity and temperature at different air outlet points, and grain temperature at various sections of the dryer. The ANN's outputs are the moisture content of the grains at the outlet and the waiting time for the discharge motors. MATLAB software was used to perform several simulations of the control system, which showed good performance. In Jin et al. [20], experimental data were directly used to train the control system, but it would be interesting to consider the idea of generating a large volume of synthetic data for ANN training through more sophisticated models of the drying process, as seen previously [21, 22] This would provide a larger training dataset without the cost of additional experiments.

Another example of AI application for drying processes can be found in Abhigna et al. [23], where a genetic algorithm (GA) is proposed for optimizing the parameters of a proportional-integral-derivative (PID) controller. PID controllers have been an industry standard for a long time; however, their performance relies on correctly tuning their parameters to avoid issues like overshooting or delay. Finding the ideal values for these parameters is not a straightforward optimization problem, and the work Abhigna et al. [23] successfully applies the GA technique for this purpose, achieving good results. GAs are inspired by the natural selection process, where successive generations of potential solutions to an optimization problem go through the processes of selection and reproduction to eventually obtain optimal or suboptimal values [24].

The pre-cleaning and drying processes are essential to ensure the uniformity of the stored grain conditions. Next, we will discuss the actual storage stage, considering the models used for the simulation and optimization of the processes necessary to maintain the grain's quality during the storage period.

Storage

Grain storage typically occurs in bulk and in large horizontal storage bins with a V-shaped floor or in vertical cylindrical silos. Ensuring the uniformity of grain conditions within appropriate physical parameters is essential to maintain grain quality for extended periods. The first process that impacts these conditions is the deposition of the grains inside the storage bin. Grains are loaded using elevators and conveyors and discharged from the top until they reach the floor. Since the pre-cleaning step does not entirely remove impurities, and because grains and impurities behave differently during free fall, regions of impurity concentration may occur within the grain mass. These impurity kernels, typically formed below the discharge point, reduce the porosity of the medium and hinder the passage of aeration airflow, posing a risk to storage [25].

During free fall, interactions occur between the particles themselves and with the fluid (air). Simultaneously simulating these interactions by coupling CFD and DEM is quite complex. As an alternative, Binelo et al. [26] proposed a drag force model for different types of seeds applied to DEM. To homogenize the distribution of impurities during the loading of the storage bin or silo, a device called a grain spreader is used, which can have different designs, including rotating ones. The aerodynamic drag model proposed by Binelo et al. in 2019 [26] was used by Binelo et al. in 2022 [27] for the 3D simulation of a grain spreader. In their work, the simulations were validated using experimental data from a scaled spreader, showing that the type of spreader studied was not entirely efficient in homogenizing the distribution of impurities. The simulations were conducted using the open-source software Yade, and the researchers hope to use the developed model for proposing optimized spreader designs.

During storage, the primary process used to maintain grain quality is aeration. Electric fans positioned at the base of storage bins and silos blow air through ducts that extend to the floor. The air then passes through the grain mass, traversing the porous medium, and exits through the roof. Aeration is carried out when ambient conditions are favorable, meaning when the ambient temperature is lower than the grain temperature, and the air humidity is appropriate [28].

Achieving homogeneous airflow within the grain mass is crucial for quality aeration. Otherwise, some grains will not receive sufficient airflow. Solving the airflow problem in the grain mass using the original Navier-Stokes model is unfeasible because it would require modeling the flow along every path between the grains using very fine meshes. Furthermore, other characteristics of the problem make the flow complex. The compaction of grains in the lower layers makes air resistance dependent on the depth of the grain layer, and the deposition of non-spherical grains results in the anisotropy of the medium, with different air resistance for vertical and horizontal flow directions. Because of this, specific models were developed for this problem. Shedd [29] was one of the first researchers to conduct experiments and develop a mathematical model for air resistance in the porous medium of the grain mass. Following his work, different researchers created airflow models in the aeration process [30, 31, 32, 33, 34, 35].

A two-dimensional model of airflow was developed in 2004 by Khatchatourian et al. [36], considering the non-uniformity of the grain mass, based on a system of partial differential equations solved by the FEM. An important aspect of this model is that it accounts for grain compaction effects, with analytical parameters defined for different types of grains through experiments. In 2008, this model was expanded to 3D by Khatchatourian et al. [37]. The model was solved using the FEM in a C++ software, and the Paraview software was used for visualization and analysis of the results. Following the development of this model, Khatchatourian et al. [38] added the anisotropy effect to the model, demonstrating that the deposition of non-spherical grains, especially rice and lentils, leads to highly anisotropic flow conditions and, consequently, non-uniformity.

The importance of the compaction effect of grains on airflow is emphasized by Nwaizu et al. [39]. The authors incorporated this effect into the CFD model used in the ANSYS software. Aeration in vertical silos is more uniform than in horizontal storage bins. Different air inlet configurations at the base of vertical metal silos through simulations using the COMSOL software were tested by Razavizadeh et al. [40].

Understanding how grain deposition and compaction affect airflow in the porous medium is essential for creating realistic models. In Liu et al. [41], DEM was applied to simulate the spaces and paths created between the grains and then used CFD to simulate airflow. The result obtained from the DEM simulation served as the basis for creating a fine mesh capable of representing the pathways of airflow between the soybeans, represented by spheres. Due to the complexity and computational cost of the process, a cylindrical domain with a diameter of 100 mm and a height of 75 mm was considered, with different levels of mechanical pressure applied to the grains to simulate compaction. While the method cannot be replicated for an entire storage bin, the study helps understand important aspects of airflow in a porous medium.

An interesting conclusion from the study is that as the grain layer compaction increases, the airflow pathways become more tortuous. The results obtained by the DEM-CFD model agreed with those obtained by the Forchheimer model, also used for calculating airflow in porous media, but it can use coarser meshes and be applied in larger domains.

The thermal state of grains in the silo is also crucial for maintaining their quality and serves as an alert signal that can indicate increased biological activity. For this reason, most silos and storage bins are equipped with temperature sensors distributed throughout the grain mass, integrated into cables attached to the ceiling and the floor. The modeling of the thermal state in vertical rice silos with aeration is presented in Khatchatourian et al. [42]. The objectives of the study were to experimentally investigate the heat transfer dynamics in rice silos, adapt four different models to calculate the thermal state of the grains, and analyze the performance of the models by comparing the results with experimental and observational data. The applied models included the homogeneous reactor model, a model based on partial differential equations, a model using the generalized homochronous number, and the moving boundary model. The experimental and observational data were collected from a grain storage plant in southern Brazil. Due to the more homogeneous nature of airflow in vertical cylindrical silos, when compared to horizontal storage bins, the study concluded that one-dimensional models are appropriate for modeling heat transfer dynamics, which predominantly occur along the vertical axis. All four models yielded satisfactory results, with each having different advantages and disadvantages.

During storage, grains must be kept dry, and this condition depends on the temperature and humidity of the air during the aeration process. Some studies address heat and mass transfer combined, as in Thorpe [43]. In this work, the CFD software ANSYS Fluent was customized with functions written in FORTRAN and C to apply heat and mass transfer models based on systems of partial differential equations suitable for the grain storage aeration problem. The evolution of computational capabilities has allowed recent studies to simulate 3D airflow, heat transfer, and mass transfer, as seen by Panigrahi et al. [44]. In this study, a CFD-based model was run using parallel computing, and various solution schemes were evaluated and validated with experimental data from an operational silo.

As seen, different models have been created with varying levels of abstraction to simulate the aeration process in silos and storage bins. The next frontier in research is to use these models to develop optimization methods for the design of new silos and storage bins. In Khatchatourian et al. [45], a new parameter called the local specific airflow rate was proposed to assess the homogeneity of airflow during aeration. This parameter applies the concept of specific airflow rate in the storage bin, measured in $m^3\ s^{-1}\ kg^{-1}$, to an infinitesimal volume at the point p(x, y, z), thus forming a three-dimensional field of specific airflow that allows for the evaluation of the amount of air at any point within the grain mass. The study then analyzes the airflow in a large storage bin and proposes an optimized aeration scheme.

In addition to the case-by-case analysis of aeration system design to find optimized schemes, it is essential to develop methods that automate this process. Binelo et al. [46] presented an optimization model of the air pressure profile for aeration in grain storage bins. In this work, the local airflow rate criterion was used to find the optimal profile of air inlet pressure for the most homogeneous airflow possible. The optimal pressure profile is defined as the pressure P_x proportional to the square of the height of the grain layer at point x_f on the storage bin floor.

This is a theoretical profile, as it assumes that the entire bottom of the storage bin is a continuous air inlet with its pressure defined at each point. Considering this limitation and the fact that real aeration systems have discrete air inlets positioned at strategic points on the bin's floor, Petravicius et al. [47] proposed a method based on a genetic algorithm to optimize the dimensions of the storage bin, the placement of the inlets, and the air pressure at the inlets. The study simulated the airflow in the storage bin using the open-source software OpenFOAM with a Darcy-Forchheimer model parameterized for airflow in soybeans. The method involved a Python-based genetic algorithm that generated possible solutions composed of parameters defining the aeration system while keeping the cross-sectional area of the storage bin and the total air flow rate of the system constant.

The problem's genotype consisted of digits representing the aeration system's parameters: lateral inlet position, flow distribution between the central and lateral inlets, inlet configurations (lateral, central, both), storage bin width, bottom bin width, and wall height. The bin's depth was calculated to maintain the cross-sectional area. The problem's phenotype involved generating a mesh definition file and boundary conditions for simulation in OpenFOAM based on the genotype's parameters. The fitness function is based on the optimization objective function, which is the homogeneity of the airflow velocity field. The developed genetic algorithm (GA) can be summarized in the following steps:

1. generate the initial population of parameters;
2. for each parameter configuration: generate mesh and boundary condition files, run simulations in OpenFOAM, and calculate the fitness function;
3. if the stopping condition is met, return the best solution found and terminate;
4. select parameter configurations with the highest fitness using the tournament method;
5. generate a new population from the crossover and mutation operators.
6. go back to step 2.

The results obtained by the GA indicated that the lateral air inlet in the storage bin of the case study was located in a position higher than necessary, and the storage bin would have a more homogeneous airflow if the lateral inlet were closer to the central inlet. Despite its high computational cost, the developed GA successfully optimized the design of the case study's storage bin. Its flexibility allows for its application in various types of storage bins and the incorporation of other parameters for optimization.

Conclusion

This chapter presented a selection of studies that illustrate the evolution and the state of the art in mathematical and computational modeling applied to the homogenization of grain storage conditions. To achieve truly homogeneous conditions, optimization of the pre-cleaning and drying stages is necessary. Various methods with different levels of abstraction have been proposed for computational simulation of these processes. Through mathematical modeling and computational simulation, several authors have proposed improvements in equipment design and operation. In addition to optimizing specific case studies, more recent research has introduced methods for automating the optimization of equipment designs, aiming for greater efficiency and suitable and homogeneous storage conditions, contributing to the fight against food waste. It is expected that over time, these scientific contributions will find their way into the tools used by engineers responsible for designing grain storage systems.

References

[1] FAO (2022). The state of food security and nutrition in the world 2022 https://www.fao.org/3/cc0639en/online/sofi-2022/food-security-nutrition-indicators.html.

[2] Anderson, J. D., & Wendt, J. (1995). *Computational fluid dynamics* (Vol. 206, p. 332). New York: McGraw-hill.

[3] Kozicki, J., & Donze, F. V. (2008). A new open-source software developed for numerical simulations using discrete modeling methods. *Computer Methods in Applied Mechanics and Engineering, 197(49-50),* 4429-4443.

[4] Badretdinov, I., Mudarisov, S., Tuktarov, M., Dick, E., & Arslanbekova, S. (2019). Mathematical modeling of the grain material separation in the pneumatic system of the grain-cleaning machine. *Journal of Applied Engineering Science*, *17(4),* 529-534.

[5] Adamchuk, V., Bulgakov, V., Ivanovs, S., Holovach, I., & Ihnatiev, Y. (2021). Theoretical study of pneumatic separation of grain mixtures in vortex flow. *Engineering for Rural Development*, *20,* 657-664.

[6] Priporov, I. E., Kurasov, V. S., Samurganov, E. E., & Shepelev, A. B. (2021). Modeling the sunflower seeds separation process in air-sieve grain-cleaning machines. In IOP *Conference Series: Materials Science and Engineering* (Vol. 1111, No. 1, p. 012048). IOP Publishing.

[7] Zhao, H., Huang, Y., Liu, Z., Liu, W., & Zheng, Z. (2021). Applications of discrete element method in the research of agricultural machinery: A review. *Agriculture*, *11(5),* 425.

[8] Cundall, P. A., & Strack, O. D. L. (1980). Discussion: A discrete numerical model for granular assemblies. *Géotechnique*, *30(3),* 331-336.

[9] Xu, Z., Li, Y., Wan, L., Ma, X., Song, J., & Huang, J. (2023). Optimising the design of ball racks to improve the sorting efficiency of vibrating screen seed cleaners using discrete element method modelling and experiment. *Biosystems Engineering, 225,* 99-117.

[10] Wang, Y., Yu, J., Yu, Y., & Zhang, Q. (2019). Validation of a coupled model of discrete element method with multibody kinematics to simulate the screening process of a swing-bar sieve. *Powder Technology*, *346,* 193-202.

[11] Bakker-Arkema, F. W., Montross, M. D., Liu, Q., & Maier, D. E. (1996, October). Analysis of continuous-flow grain dryers. In *ACIAR PROCEEDINGS* (pp. 123-131).

[12] Iroba, K. L., Mellmann, J., Weigler, F., Metzger, T., & Tsotsas, E. (2011). Particle velocity profiles and residence time distribution in mixed-flow grain dryers. *Granular Matter*, *13,* 159-168.

[13] Mellmann, J., Iroba, K. L., Metzger, T., Tsotsas, E., Mészáros, C., & Farkas, I. (2011). Moisture content and residence time distributions in mixed-flow grain dryers. *Biosystems Engineering*, *109(4),* 297-307.

[14] Khatchatourian, O. A., Binelo, M. O., & de Lima, R. F. (2014). Simulation of soya bean flow in mixed-flow dryers using DEM. *Biosystems engineering*, *123,* 68-76.

[15] Lorenzoni, R. K., Binelo, M. O., Khatchatourian, O., Ziganshin, B. G., & de FB Binelo, M. (2020). Quasi-2D simulation of soya beans flow in mixed flow dryer. *Journal of Stored Products Research*, *89,* 101727.

[16] Cao, C. W., Yang, D. Y., & Liu, Q. (2007). Research on modeling and simulation of mixed flow grain dryer. *Drying Technology*, *25(4),* 681-687.

[17] Scaar, H., Franke, G., Weigler, F., Delele, M., Tsotsas, E., & Mellmann, J. (2016). Experimental and numerical study of the airflow distribution in mixed-flow grain dryers. *Drying Technology*, *34(5),* 595-607.

[18] Khatchatourian, O. A., Vielmo, H. A., & Bortolaia, L. A. (2013). Modelling and simulation of cross flow grain dryers. *Biosystems engineering*, *116(4),* 335-345.

[19] Mellmann, J., Weigler, F., & Scaar, H. (2019). Research on procedural optimization and development of agricultural drying processes. *Drying technology*, *37(5),* 569-578.

[20] Jin, Y., Wong, K. W., Yang, D., Zhang, Z., Wu, W., & Yin, J. (2022). A neural network model used in continuous grain dryer control system. *Drying technology*, *40(9),* 1901-1922.

[21] Abhigna, R., Sharma, A., Kavya, K., Sharma, P., Salkuti, S. R., Rajesh Kumar, M., ... & Sreekeessoon, B. (2023). Genetically Optimized PID Controller for a Novel Corn Dryer. *In Soft Computing for Problem Solving: Proceedings of the SocProS 2022* (pp. 543-554). Singapore: Springer Nature Singapore.

[22] Kramer, O. (2017). *Genetic algorithms* (pp. 11-19). Springer International Publishing.

[23] Salarikia, A., Jian, F., Jayas, D. S., & Zhang, Q. (2021). Segregation of dockage and foreign materials in wheat during loading into a 10-m diameter corrugated steel bin. *Journal of Stored Products Research*, *93,* 101837.

[24] Binelo, M. O., de Lima, R. F., Khatchatourian, O. A., & Stránský, J. (2019). Modelling of the drag force of agricultural seeds applied to the discrete element method. *Biosystems Engineering*, *178,* 168-175.

[25] Binelo, M. O., Lima, R. F., Faoro, V., & Binelo, M. F. (2022). Computational modelling of a grain spreader for use in silos. *Biosystems Engineering*, *223,* 29-40.

[26] Navarro, S., Noyes, R. T. Casada, M., & Arthur, F. H. (2012). *Grain aeration. Stored product protection*. Kansas State University Kansas, USA, 121-134.

[27] Shedd, C. K. (1953). Resistance of grains and seeds to air flow. *Agricultural Engineering*, *34(9),* 616-619.

[28] Brooker, D. B. (1969). Computing air pressure and velocity distribution when air flows through a porous medium and nonlinear velocity-pressure relationships exist. *Transactions of the ASAE, 12(1),* 118-0120.

[29] Brooker, D. B., Bakker-Arkema, F. W., & Hall, C. W. (1982). Drying Cereal Grains. AVI PublishingCo. Inc., Westport, CT.

[30] Bunn, J. M., & Hukill, W. V. (1963). Pressure pattern predictions for non-linear air flow through porous media. *Transactions of the ASAE*, *6(1),* 32-0036.

[31] Pierce, R. O., & Thompson, T. L. (1975). Airflow patterns in conical-shaped piles of grain. *Transactions of the ASAE, 18(5),* 946-0949.

[32] Haque, E., Chung, D. S., & Foster, G. H. (1981). Pressure and velocity field in airflow through packed bed of corn mixed with fines under non-darcy flow conditions. *Transactions of the ASAE, 24(6),* 1595-1599.

[33] Maier, D. E., Moreira, R. G., & Bakker-Arkema, F. W. (1992). Comparison of conventional and chilled aeration of grains under Texas conditions. A*pplied Engineering in Agriculture*, *8(5),* 661-667.

[34] Khatchatourian, O. A., & Savicki, D. L. (2004). Mathematical modelling of airflow in an aerated soya bean store under non-uniform conditions. *Biosystems Engineering*, *88(2),* 201-211.

[35] Khatchatourian, O. A., & Binelo, M. O. (2008). Simulation of three-dimensional airflow in grain storage bins. *Biosystems engineering*, *101(2),* 225-238.

[36] Khatchatourian, O. A., Toniazzo, N. A., & Gortyshov, Y. F. (2009). Simulation of airflow in grain bulks under anisotropic conditions. *Biosystems Engineering*, *104(2),* 205-215.

[37] Nwaizu, C., & Zhang, Q. (2021). Computational modeling of heterogenous pore structure and airflow distribution in grain aeration system. *Computers and Electronics in Agriculture*, *188,* 106315.

[38] Razavizadeh, N., Zare, D., Nassiri, S. M., Karim, A., Eslami, M., & Nematollahi, M. A. (2023). Experimental study and numerical simulation of resistance to airflow in a storage bin of rough rice with three inlet duct configurations. *Biosystems Engineering*, *225,* 118-131.

[39] Liu, W., Chen, G., Liu, C., Zheng, D., & Ge, M. (2022). Experimental and numerical study of pressure drop characteristics of soybean grain under vertical pressure. *Applied Sciences*, *12(14),* 6830.

[40] Khatchatourian, O. A., Binelo, M. O., Neutzling, R., & Faoro, V. (2017). Models to predict the thermal state of rice stored in aerated vertical silos. *Biosystems Engineering*, *161,* 14-23.

[41] Thorpe, G. R. (2008). The application of computational fluid dynamics codes to simulate heat and moisture transfer in stored grains. *Journal of stored products research*, *44(1),* 21-31.

[42] Panigrahi, S. S., Singh, C. B., & Fielke, J. (2023). A 3D transient CFD model to predict heat and moisture transfer in on-farm stored grain silo through parallel computing using compiler directives: Impact of discretization methods on solution efficacy. *Drying Technology*, *41(7),* 1133-1147.

[43] Khatchatourian, O. A., Binelo, M. O., Faoro, V., & Toniazzo, N. A. (2016). Three-dimensional simulation and performance evaluation of air distribution in horizontal storage bins. *Biosystems Engineering*, *142,* 42-52.

[44] Binelo, M. O., Faoro, V., Khatchatourian, O. A., & Ziganshin, B. (2019). Airflow simulation and inlet pressure profile optimization of a grain storage bin aeration system. *Computers and Electronics in Agriculture*, *164,* 104923.

[45] Petravicius, D., Binelo, M. O., Binelo, M. D. F., Faoro, V., & da Silva, J. A. (2023). Genetic algorithm in the design of soybean silos for airflow homogenization. *Revista Brasileira de Engenharia Agrícola e Ambiental*, *27,* 531-538.

Chapter 5

The Effect of High-Pressure Homogenization on the Quality Characteristics of Meat and Meat Products

Arzu Imece*
Department of Gastronomy and Culinary Arts, Igdir University, Iğdır, Turkey

Abstract

Nowadays demands for high-quality foods, especially fresh-tasting and nutritious meat, have created considerable interest in the development of new preservation technologies. Although traditional methods used in preservation of meat have helped to ensure a high level of food safety and prolonged the shelf life of foods, some techniques like heat treatment may cause quality loss in product and excessive energy consumption. Since these techniques may lead to undesirable effects on meat quality, considerable effort has been made on using new preservation techniques combined with traditional methods or an alternative to traditional methods. High pressure homogenization results in microbiological inhibition, while any important degradation of vitamins, mineral salts, taste, color and appearance does not occur. Therefore, this technology has become an important alternative to heat treatments. High pressure processing can be successfully used for liquid foods as well as meat and meat products. This technology is usually utilized for inhibition of microorganisms and enzymes and tenderization and freezing-thawing processes of meat. High pressure processing has an important effect on textural properties of meat due to changes of protein structure.

High pressure homogenization (HPH) application has high potential for food preservation and development of some food properties.

* Corresponding Author's Email: arzuodunkiran@gmail.com.

In: A Closer Look at Homogenization
Editor: Gulcin Yildiz
ISBN: 979-8-89113-656-4

Generally, HPH in the range 100 to 600 MPa are applied to foods for inactivation of microorganisms and enzymes. However, HPH is used for tenderness and the control of phase transition (freezing, thawing) in meats. Because it has beneficial effects for improving protein functionality, it plays an important role in the development of textural properties of meats. It has similar effects with heat applications; however it has not decreased the nutritional quality of foods. Therefore, HPH is an important alternative for heat applications. Also, it can be used in new product development. HPH is a method of food processing that makes food safer and extends its shelf life. HPH has great potential for the meat industry. This review presents the effects of HPH on meat proteins, tenderization of meat, meat enzymes, and microbiologic quality of meat, lipid oxidation, and meat color and phase transition.

Keywords: high-pressure homogenization, meat, quality

Introduction

The main purpose of food preservation is to preserve the sensory qualities of food products including nutritional value, color, aroma and physical structure. Today, different techniques are available to protect foods from microbiological and enzymatic changes and to extend their shelf life [1]. It is known that additives and heat treatments used for preservation purposes in the food industry have some undesirable effects on foods [2]. Although heat treatment is widely used to ensure microbial inactivation in foods, it has significant undesirable effects on the sensory, nutritional, and functional properties of foods [3]. It causes some structural changes, especially in meat products [4]. Physicochemical changes that occur in the structure of meat during heat treatment causes significant changes in the components, texture, water holding capacity, color, flavor, and properties after heat treatment [5]. With all these factors, and rising demand of consumers, scientists have begun new research in the development and preservation of the properties of foods. Alternative technologies that enable the destruction of microorganisms without the need for heat treatment have come to the forefront, and the application of these technologies has recently gained great importance [6]. New preservation technologies used in meat products include high pressure application, ohmic heating, dielectric heating, high intensity pulsed light, X-rays, and electron beams. The main purpose of these technologies is to destroy pathogenic and spoilage microorganisms without damaging food and to preserve the natural appearance of food, making it safe for consumption [7].

Various additives used to improve and preserve the properties of foodsmay have negative effects on human health. However, as consumers become more aware of food safety, the goal is to safely produce food as naturally as possible without losing its nutritional value. Although various technological studies have been carried out for this purpose, recently, emphasis has been placed on high pressure homogenization (HPH). This natural method can protect foods from undesirable microbiological and enzymatic effects to a certain extent, therefore, it extends shelf life and does not cause a significant change in nutritional value, making it valuable to the food industry. In addition, it provides similar results to those desired by heat treatment. Therefore, considering the undesirable effects of heat treatment on foods, applying HPH instead of heat treatment or with low temperature is considered advantageous.

The basic principle of this method, which generally involves the application of pressure between 100 MPa and 1000 MPa, especially between 100 MPa and 600 MPa, is based on the compression of the water surrounding the material. With the compression of water, the effect of pressure is instantly transmitted equally to every point of the material, regardless of the shape and size of the material. While it can be applied at room temperature, it can be applied at different temperatures, causing differences in the effect of pressure. Therefore, in this system where high pressure is applied, the pressure and temperature must be controlled safely. During high pressure application, much less energy is consumed after the desired pressure level is reached. Once the desired pressure is reached, the pump is stopped, the valve is closed and the pressure in the container is maintained without the need for any other energy input [2]. In addition, the short application time shows that its use will be advantageous compared to heat treatment. Although gases or oils can be used in addition to water in the application of this method, water is preferred because it is economical, reliable, practical and has low volume reduction.

High pressure homogenization can be used to improve many properties of foods, especially protein modification and increasing microbiological quality in foods. Although it can be used in many foods, especially foods such as meat, fruit juice and milk, the effects of these foods need to be investigated separately due to their different properties. For example, unlike many other foods, it can also be used to crispen meat and meat products. However, it has significant effects on lipid oxidation, color and phase change in meats [8]. It is also reported that it can be used to reduce the salt level in emulsion type meat products [2].

The basis of the HPH is based on the compression of the environment surrounding the product. Although gas and oil can be used as medium, water is preferred due to its economy, reliability, practical usability and low volume reduction. The method can be worked at room temperature or at different temperatures [2]. Applying high pressure to foods causes pressure to be continuously transmitted into the product. Since the transmitted pressure is independent of the shape and geometry of the product, this pressure can also be called “isostatic pressure” [2-9].

HPH, which has become increasingly widespread in recent years; is a technology that has the mechanism of inhibiting microorganisms by denaturing the proteins in their structures, without damaging the chemical structure of foods [10-15]. Effects of HPH on microorganisms can be listed as cell membrane-related effects, cellular changes caused by pressure, biochemical changes and effects on genetic mechanisms [16]. The critical process factors in HPH are pressure, pressure application time, pressure reduction time, process heat (including adiabatic heat), initial product temperature, boiler heat distribution at pressure, product pH, product composition, product water activity and durability of the packaging material [7].

High pressure process can be successfully applied especially to liquid foods, but also to foods that are not fully fluid, such as meat and meat products [17-18]. HPH of meat was first carried out by Macfarlane [19], and following this study, many studies were conducted on the effect of high-pressure application on various properties of meat and meat products [20].

Fresh meat is a food that is highly susceptible to spoilage due to its biological composition. Many factors such as storage temperature, atmospheric oxygen, endogenous enzymes, humidity, light and microorganisms affect the freshness and shelf life of meat [21]. On the other hand, meat creates an ideal environment for the development and proliferation of spoilage microorganisms and foodborne pathogens. For these reasons, it is very important to apply sufficiently effective preservation techniques to preserve the quality and safety of meat [22].

High-Pressure Homogenization

High-pressure application is a food preservation method that allows the production of healthy and long-shelf-life products by preserving the nutritional content and functional and sensory properties of foods [23]. The basic principle of this method, which generally involves the application of pressure between 100 MPa and 1000 MPa, especially between 100 MPa and 600 MPa, is based on the compression of the water surrounding the material. With the compression of water, the pressure is instantly transmitted equally to every point of the material, regardless of its shape and size. During high pressure application, much less energy is consumed after the desired pressure level is reached [18]. Compared to traditional heat treatment methods, the shorter processing time and the reduction of damage caused by heat make this technique advantageous over other methods.

The fact that high pressure application can be used together with heat treatments, enzymatic mixtures, natural antimicrobial and antioxidant substances increases the effectiveness of this method [24]. Although high pressure application is used in many products, it has recently gained a current place in meat products as a non-thermal preservation method that preserves the appearance and taste of the product due to its effect on reducing food pathogens [25]. Studies have shown that high pressure application can affect the structural, physicochemical, morphological and textural properties of meat and cause partial color changes in fresh red meat [26]. In addition, high pressure application.

It also improves the gel formation and structural properties of emulsion type meat products. Due to freezing meat and meat products with high pressure, smaller ice crystals are formed, thus eliminating the undesirable effects of freezing and thawing. In addition, with the application of high pressure, it is possible to produce healthy, nutritionally high-quality and long-shelf-life foods with preserved sensory properties [24]. It has been determined by many researchers that high pressure applications affect meat components, especially water, protein and lipids [1].

Many physicochemical properties of water change under pressure. The volume of water decreases by around 4% at 100 MPa, 12% at 400 MPa and 15% at 600 MPa at 22°C. It has been reported that foods containing a large portion of water and a small amount of gas show a compression feature similar to the compression of water [11, 13, 16]. The compression temperature increases depending on the initial temperature of the water used in high pressure application [24]. Adiabatic compression of water causes the temperature to increase by 2-3°C for every 100 MPa [1].

Proteins are important components of meat and have a very important impact on quality. Therefore, changes that may occur in the protein structure may also affect the quality of meat and meat products [18]. Application of pressure on proteins can cause significant conformational changes that affect functional properties [8]. The effect of the applied high pressure depends on the amount of pressure, the structure of the protein and the environment in which the protein is found (temperature, pH, ionic strength) [27].

As a result of breaking many of the weak bonds of proteins with the application of high pressure, proteins are structurally affected and significant changes in their properties can be observed [16]. High pressure application can also cause changes in the volume of proteins. While it accelerates the reactions and processes that cause a decrease in the volume of the protein, it prevents the reactions that cause an increase [16]. In studies conducted on myofibrillar proteins, one of the most important proteins of meat, it was determined that high pressure application increased the solubility of these proteins and caused structural changes [28].

HPH has an effect on the polysaccharide structure of carbohydrates. However, since the only polysaccharide it affects in meat is glycogen and the amount of glycogen in the meat decreases significantly with glycolysis after slaughter, the effect of applying pressure to fresh meat on carbohydrates is insignificant [29]. It has been determined that high pressure application accelerates glycolysis. This is explained by the fact that high-pressure treated muscle has a lower pH than untreated muscle [23].

Considering lipids with nutritional value, the most important effect of high-pressure application is on lipid oxidation and hydrolysis. The main reason for the decrease in the nutritional value of foods and the change in taste is based on lipid oxidation [30]. The biggest disadvantage of high-pressure application is that it causes increased fat oxidation during aerobic storage, regardless of the presence of oxygen throughout the application. This is one of the most important factors limiting the storage time of meat and meat products [31]. However, studies have shown that the addition of ethylenediaminetetraacetic acid (EDTA), which can chelate with metal ions, reduces lipid oxidation in high-pressure meat. It is also stated that lipid oxidation can be controlled with the use of antioxidants such as rosemary extract [32].

High-pressure increases the melting point of fats and can crystallize triglycerides at room temperature [21]. Some studies have shown that high pressure application has no effect on phospholipids, triglycerides and free fatty acids in meat products. However, some studies also suggest that high pressure application increases the free fatty acid content of meat. As a result, high pressure application accelerates lipid oxidation and will also affect fatty acids as oxidation products (He et al. 2011).

Effect pf HPH Application on Meat Components

It has been determined by many researchers that high pressure applications effect meat components, especially water, protein (including enzymes) and lipids [33].

Water

Adiabatic compression of water causes the temperature to increase by 2–3°C for every 100 MPa [34].

Proteins and Enzymes

The effect of HPH on proteins depends on the molecular structure size and the composition of the environment (pH, ionic strength and temperature). Generally, the secondary structure is affected only by very high-pressure processes, while the main structure is not affected by pressure. Tertiary and quaternary structures change under 100 MPa pressure. These changes in structure may also alter the activities of endogenous meat enzymes [34]. High pressure treatment can cause inactivation and activation of enzymes depending on buffer and pressure conditions [34-36].

Proteins

Proteins are one of the most important components of meat and have a very important place on the quality of meat. Therefore, changes that may occur in the structure of proteins may also affect the quality of meat and meat products. As is known, proteins can contain covalent bonds, disulfide bonds, hydrogen bonds and other interactions in their structures. As a result of the breaking of many of these bonds, proteins are modified, and significant changes occur in their properties. Although HPH application cannot break covalent bonds, it can affect hydrogen bonds and other interactions. In general, the quaternary structure is more sensitive to HPH due to having more hydrophobic interactions. As a matter of fact, oligomeric proteins can dissociate in low pressure applications (<150-200 MPa). However, large negative pressure changes may also occur. As a matter of fact, it is reported that a volume change of 500 ml/mol occurs in the dissociation of the lactate dehydrogenase enzyme [37]. The decrease in volume (30-80 ml/mol) occurs with denaturation due to the formation or breaking of non-covalent bonds and the rearrangement of solvent molecules [27]. However, significant oligomer changes are observed above 200 MPa. While HPH applied above 200 MPa affects the dissociation of proteins, it can also enable the recombination of dissociated oligomers. However, reversible expansion of proteins can also be observed at 400-800 MPa. Therefore, this situation reveals the importance of hydrophobic interactions in HPH application [37]. Changes in the secondary structures of proteins only occur at very high pressures, as a result of the breaking of hydrogen bonds that enable the formation of the helical structure. This change in secondary structure is usually irreversible.

Proteins contain covalent, disulfide, hydrogen bonds and other interactions in their structures [27]. As a result of the breaking of many of these bonds, proteins are modified, and significant changes occur in their properties. Since YBU does not affect covalent bonds, it has no effect on the primary structure. On the other hand, very high pressure causes the hydrogen bonds that provide the formation of the helical structure to be broken, leading to changes in the secondary structure and ultimately, irreversible denaturation. HBU causes protein denaturation by rearranging and/or breaking non-covalent bonds such as ionic bonds, hydrogen bonds, and hydrophobic interactions of the tertiary structure of proteins [38]. The breaking of ionic bonds is greatly affected by the increase in pressure.

Galazka et al. [39] determined that at pressures applied to Bovine Serum Albumin from 300 to 800 MPa, a partial loss of surface hydrophobicity occurred, and this was due to the protein opening and losing its solubility under the effect of pressure. However, it is also reported that HPH improves the surface properties and gel properties of milk proteins [40]. Parés et al. [41] reported that 450 MPa pressure applied for 15 minutes at 5 and 40°C reduced the solubility of porcine blood plasma proteins. However, they reported that it did not cause a statistically significant difference in texture and water retention capacity.

The effect of pressure on hydrophobic interactions is quite complex. Significant tertiary structure changes are observed at pressures above 200 MPa. In contrast to thermal applications that affect covalent and non-covalent bonds, at room temperature HPH breaks only relatively weak chemical bonds (hydrogen bonds, hydrophobic bonds and ionic bonds) [42]. Changes in protein structure affect the functional properties of food proteins. Therefore, HPH for foods can be used to create new products [27]. While low pressures (100–300 MPa), generally applied at room temperature, result in reversible changes in the protein structure, pressures above 300 MPa cause irreversible protein denaturation [27, 38].

High pressure increases the solubility of myofibrillar proteins by depolymerizing actin and myosin. Gelatinization of myosin and actomyosin, which occurs under the influence of high pressure and temperature, increases in environments with low ionic strength (0.1-0.4 M NaCl). However, although pressure application increases actomyosin gelatinization in an environment of high ionic strength, it does not affect myosin gelatinization. Although gelation occurs when a pressure above 200 MPa is applied to myosin solutions at pH 6.0 and 0.1 M KCl at room temperature, myosin does not gel when the ionic strength is increased. It has also been reported that the gel strength of the gel formed by the effect of temperature is significantly proportional to the protein concentration [43].

Proteins can maintain their natural structure and stabilization through interactions in the protein chain and solvent interactions in the environment. HPH application affects the delicate balance of protein-solvent interactions, causing the polypeptide chain to open [27]. This situation can also provide substrates for some biochemical changes, especially enzymatic changes [44].

Although volume increase occurs in most biochemical reactions, high pressure application leads to volume decrease. With the effect of HPH, there is a decrease in molecular gaps and an increase in internal interactions. As ionization increases in aqueous systems, changes in pH occur and a decrease in volume occurs as a result of the reactions that occur [45].

Jung et al. [20] stated that applying HPH treatment to meat after hardening at low temperature causes myofibril irregularity. In research conducted with fish, the effects of high-pressure applications on the processing quality of Tilapia (Oreochromis spp.) fillets were examined. Tilapia fillets were subjected to pressure treatment in the range of 50–300 MPa at 25°C, and it was determined that the gel strength of the samples processed at 50 MPa in one hour was 1.2 times higher than the gel strength of fresh meat. However, a 25% decrease in the gel strength of samples applied to 250 MPa pressure for 1 hour was reported [46].

The gel-forming property of proteins is an important factor that determines the textural properties of many foods. Although gel formation occurs through heat treatment, heat treatment can generally negatively affect some nutritional properties of foods [18]. Examining the effects of HPH applications on proteins is very important as it provides gel formation without causing significant damage to the properties of food.

Pressure modifies the structure and function of many proteins. For example, myosin in both meat and fish is denatured by high pressure, resulting in a gel-like texture. In addition, protease activity in muscle tissue may change depending on whether the substrate is abundant or whether pressure application causes spontaneous enzyme activation. All these changes in the structure and function of the protein affect the texture of muscle tissue. In this case, the effect of pressure on the protein structure is different from the effect of temperature. The texture of heat-treated and pressure-treated meat can be very different [47].

In studies conducted on myofibrillar proteins, one of the most important proteins of meat, it was determined that HPH increased the solubility of these proteins and caused structural changes [28]. Hsu and Ko [28] determined that myosin extracted from Tilapia muscles underwent significant changes in pressure applications from 500 to 2000 atm. In the study, it was determined that myosin filaments opened, and their volume decreased after a pressure of 500 atm at 0°C, and in the application from 1000 to 1500 atm, myosins underwent aggregation and formed an insoluble structure. However, it has been determined that between 500 atm and 1000 atm, myosin turns from a viscous sol state into an elastic gel, and at 2000 atm, myosin forms irregular aggregates. Additionally, Kamiyama et al. [48] reported that the pressure applied to chicken muscle from 200 MPa to 600 MPa caused the degradation of myosin.

Since it is possible to improve the gel-forming property of muscle proteins in processed meat products and control meat hardness, many studies have been conducted on HPH in meat and meat products. It has been stated that pressure application before heat treatment increases the gelling ability of meat protein in model systems. However, it has been stated that heat treatment under pressure applied to pork and chicken meat pastes causes a weaker gel/emulsion structure. The mechanism that changes product properties is not clear, but this has been attributed to the fact that the pressure treatment partially protects the protein from thermal degradation during gelatinization. This relationship limits the formation of gel structures and affects the textural properties and water binding rates of meat emulsions [49]. Hsu and Ko [28] stated that the myosin denatured at a pressure of 500 atm, that myosin aggregated at a pressure of 1000 atm and started to form an insoluble structure, and that myosin collapsed and formed a gel at pressure applications above 1500 atm. Kamiyama et al. [48] reported that the pressure applied to chicken muscle from 200 MPa to 600 MPa caused the degradation of myosin.

The connective tissue in meat consists of endomysium and perimysium. The endomysium forms a honeycomb-like structure. Generally, a bundle of muscle fibers joins an individual sheath and the perimysium surrounds these fiber bundles. While this wavy appearance is clearly visible on the surface of the endomysium in unpressured samples, it is not seen in pressure-applied samples. While the formation of a smooth surface related to the disappearance of the wavy structure continued to be seen in the samples applied to 100 MPa pressure at 200 MPa pressure, the good texture formed on the membrane disappeared and an increase in the thickness of the membrane was observed. While rupture and roughening were observed in the membrane at 300 MPa pressure, it was determined that the honeycomb-like structure of the endomysium of meat subjected to 400 MPa pressure was disrupted in some areas. It is not clear that the structural changes caused by the pressure in the intramuscular connective tissue have a significant effect on meat crispiness [50].

The disappearance of the wavy structure on the membrane surface was observed in the sample subjected to 100 MPa pressure as well as in the sample matured for 7 and 21 days. Thus, it is clear that HPH has a different effect on the membrane structure of the intramuscular connective tissue than its maturation [50].

Jung et al. [20] reported in their study that while the myofibrils of the control group beef, to which they did not apply pressure, showed thin and thick filament ultrastructure and normal sarcomere arrangement, there was no difference compared to the control by applying 130 MPa pressure for 260 seconds at 10°C. They stated that 325 MPa pressure did not cause any change in the myofibril structure, but in the samples where 520 MPa pressure was applied, the myofibril structures completely changed.

It is thought that collagen, which is primarily stabilized by hydrogen bonds, is little affected by pressure. It is stated that contractile myofibrillar proteins are mainly responsible for the change in textural properties observed when meat is exposed to HPH [51]. Therefore, connective tissue can limit the tenderness of meat after HPH [50]. During the ripening of the meat, the meat becomes crispy. This brittleness process is thought to be due to the hydrolytic breakdown of different muscle proteins catalyzed by proteolytic enzymes. In general, degradation of cytockletal proteins causes increased embrittlement [52]. It has been reported that HPH increases the release of proteolytic enzymes in lysosomes and therefore friability increases [52]. HPH actually causes three basic changes in meat. These: enzymatic, protein modifications (mainly modification of myofibrils) and structural modifications. Meat crispiness occurs as a result of different ultrastructural changes. During ultrastructural changes, pressure leads to changes in the A-band and weakening of the gap filaments. In addition, the I-band undergoes aggregation, its integrity is disrupted, and breaks may occur in the Z-lines. Thus, HPH interferes with normal embrittlement. The main ultrastructural changes are the weakening of Z-lines with the separation of connectin and nebulin filaments, the weakening of intermediate filaments with the fragmentation of desmin molecules, and the modification of the endomysium and perimysium. However, it is stated that the contraction state of the sarcomere and the fiber diameter are also related to brittleness [20]. Jung et al. [20, 53, 54] stated that the crispness of post-rigor meat applied to pressure in the range of 100–600 MPa did not increase, but higher pressure caused ultrastructural modifications in the meat.

Jung et al. [54] stated that the Warner-Bratzler shear value of meat treated with HPH at 520 MPa for 260 s at 10 °C increased significantly compared to the control. Researchers determined that the sarcomere length of the pressure-treated samples decreased by 20% compared to the control on post-mortem day 2. They determined that in meat that was pressurized at the pre-rigor stage, there was a shortening of the sarcomere due to the increase in the concentration of Ca^{+2} ions in the sarcoplasm, the lysosomal enzyme activity of the meat increased, and the pH decreased. They also stated that cooking loss was higher in pressure-treated samples compared to meat matured for 2 and 14 days. As a result, they stated that HPH caused a significant increase in lysosomal enzyme activity in beef muscles, but the application did not improve meat tenderness and did not shorten the ripening time [54].

The gel-forming properties of proteins are an important factor determining the textural properties of many foods. Although the gel formation mechanism can be achieved by heat treatment, heat treatment can generally negatively affect some nutritional properties of foods. Heat treatment causes gel formation by affecting the weak interactions in proteins [37]. Therefore, the fact that HPH application has similar effects on proteins and does not cause significant damage to the properties of food has made it very important to investigate its effects on the gel-forming properties of proteins.

Meat Enzymes

Enzymes are a special class of proteins. Since protein denaturation is related to conformational changes, the functionality of the enzyme may also change. It is stated that low pressures (~100 MPa) activate some enzymes, but this stimulating effect is only valid for monomeric enzymes. On the other hand, higher pressure generally causes enzyme inactivation, and inactivation depends on factors such as enzyme type, pH, composition of the medium, and temperature. For example, there are significant differences in the pressure stability of lipase, which is an important enzyme in meat technology. While one study stated that lipase was stable up to 1100 MPa pressure, another study stated that inactivation was possible at ~600 MPa pressure [27].

It is generally accepted that the aroma components in muscle increase throughout ripening due to the proteolytic degradation of certain components of the muscle. Generally, after slaughter, lysosomes become free due to the breakdown of lysosomal membranes and maturation is encouraged by proteolysis of the myofibrillar structure. It is stated that cathepsins B and D are active in the degradation of myosin and actin. It has also been stated that cathepsin L breaks down myosin, actin, α-actinin, troponin T and troponin I, but has no effect on troponin C or tropomyosin. It is stated that cathepsin H has a greater specific activity against myosin than cathepsin B [55].

Applying pressure to hot meat after slaughtering causes a rapid decrease in pH and intense contraction depending on application time, meat temperature and muscle type [56]. According to the research conducted by Angsupanich and Ledward [47], while it was determined that some of the water-soluble proteins, including various protease enzymes, were resistant to 800 MPa pressure, it was also determined that there was a significant decrease in the activities of neutral proteases in applications above 200 MPa pressure and at approximately pH 6.6.

Acid phosphatase activity increased up to 500 MPa pressure applied to the muscle. It was determined that the relative activity of this enzyme was 3.5 times higher than the control group at 500 MPa. The activities of cathepsin B, D and L increased up to 400 MPa and then decreased. It was determined that cathepsin D activity was approximately 60% of the activity of the control group at 500 MPa pressure. While the activities of cathepsin H and aminopeptidase B were not different from the activities of the control group up to 100 MPa, they gradually decreased with increasing pressure above this pressure level. These results show that cathepsin B, D, L and acid phosphatase activities increase with pressure application. On the other hand, it was determined that cathepsin H and aminopeptidase B were not resistant to pressure [55]. Angsupanich and Ledward [47] stated that protease activity decreased as the pressure increased above 200 MPa. Additionally, researchers stated that pressure application in the range of 100–150 MPa increased cathepsin B1 activity.

Lipid Oxidation

One of the most important factors limiting the storage time of meat and meat products is the oxidation occurring in lipids. This situation is mostly seen in fish oils with high unsaturated fatty acid content. HPH application causes the reduced myoglobin and oxymyoglobin to transform into ferric form. Thus, fat oxidation is catalyzed under the influence of iron. Meat products contain significant amounts of iron, both in the heme complex and in the structure of other components. Therefore, the iron contained in these components can switch to the free iron form and initiate lipid oxidation with its pro-oxidant effect. However, Orlien et al. [57] reported that the reason why HPH application causes lipid oxidation is not due to the catalytic effect of metmyoglobin, that is, the effect of the iron ion released under the effect of pressure but is related to the damage to the membrane. These researchers applied 300, 400, 500, 600, 800 MPa pressure to a portion of chicken breast muscle for 10 minutes and applied only heat treatment to another portion of chicken breast muscle (10 minutes at 80 °C) and stored it at 5°C for 2 weeks. They found that the application of 800 MPa increased lipid oxidation and a similar increase occurred with the effect of heat treatment. They found that 600 and 700 MPa caused very little oxidation. However, they stated that pressure applications of 500 MPa and below do not cause rancidity and that 500 MPa is a critical value in this sense. Chevalier et al. [58] applied 100, 140, 180 and 200 MPa pressure to Turbot fillets for 15 minutes at 4°C. They reported that lipid oxidation was affected by the pressure level and duration, especially oxidation increased from 180 MPa. Beltran et al. [59] reported that HPH applied to minced chicken breast meat (30 min. 20°C, 300 and 500 MPa) caused a lower level of lipid oxidation compared to heat treatment (15 min. 90°C). However, these researchers also report that salt has a significant pro-oxidant effect on pressurized chicken mince. Cheah and Ledward [60] found that the pressure applied to ground pork meat above 300 MPa at room temperature significantly catalyzed lipid oxidation.

The biggest disadvantage of HPH is because HPH causes increased lipid oxidation during subsequent aerobic storage, regardless of the presence of oxygen throughout the application. One of the most important factors limiting the storage time of meat and meat products is the oxidation that occurs in lipids. HPH causes the reduced myoglobin and oxymyoglobin to transform into ferric (Fe^{+3}) form. Thus, lipid oxidation is catalyzed by the effect of iron [60]. Chevalier et al. [61] stated that lipid oxidation in turbot fillets, to which they applied 100, 140, 180 and 200 MPa pressure for 15 min at 4°C, was affected by the pressure level and duration, especially oxidation increased from 180 MPa pressure. Cheah and Ledward [60] found that pressure applied to ground pork meat at room temperature over 300 MPa significantly catalyzed lipid oxidation.

Fish oils are more prone to oxidation than meats. Because fish oils contain high amounts of unsaturated fat, oxidative changes caused by pressure can be much more important. Angsupanich and Ledward [47] determined little change in the initial TBA number of fish subjected to 200 MPa 20 min pressure compared to unpressurized cod. However, when the pressure was applied at 400 MPa for 20 minutes, the TBA value increased significantly. In fish packaged in a modified atmosphere (100% nitrogen), the TBA value increased significantly at 600 and 800 MPa pressures. As a result, researchers stated that lipid oxidation increased at pressures above 400 MPa due to the release of metal ions.

It has been determined that fat oxidation has a restrictive role in the application of HPH. It has been reported that the peroxide value of oils extracted by applying pressure (202, 404 and 608 MPa for 15–30 min) to cod muscles increases depending on the increase in pressure and time [62]. Oxidative changes occurring as a result of 300 MPa pressure were observed at 80°C for 15 min. It has been stated that it is no different from the changes occurring with heat application [63]. Oshima et al. [62] stated that extracted seafood oils were resistant to oxidation up to 608 MPa pressure, but if the oil was with muscle, its oxidation was more affected by the pressure application. Angsupanich and Ledward [47] reported that the oxidation of lipids in cod fish increased after pressure treatment above 400 MPa. It has been determined that the lipid in fish is more sensitive to oxidation than the lipids of other species.

Cheah and Ledward [63] stated that pressure applied up to 200 MPa did not affect lipid oxidation in ground pork, but pressures of 400 MPa and above significantly increased lipid oxidation. They stated that in samples packaged in a modified atmosphere (100% nitrogen), the pressure applied up to 800 MPa did not affect lipid oxidation much, but after 5 days of storage at 4°C, lipid oxidation increased significantly at pressures of 300 MPa and above. In conclusion, while pressure applications below 300 MPa have little effect on lipid oxidation, pressures above this level proportionally increase oxidation.

Beltran et al. [59] stated that 500 MPa pressure caused less oxidation in minced chicken breast meat than boiling it at 90°C for 15 minutes. Additionally, they stated that lipid oxidation of chicken meat samples was related to the cell membrane.

Color

Since HPH application has an effect on proteins, it can cause significant changes in color by affecting myoglobin, which is in the protein structure [18]. This color change is especially observed in beef and tuna at low temperatures such as 5-10°C [17].

The pigment that gives meat its color is myoglobin, a protein structure. Myoglobin consists of two parts: globin, which is a protein, and a heme molecule, which carries iron atoms (Fe^{+2}). While the color of the meat is dark red immediately after slaughter, myoglobin turns into oxymyoglobin in a short time under the effect of oxygenation at atmospheric pressure and the meat acquires a bright red color. The color of fresh meat changes depending on the amount of myoglobin, oxymyoglobin and metmyoglobin, and the amount of metmyoglobin increases as oxidation progresses.

After the ground beef is packaged in an oxygenated environment and/or under vacuum, it is heated at 250–350 MPa at 10 °C for 10 min. It has been reported that when pressure is applied for a long time, the Hunter "L" value increases and the ground meat turns pink. It was determined that the Hunter "a" value decreased at 400–500 MPa and the meat color turned into a grayish-brownish cooked meat color. While the total amount of myoglobin extracted decreased in the 250–500 MPa pressure range, it was determined that the metmyoglobin rate increased in the 400–500 MPa pressure range [64]. Cheah and Ledward [60] stated that 80–100 MPa pressure applied to the cattle Longissimus dorsi and Psoas major muscles for 20 minutes improved color stability, but the pressure applied 7–20 days after slaughter did not have any effect on color stability.

Since HPH is effective on proteins, it can cause significant changes in color by affecting myoglobin, which is in the protein structure. Cheah and Ledward [60] reported that 400 MPa caused irreversible denaturation on myoglobin in minced pork. Carlez et al. [64] applied HPH to vacuum-packaged ground beef for 10 minutes at 10°C and found that *L** color values increased significantly in the range of 200-300 MPa and the color turned pink. They determined that *a** values decreased in the range of 400-500 MPa and the color changed to gray brown. They found that while the oxymyoglobin ratio decreased in the range of 400-500 MPa, metmyoglobin increased. In addition, these researchers state that the denaturation of metmyoglobin under pressure is similar to heat, acid and urea denaturation and that this situation is partly due to the breaking of hydrophobic interactions. Cheah and Ledvard [60] determined that 80-100 MPa applied to Longissimus dorsi and Psoas major cattle muscles for 20 minutes affected color stability. However, they reported that the pressure they applied 7-20 days after slaughter did not provide any improvement in color stability.

In another study, it was determined that 600 MPa HPH for 20 min increased the *L** value and decreased the *a** and *b** values in sausage dough [9]. Researchers explained this color change as denaturation of myoglobin and oxidation of ferrous (Fe^{+2}) myoglobin to ferric (Fe^{+3}) metmyoglobin or haem displacement/release. Jung et al. [54] determined that the pressure applied at 520 MPa for 260 s at 10°C caused an increase in the *L**, *a** and *b** values of meat. Jung et al. [54] attributed the change in *a** value to the content of myoglobin pigments and especially metmyoglobin formation. Researchers reported that the relationship between *a** value and metmyoglobin continued even after pressure application.

Phase Change

Meat storage conditions are one of the most important factors affecting the quality of meat. For long-term preservation of meat, the freezing method is preferred. However, some problems with meat quality may occur with freezing. The volume of ice crystals formed during slow freezing can damage meat tissue. Freezing at atmospheric pressure results in an increase in volume and therefore a lower density of ice compared to water. For example, freezing at 0°C causes a 9% volume increase, and freezing at -20°C causes a 13% volume increase. Therefore, the increase in volume is not desired because it damages the tissues, and this problem can be overcome to a certain extent by the application of HPH. Because HPH application creates fast, numerous and homogeneous ice crystals, and the density of these ice crystals is higher than the density of water. Therefore, damage to tissue can be significantly reduced [24, 25]. The formation of ice crystals is closely related to the pressure level and temperature. At 200 MPa, no phase transition is observed down to -20°C. This point is critical and ice crystals form below -20°C and a phase transition is observed at this point. In pressure applications lower than or higher than 200 MPa, ice crystals may form at higher temperatures. For example, ice crystals can form at approximately 0°C at 600 MPa [25].

Under high pressure (210 MPa), the melting and freezing points of water drop to a minimum of -21°C. This occurs because the volume increases due to the ice crystals formed is prevented by the pressure. Pressure-assisted freezing can minimize the problems that arise when classical methods are applied, such as temperature changes, large ice crystals and frozen concentrated areas [43]. Therefore, it has been suggested that foods sensitive to freezing will be better preserved by applying pressure that changes the freezing point. It has been reported that the technique may be useful, especially in freezing large pieces of meat. It has been determined that since the melting point of ice decreases at elevated pressures, dissolution occurs at higher temperatures (‹-10°C) and dissolution losses decrease significantly. In addition, it has been determined that as the heat flow rate between the pressure-conducting medium and frozen meat increases, the dissolution rate and speed also increase [43].

Freezing meat is one of the most effective preservation methods. Freezing speed is very important in freezing meat and meat products. Because freezing speed is one of the factors affecting quality. Especially the volume of ice crystals formed during slow freezing can damage the tissue [65]. When water is frozen at atmospheric pressure, its volume will increase, and the density of ice will decrease compared to water. For example, freezing at 0°C causes a 9% volume increase, and freezing at -20°C causes a 13% volume increase [66]. Therefore, this increase in volume will damage the tissues during freezing. Damage to tissue can be significantly reduced with HPH [66, 67]. The formation of ice crystals is closely related to the pressure level and temperature. Phase transition down to -20°C at 200 MPa is not observed. This point is critical and ice crystals form below -20°C and a phase transition is observed at this point. In pressure applications lower than or higher than 200 MPa, ice crystals may form at higher temperatures. For example, at 600 MPa, ice crystals can form at approximately 0°C [66]. The freezing point of water decreases as the pressure is increased up to 210 MPa, at which point the ice changes phase and the freezing point increases. Therefore, the unfrozen part of the water under HPH offers a significant potential for thawing [67].

Thawing meat is a costly process as it requires a lot of time as well as freezing, and it can also cause quality problems. To solve these problems, Zhao et al. [67] reported in a study that HPH application shortened the thawing time of frozen beef and did not adversely affect the quality of the product. However, they state that the pressure level and application time may affect the dissolution rate and product quality. However, they report that product volume and freezing temperature do not affect the dissolution rate. Therefore, they state that the use of HPH application may be advantageous in thawing large pieces of meat.

The level of pressure to be applied to meat must be determined by taking into account its effects on proteins. As a matter of fact, Chevalier et al. [58]. They found that 200 MPa at -18°C did not improve lobster meat texture and caused undesirable effects on myofibrillar proteins. However, these researchers recommend that the pressure level should be reduced below 150 MPa to eliminate the negative effect on proteins. Fernández-Martín et al. [68] also reported that 200 MPa at -20°C is not suitable, the sarcomere structure is damaged at these values, and there is a change in the structure of myofibrillar proteins.

The thawing process as well as freezing of meat and meat products has a significant impact on the quality of meat. Zhao et al. [67] reported that HPH shortens the thawing time of frozen beef and may have significant advantages in thawing large pieces of meat.

Effect of HPH Application on the Microbiological Quality of Meat

One of the most important effects of HPH on the quality of meat is that it increases the microbiological quality of meat. The applied pressure level causes significant changes in the morphology, cell walls and cell membranes, genetic mechanisms and biochemical reactions of microorganisms. With the application of HPH, cell membrane permeability increases, intracellular components are broken down, energy-producing reactions in the cell are inhibited, enzymes necessary for cell growth become inactive, and the pH range required for growth decreases [45]. Thus, especially the vegetative cells of microorganisms are significantly damaged. Today, pasteurization and sterilization are mostly used to inactivate microorganisms. While this situation increases the microbiological quality of food, it can also reduce its nutritional value. Therefore, alternative sterilization practices are very important in food processes.

The amount and kinetics of inactivation of microorganisms by HPH depends on many parameters such as the type of microorganism, pressure level, processing time, temperature and pH of the food, water activity and composition. Studies have shown that the resistance of microorganisms to HPH varies. It has been reported that Gram (-) bacteria are more sensitive to pressure applications than Gram (+) bacteria, and psychrophilic bacteria are more sensitive to pressure applications than mesophilic bacteria. However, differences in resistance to pressure can also be observed between different species of the same genus. With increasing pressure, different morphological changes and metabolic activity disorders occur in microorganisms [43].

In solid foods such as meat, 300–600 MPa pressure application (10–30 min at 25°C) causes a decrease of 3–6 logarithmic units in the number of vegetative microorganisms. Although bacterial spores are more resistant to high pressure, they can be inactivated by choosing appropriate conditions [8]. Carlez et al. [69] in their study with ground meat inoculated with Pseudomonas species, they found that a certain part of the cells (0.01%) could not be inactivated by the pressure process, but were only suppressed, and that the bacteria continued to grow again after a recovery phase of 3–9 days at 3°C. However, it has been suggested that by combining high pressure and moderate temperature processes, the product can achieve excellent microbial stability. As a result of studies based on the sensitivity of various microorganisms naturally present in fresh or processed meat or subsequently contaminated to high pressure. It has been determined that the best application conditions of high pressure in the preservation of meat are 1-10 minutes at 400-600 MPa at 50-70°C [43].

Inactivation kinetics of microorganisms under high pressure depends on many factors such as pressure level, application time, temperature, pH, water activity and food ingredients [7]. The effect of HPH on microorganisms varies significantly depending on the species. Although the spores are quite resistant, Gram-positive bacteria are more resistant than Gram-negative ones. However, the resistance of bacteria decreases significantly during the growth phase. However, bacilli are more sensitive than cocci [7].

Temperature also plays an important role in microbial inactivation with HPH. At optimum growth temperatures, inactivation is less than at high and low growth temperatures. Because membrane permeability can easily deteriorate at temperatures other than the optimum development temperature [7, 70]. As a matter of fact, it has been stated that when the cell is cold shocked before HPH, the cell's resistance to pressure increases. The cell membrane, which adapts to low temperatures before HPH, causes the fatty acids in its structure to branch and decrease the chain length, increasing the resistance to high pressure. In addition, proteins subjected to cold shocking also increase the cell's resistance to high pressure [7].

The effect of HPH on the inactivation of microorganisms is significantly related to the application temperature. At optimum temperature, the resistance of microorganisms against HPH increases, and at temperature values above the optimum temperature, their resistance decreases. This situation occurs due to the increase in the permeability of the cell membrane due to the increase in temperature. Therefore, the saturation level and chain length of the fatty acid chains in the cell membrane affect the resistance of the cell membrane. As a matter of fact, when cold shocking is applied to the cell before applying HPH, the resistance of the cell against pressure increases. The cell membrane, which adapts to low temperature before HPH, causes the fatty acids in its structure to branch and decrease the chain length, increasing the resistance. In addition, proteins subjected to cold shocking also increase the cell's resistance [7]. Therefore, this situation needs to be taken into account when determining the pressure level to be applied to foods kept in the cold. In addition, temperature should be taken into account in HPH applications and the temperature level with HPH should be optimized depending on the target microorganisms, taking into account the characteristics of the food. Studies on this subject are insufficient and more studies are needed. In addition, effective results can be obtained in HPH applications with the combination of other factors such as antimicrobial substances and enzymes (Hurdle effect) as well as temperature.

Hajós et al. [9] achieved a reduction of more than 3 logarithmic units in the total number of living cells in the natural microflora of sausage dough, to which they applied 600 MPa pressure for 20 minutes. The researchers also achieved the 5 logarithmic unit reduction requested by USDAFSIS in the dough they inoculated with *L. monocytogenes*, with the specified ICU. It has been stated that in some cases, the duration of pressure application is a more important variable than the pressure level for microbial inactivation. In their study, Yuste et al. [71] stated that a 30-minute application at 350 MPa provided a greater microbial reduction than a 5 to 10-minute application at 400 MPa. Additionally, researchers reported that HPH had a significant bactericidal effect on *L. innocua.* They determined a very high reduction of 8 logarithmic units in poultry meats subjected to 400 MPa pressure for 30 minutes at 2°C and 450 MPa pressure for 10 minutes at 10°C. They also detected a reduction of 4.19 logarithmic units at 2 °C at 400, 450 and 500 MPa pressure for mesophilic bacteria.

Moerman et al. [72] stated that sterility can be achieved to a great extent at 400 MPa pressure if the pressure application temperature is minimum 65°C. Commercial sterility (12 logarithmic units) cannot be achieved if the pH of the food is high. However, it has been stated that HPH is very successful in low acid foods. As a result, it has been stated that to ensure a certain degree of commercial sterility, a pressure of at least 400 MPa must be applied at 50°C for 30 minutes. It has also been reported that pressures of 250–350 MPa at high temperatures such as 65–75°C are sufficient to achieve a certain pasteurization effect [72].

Garriga et al. [70] achieved a significant reduction of at least 4 logarithmic units in the number of aerobic, psychrotrophic and lactic acid bacteria by applying 600 MPa 6 min HPH to sliced and vacuum-packaged sirloin. They also recorded a decrease of approximately 3 logarithmic units in the number of Enterobacteriaceae. Jung et al. [73] stated that the total flora of the samples was similar to the total flora of the control group after 130 MPa pressure application, but they achieved a decrease of 2.5 logarithmic units as a result of 520 MPa pressure application. Beltran et al. [59] stated that pressure increases the effectiveness of nisin and other antimicrobial compounds by damaging bacterial cells.

Conclusion

HPH application is a new technology that commercially inactivates microorganisms in food, thus preserving food and has many advantages. As a result, high pressure has many beneficial effects in the industry, such as increasing meat crispness, microbiological inactivation, pasteurization, preventing excessive and overcooking, prolonging the storage time in the refrigerator, increasing the binding and gelatination abilities of sausages and cured meats with the help of pressure, and faster thawing of meats with less dissolution loss at low temperatures. It has been reported that high pressure, when combined with processes such as vacuum packaging, moderate heating and refrigerated storage, can be used successfully in many meat products, including cured meats, ready meals and mechanically deboned meats. This new technology does not cause environmental pollution. It will also enable the production of healthier foods that will meet the increasing demand of consumers for minimally processed foods in recent years.

The effect of high-pressure application on foods is very important, and its importance in meat technology is increasing day by day. While it can improve the quality of meat and meat products, it seems possible that it can also be used in new product development. However, studies on the application of this method are quite insufficient and more research is needed. There are many more uncertainties that need to be clarified in terms of both the advantages and disadvantages of this method. For example, there is not yet sufficient literature support as to whether it has toxic effects. The effects of the changes on proteins on human health have not been fully explained. In addition, the results of the studies are generally based on generalizations of the effects on foods. The detailed studies are limited and more literature support is needed.

Studies on high pressure technology and its practical applicability have not been carried out sufficiently due to its cost. It is necessary to make studies on this subject and make high pressure technology economically advantageous.

References

[1] Ensoy, Ü., & Cosar, B. (2006). The Effects of High Pressure Processing on the Sensorial, Physical and Biochemical Properties of Meat and Meat Products. *Journal of Agricultural Faculty of Gaziosmanpasa University, 23(2),* 1-7.

[2] Crehan, C. M., Troy, D. J., & Buckley, D. J. (2000). Effects of salt level and high hydrostatic pressure processing on frankfurters formulated with 1.5 and 2.5% salt. *Meat Science, 55(1),* 123-130.

[3] Cruz-Romero, M., Kelly, A. L., & Kerry, J. P. (2007). Effects of high-pressure and heat treatments on physical and biochemical characteristics of oysters [Crassostrea gigas]. *Innovative Food Science and Emerging Technologies, 8(1),* 30-38.

[4] Tornberg, E. (2005). Effect of heat on meat protein – Implications on structure and quality of meat products. *Meat Science, 70(3),* 493-508.

[5] Jensen, W. K., Devine, C., & Dikeman, M. (2004). *Heat effects on meat/ Physics and chemistry, Encyclopedia of meat science*, Edited by Palka K. Elsevier, Oxford, Poland, 567 p.

[6] Ross, A. I. V., Griffiths, M. W., Mittal, G. S, & Deeth, H. C. (2003). Combining nonthermal Technologies to control foodborne microorganisms. *International Journal of Food Microbiology, 89(2-3),* 125-138.

[7] Hugas, M., Garriga, M., & Monfort, J. M. (2002). New mild technologies in meat processing: high pressure as a model technology. *Meat Science, 62(3),* 359-371.

[8] Cheret, R., Chapleau, N., Delbarre-Ladrat, C., Verrez-Bagnis, V., & De Lamballerie M. (2005). Effects of high pressure on texture and microstructure of sea bass (Dicentrarchus labrax L.) fillets. *Journal of Food Science, 70(8),* 477-483.

[9] Hajós, G. Y., Polgár, M., & Farkas, J. (2004). High-Pressure Effects on IgE Immunoreactivity of Proteins in a Sausage Batter. *Innovative Food Science and Emerging Technologies, 5,* 443–449.

[10] Earnshaw, R. (1996). High Pressure Food Processing. *Nutrition & Food Science* Number 2 – March/April, 8–11.

[11] Tuboly, E., Lebovics, V. K., Gaal, Ö., Meszaros, L., & Farkas, J. (2003). Microbiological and Lipid Oxidation Studies on Mechanically Deboned Turkey Meat Treated by High Hydrostatic Pressure. *Journal of Food Engineering, 56,* 241-244.

[12] Mor-Mur, M., & Yuste, J. (2003). High Pressure Processing Applied to Cooked Sausage Manufacture:Physical Properties and Sensory Analysis. *Meat Science, 65,* 1187-1191.

[13] Cemeroğlu, B. (2004). *Fruit-Vegetable Processing Technology*, 226–227.

[14] Ramirez-Suarez, J. C., & Morrissey, M. T. (2006). Effect of High Pressure Processing (HPP) on Shelf Life of Albacore Tuna (Thunnus alalunga) Minced Muscle. *Innovative Food Science and Emerging Techologies, 7,* 19-27.

[15] Sequeira-Munoz, A., Chevalier, D., LeBail, A., Ramaswamy, H. S., & Simpson, B. K. (2006). Physicochemical Changes Induced in Carp (*Cyprinus carpio*) Fillets by High Pressure Processing at Low Temperature. *Innovative Food Science and Emerging Techonologies, 7,* 13-18.

[16] Murchie L. W., Cruz-Romero, M., Kerry, J. P., Linton, M., Patterson, M. F., Smiddy, M., & Kelly, A. L. (2005). High pressure processing of shellfish: A review of microbiological and other quality aspects. *Innovative Food Science and Emerging Technologies, 6(3),* 257- 270.

[17] Karakaya, M., Caner, C., & Sarıçoban, C. (2004). The Use of High Hydrostatic Pressure in Meat Technology. *The Journal of Food, 29(6),* 465–470.

[18] Zorba, Ö., & Kurt, S. (2005). The Effects of High Pressure Treatments on Meat and Meat Products. *Van Veterinary Journal, 16(1),* 71-76.

[19] Macfarlane, J. J. (1973). Pre-rigor pressurization of muscle: Effects on pH, shear value and taste panel assessment. *Journal of Food Science, 38,* 294–298.

[20] Jung, S., de Lamballerie-Anton, M., & Ghoul, M. (2000a). Modifications of Ultrastructure and Myofibrillar Proteins of Post-rigor Beef Treated by High Pressure. *Lebensmittel-Wissenschaft und-Technologie, 33,* 313-319.

[21] Zhou, G. H., Xu, X. L., & Liu, Y. (2010). Preservation technologies for fresh meat – a review. Meat Science, 86(1), 119-128.

[22] Aymerich, T., Picouet, P. A., & Monfort, J. M. (2008). Decontamination technologies for meat products. *Meat Science, 78(1-2*), 114-129.

[23] Ozlu, H., & Atasever, M. (2007). High Pressure Application to Foods. *Ataturk University Journal of Veterinary Sciences, 2(1),* 7-27.

[24] Ozlu, H. (2006). *High pressure applications on meat technology.* Graduate School of Natural and Applied Sciences. Master's thesis. Erzurum, Türkey.

[25] Souza, C. M., Boler, D. D., Clark, D. L., Kutzler, L. W., Holmer, S. F., Summerfield, J. W., Cannon, J. H., Smit, N. R., McKeith, F. K., & Killefer, C. (2011). The effects of high pressure processing on pork quality, palatability and further processed products. *Meat Science, 87(4),* 419-427.

[26] Kruk, Z. A, Yun, H., Rutley, D. L., Lee, E. J., Kim, Y. J., & Jo, C. (2011). The effect of high pressure on microbial population, meat quality and sensory characteristics of chicken breast filet. *Food Control, 22(1),* 6-12.

[27] Hendrickx, M., Ludikhuyze, L., Van den Broek, I., & Weemaes, C., (1998). Effects of high pressure on enzymes related to food quality. *Trends in Food Science & Technology, 9(5),* 197-203.

[28] Hsu, K. C., & Ko, W. C. (2001). Effect of hydrostatic pressure on aggregation and viscoelastic properties of tilapia [Orechromis niloticus] myosin. *Journal of Food Science, 66(8),* 1158-1162.

[29] Kerry, J., Kerry, J., & Ledward, D. (2002). Meat processing: improving quality.2nd Edition, CRC Press, Cambridge, England, 464 p.

[30] Henry, C. J. K., & Chapman, C. (2002). *The nutrition handbook for food processors.* 2nd Edition, CRC Press, Cambridge, England, 483 p.

[31] Oz, F., & Kaya, M. (2006). The Effect of High Pressure Treatment on Meat Quality. *Ataturk University Journal of Agricultural Faculty, 37(2),* 249-255.

[32] Bajovic, B., Bolumar, T., & Heinz, V. (2012). Quality consideration with high pressure processing of fresh and value-added meat products. *Meat Science, 92(3),* 280-289.

[33] He, Z., Haung, Y., Li, H., Qin, G., Wang, T., & Yang, J. (2011). Effect of high-pressure treatment on the fatty acid composition of intramuscular lipid in pork. *Meat Science, 90(1),* 170-175.

[34] Lamballerie-Anton, M., Enitiaa, N., Taylor, R. G., Culioli, J., & Theix, I. (2002). *High Pressure Processing of Meat. "Meat Processing: Improving Quality"* Ed. Kerry, J., Kerry, J. and Ledward, D. S, 313-331.

[35] Serra, X., Sarraga, C., Grebol, N., Guardia, M. D., Guerrero, L., Gou, P., Masoliver, P., Gassiot, M., Monfort, J. M., & Arnau, J. (2007a). High Pressure Applied to frozen Ham at Different Process Stages. 1. Effect on The Final Physicochemical Parameters and on The Antioxidant and Proteolytic Enzyme Activities of Dry-Cured Ham. *Meat Science, 75,* 12-20.

[36] Serra, X., Grebol, N., Guardia, M. D., Guerrero, L., Gou, P., Masoliver, P., Gassiot, M., Sarraga, C., Monfort, J. M., & Arnau, J. (2007b). High Pressure Applied to Frozen Ham at Different Process Stages. 2. Effect on The Sensory Attributes and on The Colour Characteristics of Dry-Cured Ham. *Meat Science, 75,* 21-28.

[37] Lullien-Pellerin, V., & Balny, C. (2002). High-Pressure as a Tool to Study Some Proteins' Properties; Conformational Modification, Activity and Oligomeric Dissociation. *Innovative Food Science & Emerging Technologies, 3,* 209-221.

[38] Ko, W. C., Jao, C. L., Hwang, J. S., & Hsu, K. C. (2006). Effect of High-Pressure Treatment on Processing Quality of Tilapia Meat Fillets. *Journal of Food Engineering, 77,* 1007-1011.

[39] Galazka, V. B., Ledward, D. A., Summer, I. G., & Dickinson, E. (1997). Influence of high pressure on bovine serum albumin and its complex with dextran sulphate. *Journal of Agricultural and Food Chemistry, 45,* 3465-3471.

[40] Johnston, D. E., Austin, B. A., & Murphy, R. J. (1993). Properties acid-set gels prepared from high pressure treated skim milk. *Milchwissenschaft, 49(12),* 206-209.

[41] Parès, D., Saguer, E., Toldrà, M., & Carretero, C. (2000). Effect of High Pressure Processing at Different Temperatures on Protein Functionality of Porcine Blood Plasma. *Journal of Food Science, 65(3),* 486-490.

[42] Jiménez Colmenero, F. (2002). Muscle Protein Gelation by Combined Use of High Pressure/Temperature. *Trends in Food Science & Technology, 13,* 22–30.

[43] Yetim, H., Kesmen, Z., & Kayacier, A. (2003). *High Pressure Applications in the Meat Industry.* 3rd Food Engineering Congress. Ankara, Turkey.

[44] Kurt, Ş. & Zorba, Ö. (2004). Transglutaminase and its use for protein modification. *The Journal of Food, 29(5),* 357-364.

[45] Şanal, İ. S., & Çalımlı, A. (2000). Application of High Hydrostatic Pressure Technology in the Food Industry. *The Journal of Food, 25(3),* 193 – 201.

[46] Ko, W. C., Jao, C. L., Hwang, J. S., & Hsu, K. C. (2005). Effect of High Pressure Treatment on Processing Quality of Tilapia Meat Fillets. *Journal of Food Engineering, 85,* 295-350.

[47] Angsupanich, K., & Ledward, D. A. (1998). High Pressure Treatment Effects on Cod (*Gadus Morhua*) Muscle. *Food Chemistry, 63(1),* 39–50.

[48] Kamiyama, K., Ikeuchi, Y., Suzuki, A., Kim, K., Hayashi, T., & Ito, T. (2001). An Immunological Assessment of Myosin Degradation in Pressurized Chicken Muscle. *Journal of Food Science, 66,* 1126–1129.

[49] Jiménez-Colmenero, F., Cofrades, S., Carballo, J., Fernández, P., & Fernández-Martín, F. (1998). Heating of Chicken and Pork Meat Batters under Pressure Conditions: Protein Interactions. *Journal of Agricultural and Food Chemistry, 46,* 4706-4711.

[50] Ueno, Y., Ikeuchi, Y., & Suzuki, A. (1999). Effects of High-Pressure Treatments on Intramuscular Connective Tissue. *Meat Science, 52,* 143–150.

[51] Ma, H.-J., & Ledward, D. A. (2004). High Pressure/Thermal Treatment Effects on the Texture of Beef Muscle. *Meat Science, 68,* 347–355.

[52] Bertram, H. C., Whittaker, A. K., Shorthose, W. R., Andersen, H. J., & Karlsson, A. H. (2004). Water Characteristics in Cooked Beef as Influenced by Ageing and High-Pressure Treatment-an NMR Micro Imaging Study. *Meat Science, 66,* 301–306.

[53] Jung, S., de Lamballerie-Anton, M., Taylor, R. G., & Ghoul, M. (2000b). High-Pressure Effects on Lysosome Integrity and Lysosomal Enzyme Activity in Bovine Muscle. *Journal of Agricultural and Food Chemistry, 48,* 2467-2471.

[54] Jung, S., Ghoul, M., & de Lamballerie-Anton, M. (2000c). Changes in Lysosomal Enzyme Activities and Shear Values of High Pressure Treated Meat During Ageing. *Meat Science, 56,* 239–246.

[55] Homma, N., Ikeuchi, Y., & Suzuki, A. (1994). Effects of High-Pressure Treatment on the Proteolytic Enzymes in Meat. *Meat Science, 38,* 219–228.

[56] Des Champs. O., Cottin, P., Largeteau, A., Demazeav, G., & Ducastaing, A. (1992). In *High Pressure and Biotechnology* Eds. C. Balny, R., Hayashi, K. Heremans and P. Masson. coll. INSERM, 175–224, Paris.

[57] Orlien, V., Hansen, E., & Skibsten, L. H. (2000). Lipid Oksidation in High-pressure Processed Chicken Breast Muscle During Chill Storage: Critical Working Pressure in Relation to Oxidation Mechanism. *European Food Research and Technology, 211,* 99-104.

[58] Chevalier, D., Sentissi, M., Havet, M., Le Bail, A. (2000). Comparison Of Air-blast and Pressure Shift Freezing on Norway Lobster Quality. *Journal of Food Science, 65(2),* 329-333.

[59] Beltran, E., Pla, R., Yuste, J., & Mor-Mur, M. (2003). Lipid Oxidation of Pressurized and Cooked Chicken: Role of Sodium Chloride and Mechanical Processing on TBARS and Hexanal Values. *Meat Science, 64*, 19–25.

[60] Cheah, P. B., & Ledward, D. A. (1997). Inhibition of Metmyoglobin Formation in Fresh Beef by Pressure Treatment. *Meat Science, 45(3),* 411-418.

[61] Chevalier, D., Le Bail, A., & Ghoul, M. (2001). Effects of High Pressure Treatment (100-200 MPa) at Low Temperature on Turbot (Scophtalmus maximus) Muscle. *Food Research International, 34,* 425–429.

[62] Oshima, T., Ushio, H., & Kizumic, J. (1993). High Pressure Processing of Fish and Fish Products. *Trends in Food Science & Technology, 4,* 370.

[63] Cheah, P. B., & Ledward, D. A. (1996). High Pressure Effects on Lipid Oxidation in Minced Pork. *Meat Science, 43(3),* 123-134.

[64] Carlez, A., Veciana-Nogues, T., & Cheftel, J. C. (1995). Changes in Colour and Myoglobin of Minced Beef Meat Due to High Pressure Processing. *Lebensmittel-Wissenschaft und-Technologie*, *28,* 528-538.

[65] Gökalp, H. Y., Kaya., M., & Zorba., O. (2004). Engineering of meat products processing (5th ed.). Atatürk Univ. Publ. No: 786, *Faculty of Agriculture*, No: 320. Erzurum, Turkey,

[66] Li, B., & Sun, D. W. (2002). Novel Methods for Rapid Freezing and Thawing of Foods: A Review. *Journal of Food Engineering, 54,* 175-182.

[67] Zhao, Y., Flores, R. A., & Olson, D. G. (1998). High Hydrostatic Pressure Effects on Rapid Thawing of Frozen Beef. *Journal of Food Science, 63,* 272–275.

[68] Fernandez-Martin, F., Otero, L., Solas, M. T., & Sanz, P. D. (2000). Protein Denaturation and Structural Damage During-High-Pressure Schift Freezing of Porcine and Bovine Muscle. *Journal of Food Science, 65(6*), 1002-1008.

[69] Carlez, A. Rosec., J. P., Richard, N., & Cheftel, J.C. (1994). Bacterial Growth During Chilled Storage of Pressure Treated Minced Meat. *Lebensmittel-Wissenschaft und-Technologie*, *27,* 48–54.

[70] Garriga, M., Grèbol, N., Aymerich, M. T., Monfort, J. M., & Hugas, M., (2004). Microbial Inactivation after High-Pressure Processing at 600 MPa in Commercial Meat Products over its Shelf Life. *Innovative Food Science and Emerging Technologies, 5,* 451–457.

[71] Yuste, J., Mor-Mur, M., Capellas, M., & Pla, R. (1999). Listeria innocua and Aerobic Mesophiles during Chill Storage of Inoculated Mechanically Recovered Poultry Meat Treated with High Hydrostatic Pressure. *Meat Science, 53,* 251–257.

[72] Moerman, F., Mertens, B., Demey, L., & Huyghebaert, A. (2001). Reduction of Bacillus subtilis, Bacillus stearothermophilus and Streptococcus faecalis in Meat Batters by Temperature- High Hydrostatic Pressure Pasteurizaiton. *Meat Science, 59,* 115–125.

[73] Jung, S., Ghoul, M., & de Lamballerie-Anton, M. (2003). Influence of High Pressure on the Color and Microbial Quality of Beef Meat. *Lebensmittel-Wissenschaft und-Technologie*, *36,* 625-631.

Chapter 6

The Impact of High-Pressure Homogenization on Milk and Dairy Products

Suzan Sever
and Gulcin Yildiz*
Department of Food Engineering, Igdir University, Iğdır, Turkey

Abstract

A non-thermal food preservation technique called high pressure homogenization (HPH) allows foods to retain their nutritional value, as well as their functional and sensory qualities, for a long shelf life. The application of HPH is effective because it inactivates microorganisms, activates and/or deactivates food enzymes, preserves food freshness, and develops the texture, color, natural taste, and flavor characteristics of food products via the formation of covalent bonds. It also causes structural changes in proteins and improves their ability to bind water. The efficiency of non-thermal techniques in the food industry is increased by their availability for use in heat processing, enzymatic mixes, and natural antibacterial and antioxidant compounds. HPH is significantly more favorable than the other thermal processing techniques since it requires less time and reduces the damage caused by heat application. This chapter reviews the impacts of HPH on the components and quality attributes of milk and dairy products.

Keywords: high pressure homogenization, milk, dairy products

* Corresponding Author's Email: gulcn86@gmail.com.

In: A Closer Look at Homogenization
Editor: Gulcin Yildiz
ISBN: 979-8-89113-656-4

Introduction

Dairy products and milk have unique sensory qualities and are high in nutrients, making them valuable components of the human diet. Homogenization occurs during the processing of milk, resulting in physicochemical alterations that shape the texture and flavor of dairy products. Homogenization is a mechanical procedure that lessens the separation of the creamy portion of the product and uses pressure to shrink fat globules in size. A temperature greater than 45°C (113°F) is required for homogenization because many microbial and milk lipases become inactive at this level. Nevertheless, at 50°C, pressure begins to affect the size and distribution of fat globules. Because of the durability of the fat particle membrane, no modifications have occurred below this temperature.

A new preservation method known as high pressure homogenization (HPH) uses significantly higher pressures (>100 MPa) to enhance the quality of dairy products. The functional properties of milk fat globules, protein, and polysaccharides are significantly altered by HPH. During acidification and gel formation, the small fat globules produced by HPH typically incorporate into the milk proteins casein and denatured whey. The practicality of probiotics microorganisms is improved by expanding the homogenization strain due to delivering peptides and free amino acids which are essential for their development and feasibility during capacity.

The basic principle of this method, which generally involves the application of pressure between 100 MPa and 1000 MPa, especially between 100 MPa and 600 MPa, is based on the compression of the water surrounding the material. With the compression of water, the effect of pressure is instantly transmitted equally to every point of the material, regardless of the shape and size of the material. While it can be applied at room temperature, it can be applied at different temperatures, causing differences in the effect of pressure. Therefore, in this system where high pressure is applied, the pressure and temperature must be controlled safely. During high pressure application, much less energy is consumed after the desired pressure level is reached. Once the desired pressure is reached, the pump is stopped, the valve is closed and the pressure in the container is maintained without the need for any other energy input [1]. In addition, the short application time shows that its use will be advantageous compared to heat treatment. Although gases or oils can be used in addition to water in the application of this method, water is preferred because it is economical, reliable, practical and has low volume reduction.

Use of High-Pressure Homogenization in the Food Industry

Many food preservation methods have been developed to prevent food spoilage and inhibit pathogenic microorganisms [2]. One of these developed techniques is the application of high-pressure homogenization.

The first successful studies on the use of HPH application in foods were carried out by a scientist named Hite [3-6]. Hite proved that the shelf life of raw milk can be extended up to four days by exposing it to 600 MPa pressure for one hour at room temperature [6, 7].

Today, the main countries where HPH applications are used the most in the food industry and continue intensive research activities on this subject are Japan (juice, jam, ham, fish, sake (rice liquor), rice products), USA (juice and oysters), Mexico (juice), France (especially apple juice), and Spain (ham and some other meat products) [8-12].

High pressure application is a method that does not require the application of heat in food technology today [13], providing extraction and degradation by modification of biopolymers such as inactivation of microorganisms, protein denaturation, enzyme activation or inactivation, gel formation, and sensory effects such as taste, smell, and color. It is widely used in the food industry to preserve quality elements and change density, freezing and melting points, or textural properties [14-24].

HPH is applied as a non-thermal processing technology that can reduce the number of microorganisms and extend the shelf life by improving microbial safety in many foods such as fruits and vegetables, ready-to-eat meats, and seafood [25].

HPH applications are used in the pasteurization of acidic foods such as marmalades, fruit jellies, citrus family fruit juices, fruit yogurts, pork meat and products, oysters, as well as increasing the crispness of meats, freezing and thawing foods, hardening chocolate, gelatinization of starches and proteins, peeling of vegetables, It is used in the coagulation and texturing of fish and ground meat [26, 27].

Mineral Balance of Milk

High-pressure homogenization (HPH) has an effect on two basic features of the mineral balance of milk. One of them is effective by changing the amount of ionized minerals (especially calcium), and the second is by changing the mineral distribution (calcium and phosphate) between the serum phase and micelles of the milk [28]. It has been reported that 100–600 MPa high-pressure homogenization application causes an increase in the ionic calcium concentration in milk [29]. It was stated that 300 MPa high-pressure homogenization application caused a significant increase in the phosphate, magnesium and calcium contents in the serum phase of milk [30, 31]. However, it has been suggested that the increase in the phosphate concentration of the serum phase of milk also causes an increase in the pH of the milk [28].

It was emphasized that the increase in calcium (12%) and pH (0.08) in HPH-treated milk was restored when the milk was stored at 20°C for 4 hours but was almost irreversible after 48 hours of storage at 4°C [29]. In a study conducted on buffalo milk, while 100 MPa HPH application did not cause a significant change in the level of non-sediment calcium in milk, it caused a 26-30% increase in the amount of non-sediment calcium at 400–800 MPa pressures and the lowest at 600 MPa. It was also observed that pressure application caused an increase in pH of 0.06–0.08 units, but there was no significant increase in pH despite increasing the pressure [32].

High-pressure homogenization process causes minimal changes in the volatile components of milk at a low temperature (25°C). It has been stated that the effect of HPH treatment applied at different pressures (482, 586 and 620 MPa) and high temperature (60°C) on the formation of volatile components is different from atmospheric pressure conditions. While high temperature applications increase the formation of methyl ketones, methyl ethanol, hydrogen sulfide and aldehydes, it has been suggested that high-pressure homogenization application together with high temperature promotes the formation of aldehydes and hydrogen sulfide [33]. It has also been reported that increasing the HPH level and application time increases the level of proteose-peptone nitrogen and non-protein nitrogenous components [34].

Casein Micelles

Changes in casein micelles as a result of HPH application depend on the size and amount of casein micelles [28]. While 100–200 MPa pressure applied at room temperature causes very little change in the size of the average casein micelles [35-37], at pressures above 300 MPa, the casein micelles deteriorate, causing the micelles to lose their original size. It has been suggested that their size decreases by approximately 50% [38-44]. It has been reported that with the application of a pressure of up to 400 MPa, casein micelles break down into smaller particles called submicelles. The separation of casein micelles under the influence of HPH can be explained by the weakening of hydrophobic and electrostatic interactions between submicelles and the dissolution of colloidal calcium phosphate outside the micellar network [44]. When the application is made for longer than 15 d, at higher temperature [36-41] and at higher pH [36], the size of the micelles is reduced by 20%. It was observed that there was an increase of 30%. It is thought that the increase in the size of the micelles is probably due to the formation of large aggregations of casein micelles, which are degraded by the effect of HPH [36, 40, 42].

The dissolution of casein fractions under the influence of HPH is in the order of β-casein, κ-casein, αs1-casein, αs2-casein, from highest to lowest. It has been reported that the applied pressure of 250–300 MPa causes a maximum increase in the level of dissolved αs1-casein and β-casein [45, 46]. It has been stated that the dissolution of caseins under the effect of HPH is a possible result of the dissolution of colloidal calcium phosphate and the disruption of hydrophobic interactions. Although the dissolved casein level increased by HPH application quickly returned to its previous state when stored at 20°C, it was found to be almost irreversible when stored at 5°C [46]. In addition, the application of 800–1000 MPa HPH causes an increase in the amount of dissolved casein, which constitutes approximately 1/3 of the total casein [34].

It was observed that a tighter gel was formed by applying 600 MPa pressure to the prepared 7% casein solution, while a weak gel with a rough structure was formed at 200 MPa pressure. On the other hand, in 5% casein solution, casein aggregates consisting of smaller casein micelles with larger volume and larger hydrodynamic diameter are seen at 600 MPa, while at 200 MPa, aggregates with smaller volume and smaller hydrodynamic diameter but the same shape and diameter have been observed [44].

It is not possible to directly measure the number of casein micelles per unit volume of HPH-treated milk. For this reason, changes in the light scattering properties of milk are used to determine the number of casein micelles [28].

Color of Milk

HPH application disrupts the micellar integrity of milk and causes some changes in the color of milk. While applying 100–200 MPa HP to milk had little effect on the light scattering of milk, it was reported that there was a sharp decrease after 200–400 MPa pressure application and decreased from approximately 78% to 42% in 30 d HPH application at 600 MPa [40, 41]. Gervilla et al. [47] reported that after the application of 100–500 MPa HPH to sheep milk at 4–25–50°C for 10 days, there was a decrease in the *L** value and an increase in the *a** and *b** values. In addition, it was stated that HPH applied at 4°C decreased the *L** value of milk to the minimum level and increased the *a** and *b** values to the maximum level. Huppertz et al. [40] stated in their study that the decrease in the *L** value of milk with HPH application was irreversible, however, this decrease was completely reversible with heat application after HPH treatment. However, it was observed that this decrease in the *L** value did not change when the milk was stored at 5°C, but increased when it was stored at 37°C. Although there was an increase in the size of casein micelles by applying 200–250 MPa pressure to milk, it was determined that it caused a significant decrease in the *L** value and milk viscosity [28], turbidity and the number of light scattering particles [28, 32]. However, it was observed that there was an increase in turbidity by increasing the temperature (40°C, 250–300 MPa) [48]. Moreover, the reduction in light scattering appears to be broader than that explained by a decrease in micelle size alone at pressures greater than 300 MPa and is suggested to be accompanied by a decrease in the number of casein micelles [28].

Whey Proteins

β-lactoglobulin, α-lactoalbumin and serum albumin are the whey proteins of milk. It has been reported that the denaturation of whey proteins in pressure-treated milk is similar to the application of heat and the level of denaturation increases with increasing the level and duration of pressure [49].

The proteins are known as milk proteins that can be dissolved at 20°C and pH 4.6 [50]. β-lactoglobulin, one of the whey proteins, is very sensitive to HPH (Hinrichs and Rademacher, 2005) and denatures at pressures above 100 MPa at 25°C (loss of solubility at pH 4.6), whereas α-lactoalbumin and serum albumin found to be resistant to pressure up to 400 MPa [49, 50].

The magnitude of the denaturation of α-lactoalbumin and β-lactoglobulin under the influence of HPH depends on the applied pressure time [51], application temperature [52], milk pH and micellar calcium phosphate level in milk [53].

It is known that β-lactoglobulin, which is the dominant whey protein, is significantly responsible for the physicochemical characteristics of whey, such as increasing binding properties, dissolution, gelatinization, foaming and emulsification. It has been stated that the application of pressure up to 100 MPa causes reversible denaturation of β-lactoglobulin in milk, while it does not cause a change in the β-lactoglobulin core, but causes significant denaturation at higher pressures [49, 50]. It is known that β-lactoglobulin denization occurs through two reactions: β-lactoglobulin unfolding and aggregation. It has been reported that if the pressure is applied up to 200 MPa, the opening of β-lactoglobulin is limited, and at higher pressures, the rate of β-lactoglobulin aggregate formation is limited [54]. It has been suggested that the exposure of hidden hydrophobic groups as a result of the opening of β-lactoglobulin with HPH application causes an increase in the hydrophobic surface area, leading to protein aggregation [50]. It has been suggested that increasing the pressure to 400 MPa, reactions of sulfhydryl groups (pH 6.8) lead to irreversible denaturation of β-lactoglobulin. It has also been stated that most of the S-S (disulfide) bonds are formed as a result of SH/S-S exchange reactions rather than oxidation of SH (sulfhydryl) groups (450 MPa, 25°C, pH 7.0) [49].

As a result of the analysis made with measurement techniques such as Fourier Transform Infrared Spectrometry and Small-Angle X-ray Scattering as a result of applying a pressure of approximately 130 MPa to milk, the α-helix structure of β-lactoglobulin and the amount of β-sheet decreased. It was observed that the disordered structure of β-lactoglobulin increased, indicating the beginning of its unfolding. It was determined that this protein was not completely opened even when the pressure was increased to 1000 MPa [49, 50].

It was reported by Huppertz et al. [32] that 400–800 MPa pressure application denatured almost all of the total β-lactoglobulin in buffalo milk. It was observed that 20% of β-lactoglobulin was denatured with the application of 200 MPa pressure, and at higher pressures and increasing the application time, denatation increased to 80-90% [34].

Denaturation of α-lactoalbumin occurs at pressures above 400 MPa. It has been suggested that approximately 10% of α-lactoalbumin is denatured by applying a pressure of 400 MPa [34], more than 50% is denatured at a pressure of 600 MPa, and 90% is denatured at a pressure of 800 MPa. [32] .It has been reported that α-lactoalbumin is more resistant to pressure than β-lactoglobulin due to the presence of four disulfide bonds [34] and the lack of free sulfhydryl groups in α-lactoalbumin [53].

In order to prevent the denaturation of whey proteins under the influence of HPH, it was almost completely prevented by adding the sulfhydryl-blocking agent N-ethylmaleimide to the milk before the application of HPH. However, the addition of the sulfhydryl oxidizing agent potassium iodate (KIO_3) increased the denaturation of β-lactoglobulin and α-lactoalbumin. It has been suggested that it has decreased [53]. It has been reported that immunoglobulin is denatured by 17% with the application of a pressure of 200 MPa, and denaturation increases up to 36% by increasing the pressure and application time [34].

Use of HPH in Cheese Production

HPH effect is effective on variables such as the magnitude of the applied pressure, duration, temperature and milk pH in the changes occurring in milk. As a result of the effect of HPH application on the components of milk; Enzyme coagulation time (RCT), gel formation rate, density of the gel, water retention capacity of the gel and network structure may also vary.

Although microbial quality and safe products are obtained by pasteurizing milk (15 seconds at 72-74°C), cheeses made from raw milk have a stronger taste and faster ripening. It allows the production of cheese with the desired sensory properties by applying HPH to the milk without applying high temperatures. As a result of the effect of HPH application on the components of milk, RCT is effective on the gel formation rate, gel firmness, water retention capacity and network structure of the gel [55, 56].

It has been reported that applying 150 MPa pressure to skim milk has no effect on the RCT of milk, but RCT decreases at pressures of 150–300 MPa, does not change much at 400 MPa, and increases by 15% at 600 MPa [57].

While the application of pressure at 200 MPa for 10, 20, 30 or 60 d reduces the RCT, the application of 400 MPa pressure for 10 d increases the RCT, and in the applications of 20, 30 or 60 d at the same pressure, the RCT gradually increases [28].

It is stated that milk treated with HPH at temperatures above ambient temperature has a longer RCT than milk applied at the same pressure at ambient temperature [58], and RCT decreases at pressures below ambient temperature [59]. When the milk was heated at 90°C for 10 minutes and then 100 MPa pressure was applied, it did not coagulate with rennet, but when the milk heated at the same temperature was applied for 0–30 d and 250–600 MPa HPH, the RCT was equal to or lower than that of milk without heat and pressure, and the clot strength was higher. It was observed that by applying 600 MPa pressure to heat-treated milk for 30 d, RCT was reduced by approximately 50% (compared to milk without heat and pressure). It has also been stated that increasing the time in temperature-HP application also causes a decrease in RCT [32].

Whey proteins, which are denatured by the application of HPH to milk, cause an increase in the pH of the milk. For this reason, readjusting the pH of milk after the application of HPH allows the changes that may occur due to the effect of HPH in the RCT to be significantly eliminated [29].

Two separate mechanisms are responsible for the changes in the properties of milk due to the effect of HPH in rennet coagulation:

- Denaturation of β-lactoglobulin and its association with casein micelles may affect the hydrolysis of κ-casein by rennet and prevent the assembly of caseins during the second phase of rennet coagulation.
- Dispersion of casein micelles at high pressures can increase the surface area for intramicellar interactions and thus promote the aggregation of paracasein micelles [29].

Under the influence of high pressure, β-lactoglobulin is denatured and its combination with casein micelles causes an increase in the bound β-lactoglobulin in the curd made from HPH-treated milk. Thus, the water release of the curd will decrease and the moisture content of the curd will increase; As a result, it is understood that the yield of cheese will increase [28]. It has also been reported that curd yield is 15% higher and whey protein is 30% lower with HPH application [32]. In another study [60], it was determined that HPH applied at 600 or 800 MPa increased the yield of curd by up to 25%. It was observed that the hardness of curd made from milk subjected to 10-60 d pressure at 200 MPa increased; It was observed that curd hardness increased after 10 d of application at 400 MPa, but gradually decreased after 20, 30 or 60 d of application [30].

The effect of HPH application on the gel formation rate is similar to its effect on RCT. It has been reported that 100–300 MPa pressure application to milk increases the gel formation rate, this effect is maximum at 200 MPa, and 400 MPa application has little effect on these properties [30]. However, it has been stated that there is an increase in the gel formation rate with the application of 100–600 MPa pressure, but the maximum effect is achieved at 200–400 MPa. It has been reported that the firmness of the milk gel formed by the effect of rennet increases with the application of 100–600 MPa and reaches its maximum at 250–400 MPa [29, 35].

It has been reported that potassium iodate (KIO_3) added to milk before HEA application (400–600 MPa) prevents the association of casein micelles with β-lactoglobulin, which is denatured by the effect of HPH, and that stronger clots form than milk without added potassium iodate (KIO_3). It was also observed that milk treated with HP at 250 MPa formed stronger clots than milk without HPH. It has been suggested that the reason for this is that the disruption of casein micelles by the effect of HPH is more effective in clot formation rather than the combination of denatured β-lactoglobulin with casein micelles [29].

Applying 200–500 MPa pressure on the first and 15th day after the production of semi-hard sheep cheese caused an increase in the water holding capacity and pH of the cheese. It has been suggested that they are cheeses with less crunchy properties [61]. It has been determined that the ripening of cheese can be accelerated, stopped or activated by applying HPH to cheese, and changes in the taste of cheese can be made by activating or inactivating proteolytic enzymes with reduced starter culture production [62].

It has been reported that applying pressure of 483 and 676 MPa at 10 °C to the milk used in the production of Cheddar cheese causes an increase in the protein content and moisture content of the cheese, and as a result, its yield increases. However, it has been determined that the application of 483 MPa pressure at 40°C does not allow the formation of a suitable gel for cheese making [63].

It has been stated that proteolysis is higher in goat cheese made from milk treated with HPH (500 MPa at 15 d, 20°C) as a result of differences in free amino acid and small peptide content, and as a result, the cheese ripens faster and has a positive effect on the taste of the cheese [64]. At the stage when the cheese begins to ripen; It was determined that cheese ripening was accelerated due to the increase in enzyme activity of starter cultures with 5 d HPH application at 400 MPa, and also increased humidity, pH [65, 66] and proteolysis [66]. It was observed that, due to the high moisture content during the ripening stage of the cheese, the L^* value decreased and the a^* and b^* values increased, but reached values close to the control samples after 60–75 days of storage [67, 68].

It has been determined that while HPH does not affect the free water in the moisture content of cheeses treated with HPH, it causes the bound water to be higher [69]. In addition, HPH leads to decreased lipolysis and lower free fat content and volatile components [70]. It has been determined that friability decreases, and elasticity increases in HPH-treated cheeses [65]. It was stated that when 300–400 MPa pressure was applied to Le Serena cheese made from raw milk during the ripening stage, the desired flavor formation did not occur because the pressure applied on the 2nd day accelerated the secondary proteolysis and prevented the primary proteolysis, but the HPH applied on the 50th day had a positive effect on both ripening and microbial safety [71].

Harmful and Pathogenic Microorganisms in Cheese Inactivation

Most of the high-pressure homogenization applications in cheese focus on the inactivation of pathogenic microorganisms that cause spoilage in cheese. In order to inactivate microorganisms, milk can either be subjected to high pressure homogenization and then cheese can be produced, or high pressure can be applied after the cheese is packaged. Applying high pressure homogenization to the cheese after packaging will eliminate any possible contamination during production, resulting in a completely safe product. Studies targeting microbial inactivation aim to extend the shelf life of cheese and make it safer. However, the possibility of producing safe cheese from raw milk has not been ignored and high-pressure technology has been shown as an alternative to pasteurization.

Research has focused on the inactivation of *Listeria monocytogenes, Staphylococcus aureus* and *E. coli* with high pressure homogenization because they cause cheese-related epidemics. Studies on pathogens were carried out by inoculating the microorganism into milk or cheese. However, it should not be forgotten that the microbial inactivation efficiencies of such technologies vary depending on whether the microorganism is already present in the food or is subsequently inoculated. Szczawinski et al. [72] detected a 5.38 to 6.18 log reduction in the number of *L. monocytogenes* in three different hard cheese varieties by applying 500 MPa pressure for 15 minutes. Lopez-Pedemonte et al. [58] determined that the application of 400 and 500 MPa pressure at 5 and 20°C affected the inactivation of *L. monocytogenes* strains at different levels. Additionally, *L. monocytogenes* reached undetectable levels on the 30th day of ripening in cheese samples subjected to 500 MPa pressure. A pressure of 500 MPa and 10 min at 20°C was applied to raw milk inoculated with 2 and 4 log colonies/mL. *L. monocytogenes,* was not detected in the resulting Camembert type cheese [73]. Application of 600 MPa pressure to white cheese reduced the number of *L. monocytogenes* by 4.9 log [24]. Capellas et al. [74] achieved a 7-log decrease in the number of *E. coli* in Mato cheese (fresh goat cheese) as a result of applying 400-500 MPa pressure for 5-15 minutes at room temperature and refrigerator temperature. The same researchers also discussed the inactivation of cocci (*Staphylococcus carnosus*) and spores (*Bacillus subtilis*) in fresh goat cheese by the effect of pressure. The high-pressure process, which completely inactivated *E. coli,* could only cause a 2-log reduction in the number of Staphylococcus carnosus. 500 MPa pressure was applied to germinated *Bacillus subtilis* spores at 40ºC for 15 minutes and inactivation was achieved at 4.9 log level [75]. Studies on the effect of pressure on *S. aureus* inactivation have shown that the inactivation level increases with increasing pressure. However, not only *S. aureus* but also the enterotoxins produced by this microorganism are important in cheese. However, as a result of the research, *Staphylococcus enterotoxin* was still found positive in the cheese after 500 MPa pressure application and 30 days of ripening. On the other hand, it is not clear whether these enterotoxins maintain their toxic activity as a result of high-pressure application or whether they are present at sufficient levels to cause food poisoning [52].

After high pressure application, cells appear in three shapes.

1. *Living cells:* They form colonies on selective and non-selective media.
2. *Primary injury*: Although they form colonies on non-selective media, they are not seen on selective media, but they also form colonies on this media during storage.
3. *Secondary injury*: They do not form colonies on either medium, but during storage, they first form colonies on the non-selective and then on the selective medium. For this reason, to be sure that high-pressure food is completely microbiologically safe, a shelf-life study must be carried out [76]. In studies investigating the effect of high-pressure application in cheese on *Salmonella enterica* (*Salmonella enteridis and Salmonella typhimurium*) [77, 78].

In addition to pathogenic microorganisms, microorganisms that cause spoilage in cheese are also important. It has been stated that in fresh cheese, applying pressure higher than 300 MPa can control yeasts for 6-8 weeks [79]. Yeast and molds were completely inactivated by applying pressure of 200 MPa in Gouda and Kurpiowski cheese, 400 MPa in Camembert cheese [80], and 300 MPa and 10 minutes of pressure in White cheese made from raw milk.

It has also been shown that for a more effective microbial inactivation, increasing the process temperature, applying cyclic pressure treatment and using bacteriocin or bacteriocin-producing lactic acid bacteria are beneficial [81]. As a matter of fact, 500 MPa, 50°C and 5 minutes of pressure application increased the Staphylococcus carnosus inactivation level to 7 log level. Application of cyclic pressure of 500 MPa between 15 and 30 minutes at 10°C increased the inactivation. The combination of 500 MPa pressure application and nisin gave the best results [75]. Similarly, it has been stated that temperatures higher than room temperature (30°C) and the use of nisin are beneficial in the inactivation of *Bacillus cereus* spores inoculated into soft cheese curd produced from raw milk. However, no significant effect of lysozyme use was found [58].

Accelerating or Controlling Cheese Ripening

Cheese ripening is a very complex process that includes biochemical reactions such as proteolysis, lipolysis and glycolysis, and includes flavor and texture development. It also requires high storage costs. For this reason, studies on the use of high-pressure homogenization technology in cheese have focused on accelerating ripening in cheese as well as microbial inactivation.

In studies conducted on the ripening of cheese under the influence of high-pressure homogenization, either the process was applied for a long time at low pressure values, or for a short time at high pressure values, or first for a short time at high pressure values, and then for a long time at low pressure values. Additionally, studies have been conducted on cyclic pressure applications between 100-1000 MPa. It is known that the most important biochemical mechanism in the ripening process of cheese is proteolysis. Therefore, studies on this subject have mostly focused on the proteolysis changes in cheese caused by high pressure homogenization.

Proteolysis in cheese can be handled in two stages:

- The primary proteolysis stage, where casein is broken down by chymosin and plasmin into large and medium molecule peptides, and
- Secondary proteolysis phase, where proteinase and peptidase enzymes of starter bacteria, non-starter lactic acid bacteria and secondary starter flora break down peptides into smaller molecule peptides and amino acids [82].

The effect of high-pressure homogenization on proteolysis is shaped by its effect on proteolytic enzymes such as chymosin and plasmin, which shape these two stages, and on the bacteria that play a role in maturation and the enzymes secreted by these bacteria. Most starter culture enzymes are generally intracellular enzymes. Therefore, it is stated that maturation will accelerate when intracellular enzymes are released by autolysis of the starter bacterium by any agent [83]. It has been reported that high pressure homogenization damages the cell membrane and increases cell permeability [84] and promotes the release of intracellular enzymes such as peptidase [64]. On the other hand, it has been stated that high pressure treatment may facilitate the degradation of proteins by enzymes by promoting the release of Ca and P and conformational changes in the casein structure [85]. The effects of high-pressure homogenization on these stages vary depending on the cheese variety and the applied pressure conditions.

On the acceleration of cheese ripening by high pressure homogenization, a patent was published by Yokohama et al. [86]. In this patent study, high pressure between 0.1-300 MPa was applied to Cheddar cheese for 72 hours at 25ºC. It was stated that after applying 50 MPa pressure, values comparable to the amino acid level and taste of the control cheese ripened for 6 months were obtained in the cheese. Although the average amino acid value was found to be 21.3 mg/g in cheese ripened for 6 months, the average amino acid value in cheese subjected to high pressure was found to be 26.5 mg/g. In this research, a higher number of lactobacilli was added as a starter culture and pressure was applied at a temperature higher than the ripening temperature of commercially available cheeses. This situation should not be ignored in terms of maturation. Similarly, in the same patent study, the high-pressure process applied to Cheddar cheese was applied to Parmesan cheese with added lipase and protease, and the amount of free amino acids and taste development equivalent to commercial cheeses were determined [86]. In later studies using the same conditions, although it was demonstrated that maturation accelerated, results as ambitious as Yokohama et al. [86] were not obtained.

O'Reilly et al. [87] applied high pressure homogenization to Cheddar cheese using the same conditions. A sudden increase in soluble nitrogen values and free amino acid amounts at pH 4.6 was detected in the first two days of ripening, but this effect decreased as the cheese ripened.

Juan et al. [85] stated that applying 300 MPa pressure to sheep cheese on the first day of ripening increased peptidolytic activity, and therefore secondary proteolysis. Application of 500 MPa pressure slowed down proteolysis due to enzyme inactivation and microbial inactivation. Malone et al. [62] stated that chymosin activity did not change up to 400 MPa pressure values, while plasmin activity was not affected even by 800 MPa pressure values.

Similarly, an increase in the level of proteolysis was detected in surface ripened cheese [88], surface mold cheese [89] and goat cheese obtained from high-pressure milk [55]. Voigt et al. [82] found that in 42-day-old ripe, blue-veined cheeses subjected to 400 and 600 MPa pressure, the nitrogen dissolved at pH 4.6 and the nitrogen dissolved in 12% trichloroacetic acid increased after high pressure application, but after 28 days of storage, these values were lower than the control sample. They have determined. O'Reilly et al. [90] stated that only the primary phase of proteolysis was accelerated in Cheddar cheese.

There are also studies that do not support that high pressure accelerates ripening. For example, Messens et al. [91] stated that high pressure applications of 50-400 MPa and 20-100 minutes did not affect proteolysis in Gouda cheese, and Iwanczak and Wisniewska [92] stated that 200 and 400 MPa pressure applications did not affect proteolysis in Edam cheese. It has been reported that high pressure technology can be used to stop or slow down cheese ripening, on the contrary to accelerate it. Indeed, Wick et al. [93] stated that the application of 400 MPa and higher pressure in Cheddar cheese reduces the amount of free amino acids due to enzyme inactivation, less live lactic acid bacteria number and changes in the cheese, and these applications can be used to control ripening.

Studies on the effect of high-pressure homogenization on lipolysis and volatile flavor compounds are quite limited. Lower amounts of free fatty acids were detected in goat cheese [94], and lower concentrations of n-butanoic acid and 3-hydroxy-2-butanone were detected in Gouda cheese as a result of 50-400 MPa pressure application [95]. On the other hand, it was reported that the application of 300-400 MPa pressure on the 2nd day had a negative effect on the odor and flavor of La Serena cheese made from sheep's milk, but the application on the 50th day did not change these properties [96]. Similarly, no significant change in flavor components was detected in blue mold cheese [82]. In contrast, Juan et al. [61] reported that applying 300 MPa pressure to sheep cheese increased lipolytic activity on the 1st day of ripening, while 400 MPa and higher pressures had the opposite effect. Increased lipolysis has been associated with cell lysis and increased interaction of microbial lipase with fat. Increasing pressure values in fresh cheese reduced lipid oxidation and suppressed bitterness formation during storage [48].

Use of HPH in Yogurt Production

The quality of yogurts made from HPH-treated milk improves. It has been reported that yogurt made from HPH-treated milk coagulates at higher pH, exhibits less water release (syneresis), and increases the incorporation of whey proteins, gel strength, and water retention capacity when compared to unpressurized milk [97]. In addition, when the casein micelles of yogurts made from milk treated with 30 d 676 MPa HPH were examined using TEM, it was observed that small round micelles [97] formed aggregates in the form of smooth-surfaced particles consisting of dense thread packages, while in yogurts made from milk treated with heat, the micelles were separated as dense fibrous protrusions [35].

The acidification rate during yogurt production from HPH treated milk is higher than that from milk treated at high temperature (20 d at 85°C). In a study on the acidification of milk under pressure (reducing pH to 5.5 by adding glucono-lactone), it was observed that the milk coagulated when a pressure of 100 MPa was applied for more than 40 d or 200 MPa for a period of more than 10 d. In the microstructural analysis of coagulated milk, it was observed that although large flocs were formed in yogurts made from milk that was subjected to 100 MPa pressure for more than 40 d, a well-threaded, adhesive coagulated structure was formed in yogurts that were subjected to 200 MPa pressure for 10-40 d [28].

It has been observed that the irregular casein micelles of yogurts made from only heat-treated milk consist of less connected chains, while in yogurts made from HPH-treated milk, protein aggregates with small particle sizes cluster more and show a softer, regular surface and a more uniform distribution. However, it was observed that larger casein micelle clusters and compact yogurt gel were formed in yogurts produced by applying HPH and heat together, and also a higher degree of cross-linking was formed. Thus, the formation of small irregular aggregates around the micelles and their tendency to come together enabled the development of gel texture and viscosity [13]. In a study [43], it was reported that 400 MPa pressure application shortened the gel formation time of milk, increased the gelation pH and caused an increase in final G (storage modulus) similar to temperature. It has also been suggested that the G of the acid gel is higher when transglutaminase is applied to milk together with HPH, and this is due to the increase in whey protein cross-links, especially β-lactoglobulin, or cross-links between casein and whey proteins.

High Pressure Applications and Nutrient Losses in Milk and Dairy Products

- At the end of the high-pressure treatment (108-309 MPa, 60-180 min., 66-71°C) and subsequent storage period (2 weeks), a cooked taste can be observed in some milk samples [98].
- It has been reported that the viscosity of fat particles does not change significantly when 450 MPa pressure is applied to milk cream (15-30 min at 25°C and 30 min at 10°C), and there is no increase in acidity during storage for up to 8 days [99].
- In another study, the effect of high-pressure treatment (400 MPa, 25-60°C for 15 min) on the micelle structure, proteolytic activity and sensory properties of milk was investigated and it was found that pressure application at 25°C significantly reduced the micelle size, whereas pressure application at higher temperatures reduced the micelle size significantly. It was observed that micelle sizes gradually increased. In the same study, it was reported that pressure application at high temperatures reduced the proteolytic activity and increased the organoleptic properties of milk when compared to the pressure effect applied at 25°C and the pressure effect applied at temperatures between 40 and 60 °C. It may be recommended to use this combined method to produce drinking milk with long shelf life and good sensory properties [100].
- Sierra et al. [101] tested milk at 25°C, 30 min and 2.5 MPa/milk. They reported that there were significant losses in vitamins B_1 and B_6 with the application of 400 MPa high pressure at s speed.

Conclusion

HPH application, which has found widespread application in foods, has become one of the important preservation methods in the food industry. HPH application is effective on the flavor, color, and texture of foods and its importance in the food industry is increasing day by day as an alternative technique to heat treatment techniques.

When HPH-treated milk is used in cheese making, the enzyme coagulation time of the milk is shortened, and the water retention capacity and yield of the cheese are increased. In addition, the water release of yogurt decreases and the gel strength increases.

It causes an increase in the nutritional values and sensory quality of products pasteurized by applying HPH. Additionally, it is suggested that HPH application is a more suitable method for pasteurization of heat-sensitive animal food products. It is thought that it will reduce the product cost since there is no need for heat-resistant packaging material during HPH application.

It can be easily applied to heat-sensitive ingredients, it does not affect free amino acids, it does not affect or improve the natural aroma of the product, it does not affect the vitamin content of the product, there is no need to use additional preservatives in the products, the pressure application time does not depend on the mass and shape of the product, it can be applied to packaged products and the amount of energy used is low. It is an advantage of high pressure applications.

Its main disadvantages are that the installation cost is high and process control is difficult, the system operates at low capacity, it causes changes in the texture of foods at high pressures, and it causes a rapid pH drop due to increased ionic dissociations.

References

[1] Crehan, C. M., Troy, D. J., & Buckley, D. J. (2000). Effects of Salt Level and High Hidrostatic Pressure Processing on Frankfurters Formulated with 1.5 and 2.5% Salt. *Meat Science, 55,* 125-130.

[2] Gómez-Estaca, J., López de Lacey, A., López-Caballero, M. E., Gómez-Guillén, M. C., & Montero, P. (2010). Biodegradable gelatinechitosan films incorporated with essential oilsas antimicrobial agents for fish preservation. *Food Microbiology,* 27, 889-896.

[3] Hoover, D. G., Metrick, C., Papineau, A. M., Furkas, D. F., & Knorr, D. (1989). *Food Technology,* March 99-107.

[4] Gökmen, V., & Acar, J. (1995). Applications of High-pressure Technology in the Food Industry. *The Journal of Food, 20 (3),* 167-172.

[5] Knorr, D. (1999). Novel approaches in food-processing technology: new Technologies for preserving foods and modifying function, *Current Opinion in Biotechnology, 10,* 485-491.

[6] Özcan, D., & Obuz, E. (2006). Use of High Pressure Application in the Food Industry. Turkey 9th Food Congress, 675-678. Bolu, Turkey.

[7] Arıcı, M. (2006). The Effect of High Hydrostatic Pressure on Microorganisms in Food Preservation. *Journal of Tekirdag Agricultural Faculty, 3(1),* 41-49.

[8] Angsupanich, K., & Ledward, D. A. (1998). High Pressure Treatment Effects on Cod (*Gadus morhua*) Muscle. *Food Chemistry, 63,* 39-50.

[9] Hugas, M., Garriga, M., & Monfort, J. M. (2002). New Mild Technologies in Meat Processing: High Pressure as a Model Technology. *Meat Science, 62,* 359-371.

[10] Zorba, Ö., & Kurt, Ş. (2005). The Effects of High Pressure Treatments on Meat and Meat Products. *Van Veterinary Journal, 16(1),* 71-76.

[11] Öz, F., & Kaya, M. (2006). The Effect of High Pressure Treatment on Meat Quality. *Ataturk University Journal of Agricultural Faculty, 37(2),* 249-255.

[12] Özlü, H., & Atasever, M. (2007). High Pressure Application to Foods. *Ataturk University Journal of Veterinary Sciences, 2(1),* 7-27.

[13] Penna, A. L. B., Subbarao-Gurram, & Barbosa-Cánovas, G. V. (2007). High hydrostatic pressure processing on microstructure of probiotic low-fat yogurt. *Food Research International, 40,* 510–519.

[14] Popper, L. & Knorr, D. (1990). Applications oh high pressure homogenization for food preservation, *Food Technology,* 84-89.

[15] Watanabe, M., Makino, T., Kumeno, K., & Arai, S. (1991). High-pressure sterilizationof ice nucleation-active bacterial cells. *Agricultural and Biological Chemistry, 55(1),* 291-292.

[16] Mertens, B., & Knorr, D. (1992). Developments of non-thermal processes for food preservation, *Food Technology,* 124-133.

[17] Zimmerman, F., & Bergman, C. (1993). Isostatic high-pressure equipment for food preservation, *Food Technology*, 162-163.

[18] Earnshaw, R. (1996). High Pressure Food Processing. *Nutrition & Food Science, 2,* 8-11.

[19] Cheftel, J. C., & Culioli, J. (1997). Effects of High Pressure on Meat: A Review, *Meat Science, 46(3),* 211-236.

[20] Mor-Mur, M., & Yuste, J. (2003). High Pressure Processing Applied to Cooked Sausage Manufacture: Physical Properties and Sensory Analysis, *Meat Science, 65,* 1187-1191.

[21] Cemeroğlu, B. (2004). *Fruit-Vegetable processing Technology*. 1, 226-227.

[22] Ramirez-Suarez, J. C., & Morrissey, M. T. (2006). Effect of High Pressure Processing (HPP) on Shelf Life of Albacore Tuna (*Thunnus alalunga*) Minced Muscle. *Innovative Food Science and Emerging Tec*hnology, *7,* 19-27.

[23] Sequeira-Munoz, A., Chevalier, D., LeBail, A., Ramaswamy, H. S., & Simpson, B. K. (2006). Physicochemical changes induced in carp (*Cyprinus carpio*) fillets by high pressure processing at low temperature, *Innovative Food Science and Emerging Technologies, 7,* 13-18.

[24] Akdemir Evrendilek, G., Çağrı Mehmetoğlu, A., Çoşansu, S., & Erkmen, O. (2010). Preservation of food with innovative technologies, 307-344. (O. Erkmen ed. Food Microbilogy). Efil Publishing House. Ankara, Turkey.

[25] Uçak, İ. & Gökoğlu, N. (2012). *High hydrostatic pressure applications in seafood preservation.* Turkey 11th Food Congress. Hatay, Turkey.

[26] Ohlsson, T. Bengtsson, N. (2002). Minimal Processing Technologies In The Food Industry. Woodhead publishing limited. 41 s, England.

[27] Yemencioğlu, A., & Özkan, M. (2004). Methods of Preservation of Fruits and Vegetables and Products Obtained from These. *Fruit-Vegetable Processing Techniques*. 226-227, Ankara, Turkey.

[28] Huppertz, T., Smiddy, M. A., Upadhyay, V. K., & Kelly, A. L. (2006). High-pressure induced changes in bovine milk: a review. *International Journal of Dairy Technology, 59,* 2, 58-66.

[29] Zobrist, M. R., Huppertz, T., Uniacke, T., Fox, P. F., & Kelly, A. L. (2005). High-pressure induced changes in rennet-coagulation properties of bovine milk. *International Dairy Journal, 15,* 655–662.

[30] López-Fandiño, R. (2006a). High pressure-induced changes in milk proteins and possible applications in dairy technology. *International Dairy Journal, 16,* 1119–1131.

[31] López-Fandiño, R. (2006b). Functional improvements of milk whey proteins induced by high hydrostatic pressure. *Critical Reviews in Food Science and Nutrition, 46,* 351–363.

[32] Huppertz, T., Hinz, K., Zobrist, M. R., Uniacke, T., Kelly, A. L., & Fox, P. F. (2005). Effects of high-pressure treatment on the rennet coagulation and cheese-making properties of heated milk. *Innovative Food Science and Emerging Technologies, 6,* 279–285.

[33] Vazquez-Landaverde, P. A., Torres, J. A., & Qian, M. C. (2006). Effect of high-pressure moderate-temperature processing on the volatile profile of milk. *Journal of Agricultural and Food Chemistry, 54,* 9184–9192.

[34] Kielczewska K., Czerniexicz M., Michalak J., & Brandt W. (2004). The effect of high pressure on nitrogen compounds of milk. *Journal of Physics: Condensed Matter, 16,* 1067–1070.

[35] Needs, E. C., Capellas, M., Bland, A. P., Manoj, P., MacDougal, D., & Paul, G. (2000b). Comparison of heat and pressure treatments of skim milk, fortified with whey protein concentrate, for set yogurt preparation: effects on milk proteins and gel structure. *Journal of Dairy Research, 67,* 329–348.

[36] Huppertz, T., Fox, P. F., & Kelly, A. L. (2004a). High-pressure treatment of bovine milk: effects on casein micelles and whey proteins. *Journal of Dairy Research, 71,* 97–106.

[37] Anema, S. G., Lowe, E. K., & Stockmann, R. (2005a). Particle size changes and casein solubilisation in high-pressure-treated skim milk. *Food Hydrocolloids, 19,* 257–267.

[38] Gaucheron, F., Famelart, M. H., Mariette, F., Raulot, K., Michel, F., & Le Graet, Y. (1997). Combined effects of temperature and high-pressure treatments on physicochemical characteristics of skim milk. *Food Chemistry, 59,* 439–447.

[39] Huppertz, T., Fox P. F., & Kelly A. L. (2004b). Plasmin activity and proteolysis in high-pressure- treated bovine milk. *Lait, 84,* 297–304.

[40] Huppertz, T., Fox, P. F., & Kelly, A. L. (2004c). Properties of casein micelles in highpressure-treated bovine milk. *Food Chemistry, 87,* 103–110.

[41] Huppertz, T., Grosman, S., Fox, P. F., & and Kelly, A. L. (2004d). Heat and ethanol stabilities of high-pressure-treated bovine milk. *International Dairy Journal, 14,* 125–133.

[42] Regnault, S., Thiebaud, M., Dumay, E., & Cheftel, J. C. (2004). Pressurisation of raw skim milk and of a dispersion of phosphocaseinate at 9°C or 20°C: effects on casein micelle size distribution. *International Dairy Journal, 14,* 55–68.

[43] Anema, S. G., Lauber, S., Lee, S. K., Henle, T., & Klostermeyer, H. (2005b). Rheological properties of acid gels prepared from pressure and transglutaminase-treated skim milk. *Food Hydrocolloids, 19,* 879–887.

[44] Merel-Rausch, E. (2006). Hydrostatic high-pressure treatment of casein to generate defined particle and gel structures. Doctora Thesis. Fakultät Naturwissenschaften Universität Hohenheim.

[45] López-Fandiño, R., & Olano, A. (1998). Effects of high pressures combined with moderate temperatures on the rennet coagulation properties of milk. *International Dairy Journal, 8,* 623–627.

[46] Huppertz, T., Fox, P. F., & Kelly, A. L. (2004e). High-pressure induced dissociation of caseins from micelles. *International Dairy Journal, 14,* 575–580.

[47] Gervilla, R., Ferragut, V., & Guamis, B. (2001). High hydrostatic pressure effects on color and milk-fat globule of ewe's milk. *Journal of Food Science, 66(6),* 880–885.

[48] Okpala, C. O. R., Piggott, J. R., & Schaschke, C. J. (2010). Influence of high pressure processing on physico-chemical properties of fresh cheese. *Innovative Food Science and Emerging Technologies, 11,* 61-67.

[49] Considine, T., Patel, H. A., Anema, S. G., Singh, H., & Creamer, L. K. (2007). Interactions of milk proteins during heat and high hydrostatic pressure treatments-a review. *Innovative Food Science and Emerging Technologies, 8,* 1–23.

[50] López-Fandiño, R., Carrascosa, A. V., & Olano, A. (1996). The effects of high pressure on whey protein denaturation and cheesemaking properties of raw milk. *Journal of Dairy Science, 79,* 929–936.

[51] Scollard, P. G., Beresford, T. P., Needs, E. C., Murphy, P. M., & Kelly, A. L. (2000). Plasmin activity, ß-lactoglobulin denaturation and proteolysis in high pressure treated milk. *International Dairy Journal, 10,* 835–841.

[52] López-Pedemonte, T. J, Roig-Sagues, A. X., de Lamo S, Hernandez-Herrero, M., & Guamis, B. (2007). Reduction of counts of Listeria monocytogenes in cheese by means of high hydrostatic pressure. *Food Microbiology, 24,* 59-66.

[53] Huppertz, T., Fox, P. F., & Kelly, A. L. (2004f). High-pressure induced denaturation of α-lactalbumin and β-lactoglobulin in bovine milk and whey: a possible mechanism. *Journal of Dairy Research 71,* 489–495.

[54] Hinrichs, J., & Rademacher, B. (2005). Kinetics of combined thermal and pressure-induced whey protein denaturation in bovine skim milk. *International Dairy Journal, 15,* 315–323.

[55] Trujillo, A. J., Royo, C., Ferragut, V., & Guamis, B. (1999). Influence of pressurization on goat milk and cheese composition and yield. *Milchwissenschaft 54,* 1977-1979.

[56] Trujillo, A. J., Capellas, M., Saldo, J., Gervilla, R., & Guamis, B. (2002). Applications of high-hydrostatic pressure on milk and dairy products: a review. *Innovative Food Science and Emerging Technology, 3(4)*, 295-307.

[57] Needs, E. C., Stenning, R. A., Gill, A. L., Ferragut, V., & Rich, G. T. (2000a). High-pressure treatment of milk: Effects on casein micelle structure and on enzymic coagulation. *Journal Dairy Research, 67,* 31–42.

[58] López-Pedemonte, T. J, Roig-Sagues, A. X., Trujilla, A. J., Capellas, M., & Guamis, B. (2003). Inactivation of spores of Bacillus cereus in cheese by high hydrostatic pressure with the addition of nisin or lysozym. *Journal of Dairy Science, 86,* 3075-3081.

[59] Pandey, P. K., Ramaswamy, H. S., & St-Gelais, D. (2003). Effect of high-pressure processing on rennet coagulation properties of milk. *Innovative Food Science and Emerging Technologies, 4,* 245–256.

[60] Huppertz, T., Fox, P. F., & Kelly, A. L. (2004g). Effect of high pressure on the yield of cheese curd from bovine milk. *Innovative Food Science and Emerging Technologies, 5,* 1–8.

[61] Juan, B., Ferragut, V., Bufa, M., Guamis, B., & Trujillo, A. J. (2007a). Effects of high-pressure treatment on free fatty acids release during ripening of ewes' milk cheese. *Journal of Dairy Research, 74,* 438–445.

[62] Malone, S. A., Wick, C., Shellhammer, T. H., & Courtney, P. D. (2003). High pressure effects on proteolytic and glycolytic enzymes involved in cheese manufacturing. *Journal Dairy Science, 86,* 1139–1146.

[63] San Martín-González, M. F., Rodríguez, J. J., Gurram, S., Clark, S., Swanson, B. G., & Barbosa-Cánovas, G. V. (2007). Yield, composition and rheological characteristics of cheddar cheese made with high pressure processed milk. *Lebensmittel- Wissenschaft und-Technologie, 40,* 697–705.

[64] Trujillo, A. J., Capellas, M., Buffa, M., Royo, C., Gervilla, R., Felipe, X., Sendra, E., Saldo, J., Ferragut, V., & Guamis. B. (2000). Application of high-pressure treatment for cheese production. *Food Research International, 33,* 311–316.

[65] Saldo, J., Sendra, E., & Guamis, B. (2000). High hydrostatic pressure for accelerating ripening of goat's milk cheese: Proteolysis and Texture. *Journal of Food Science, 65(4),* 636– 640.

[66] Saldo, J., McSweeney, P. L. H., Sendra, E., Kelly, A. L., & Guamis, B. (2002a). Proteolysis in caprine milk cheese treated by high pressure to accelerate cheese ripening, *International Dairy Journal, 12,* 35–44.

[67] Saldo, J., Sendra, E., & Guamis, B. (2002b). Colour changes during ripening of high pressure treated hard carpine cheese. *High Pressure Research, 22 (3–4),* 659–663.

[68] Sheehan, J. J., Huppertz, T., Hayes, M. G., Kelly, A. L., Beresford, T. P., & Guinee, T. P. (2005). High pressure treatment of reduced-fat Mozzarella cheese: effects on functional and rheological properties. *Innovative Food Science and Emerging Technologies, 6,* 73–81.

[69] Saldo, J., Sendra, E., & Guamis, B. (2002c). Changes in water binding in high-pressure treated cheese, measured by TGA (thermogravimetrical analysis. *Innovative Food Science and Emerging Technologies, 3(3),* 203-207.

[70] Saldo, J., Fernandez, A., Sendra, E., Butz, P., Tauscher, B., & Guamis, B. (2003). Analysis of the volatile fraction of high pressure treated caprine cheese by

simultaneous distillation-extraction and gas chromatography. *Food Research International, 36(9-10),* 1061-1068.

[71] Garde, S., Arqués, J. L., Gaya, P., Medina, M., & Nuñez, M. (2007). Effect of high-pressure treatments on proteolysis and texture of ewes' raw milk La Serena cheese. *International Dairy Journal,* 1–31.

[72] Szczawinski, J., Stanczack, B., Szczawinska, M., Fonberg-Broczek, M., Arabas, J., & Szczepek, J. (1997). Effect of high pressure on survival of Listeria monocytogenes in ripened, sliced cheeses at ambient temperature (pp: 295-298). In: K. Heremans (Ed.), *High Pressure Research in the Biosciences and Biotechnology.* Leuven, Belgium, Leuven University Press.

[73] Linton, M., Mackle, A. B., Upadhyay, V. K., Kelly, A. L., & Patterson, M. F. (2008). The fate of Listeria monocytogenes during the manufacture of Camembert-type cheese: A comparison between raw milk and milk treated with high hydrostatic pressure. *Innovative Food Science and Emerging Technologies, 9,* 423–428.

[74] Capellas, M., Mor-Mur. M, Sendra, E., Pla, R., & Guamis, B. (1996). Populations of aerobic mesophilic and inoculated *E. coli* during storage of fresh goat's milk fresh cheese treated with high pressure. *Journal of Food Protection, 59(6),* 582-587.

[75] Capellas, M., Mor-Mur, M., Gervilla, R. Yuste, J., & Guamis B. (2000). Effect of high pressure combined with mild heat or nisin on inoculated bacteria andmesophiles of goa'smilk fresh cheese. *Food Microbiology, 17,* 633-641.

[76] Bozoğlu, F., Alpas, H., & Kaletunc, G. (2004). Injury recovery of foodborne pathogens in high hydrostatic pressure treated milk during storage. FEMS Immunol. *Medical Microbiology, 40,* 243-247.

[77] de Lamo-Castellvi, S., Capellas, M., Roig-Sagues, A. X., Lopez-Pedemonte, T., Hernandez-Herrero, M. M., & Guamis, B. (2006). Fate of Escherichia coli strains inoculated in model cheese elaborated with or without starter and treated by high hydrostatic pressure. *Journal Food Protection, 69,* 2856-2864.

[78] de Lamo-Castellví, S., Roig-Sagués, A. X., López-Pedemonte, T., Hernández-Herrero, M. M., Guamis, B., & Capellas, M. (2007). Response of two Salmonella enterica strains inoculated in model cheese treated with high hydrostatic pressure. *Journal of Dairy Science, 90,* 99-109.

[79] Daryaei, H., Coventry, M. J., Versteeg, C., & Sherkat, F. (2008). Effect of high pressure treatment on starter bacteria and spoilage yeasts in fresh lactic curd cheese of bovine milk. *Innovative Food Science and Emerging Technologies, 9,* 201–205.

[80] Reps, A., Kolakowski, P., & Dajnowiec, F. (1998). In: Isaacs, N. S. (Ed.), *High Pressure Food Science, Bioscience and Chemistry.* Royal Society of Chemistry, Cambridge, UK, p. 265.

[81] Arques, J. L., Rodriguez, E., Gaya, P., Medina, M., & Nunez, M. (2005). Inactivation of Staphylococcus aereus in raw milk cheese by combinations of high-pressure treatments and bacteriocin-producing lactic acid bacteria. *Journal of Applied Microbiology, 98,* 254-260.

[82] Voigt, D. D., Chevalier, F., Qian, M. C., & Kelly, A. L. (2010). Effect of high-pressure treatment on microbiology, proteolysis, lipolysis and levels of flavour

compounds in mature blue-veined cheese. *Innovative Food Science and Emerging Technologies, 11,* 68-77.

[83] Juan, B., Ferragut, V., Bufa, M., Guamis, B., & Trujillo, A. J. (2007b). Effects of high pressure on proteolytic enzymes in cheese: Relationship with the proteolysis of ewe milk cheese. *Journal of Dairy Science, 90,* 2113–2125

[84] Cheftel, J. C. (1995). High-pressure, microbial inactivation and food preservation. *Food Science and Technology International, 1(2-3),* 75-90.

[85] Juan, B., Ferragut, V., Guamis, B., & Trujillo, A. J. (2008). The effect of high-pressure treatment at 300 MPa on ripening of ewes' milk cheese. *International Dairy Journal, 18,* 129-138

[86] Yokoyama, H., Sawamura, N., & Motobayashi, N. (1992). Method for accelerating cheese ripening. *Eur Pat Applic* No 00469 857 A1.

[87] O'Reilly, C. E., O'Connor, P. M., Murphy, P. M., Kelly, A. L., & Beresford, T. P. (2000). The effect of exposure to pressure of 50 MPa on Cheddar cheese ripening. *Innovative Food Science and Emerging Technologies, 1,* 109-117.

[88] Messens, W., Foubert, I., Dewettinck., K., & Huyghebaert, A. (2000). Proteolysis of a high-pressure treated mould-ripened cheese. *Milchwissenschaft, 56,* 201-204.

[89] Messens, W., Foubert, I., Dewettinck., K., & Huyghebaert, A. (2001). Proteolysis of a high-pressure treated smear-ripened cheese. *Milchwissenschaft, 55(6),* 328–332.

[90] O'Reilly, C. E., Kelly, A. L., Oliveria, J. C., Murphy, P. M., Auty, M. A. E., & Beresford, T. P. (2003). Effect of varying highprssure treatment conditiins on accelaratıon of ripening of Cheddar cheese. *Innovative Food Science and Emerging Technologies, 4, 277*-284.

[91] Messens, W., Dewettinck, K., & Huyghebaert, A. (1999). Transport of sodium chloride and water in Gouda cheese as affected by high-pressure brining. *International Dairy Journal, 9,* 569-576.

[92] Iwanczak, M., & Wisniewska, K. (2005). Effect of high pressures on the process of Edam cheese proteolysis. *High Pressure Research, 25,* 43–50.

[93] Wick, C., Nienaber, U., Anggraeni, O., Shelhammer, T. H., & Courtney, P. D. (2004). Texture, proteolysis and viable lactic acid bacteria in commercial Cheddar cheeses treated with high pressure. *Journal of Dairy Research, 71,* 107-115.

[94] Saldo, J., Fernandez, A., Sendra, E., Butz, P., Tauscher, B., & Guamis, B. (2003). High pressure treatment decelerates the lipolysis in a caprine cheese. *Food Research International, 36,* 1061–1068.

[95] Butz, P., Fernandez, A., Koller, W. D., Messens, W., & Tauscher. B. (2000). Effects of high-pressure treatment on fermentation process during ripening of Gouda cheese. *High Pressure Research, 19,* 37–41.

[96] Arques, J. L., Garde, S., Fernandez-Garcia, E., Gaya, P., & Nunez, M. (2007). Volatile compounds, odor, and aroma of La Serena cheese high-pressure treated at two different stages of ripening *Journal of Dairy Science, 90,* 3627–3639.

[97] Harte, F., Amonte, M., Luedecke, L., Swanson, B. G., & Barbosa-. Canovas, G. V., 2002. Yield stress and microstructure of set yogurt made from high hydrostatic pressure-treated full fat milk. *Journal Food Science, 67(6),* 2245–2250.

[98] Hite, B. H. (1899). The Effect of Pressure in the Preservation of Milk, A Preliminary Report. West Virginia Agricultural Experiment Station, Morgantown, W. VA. Bulletin 58.

[99] Dumay, E. M., Lambert, C., Funtenberger, S., & Cheftel, J. C. (1996). Effects of high pressure on the physicochemical characteristics of dairy cream and model oil/water emusion. *Lebensmittel Wissenschaft und Technologie, 2*, 606-625.

[100] Garcia-Risco, M. R., Olano, A., Ramos, M., & Lopez-Fandino, R. (2000). Micellar Changes Induced by High Pressure. Influence in the Proteolytic Activity and Organoleptic Properties of Milk. *Journal of Dairy Science, 83(10)*, 2184-2189.

[101] Sierra, I., Vidal Valverde, C., & Lopez-Fandio, R. (2000). Effect of high pressure on the vitamin B1 and B6 content in milk. *Milchwissenschaft, 55(7),* 365-367.

Chapter 7

The Importance of Homogenization on Plant-Based Oils and Fats

Gökçen Yıldız*
Department of Food Engineering, Bursa Technical University, Bursa, Turkey

Abstract

Oils, which have an important place in human nutrition and are among the basic food products, are divided into two groups: animal-based and plant-based. Plant-based oils, which are beneficial for human health due to their low saturated fat content and free fatty acids, are obtained from avocados, canola seeds, coconuts, peanuts, soybeans, corn kernels, sesame seeds, and olives. Plant-based oils are fatty oils obtained from various parts of oil plants, unlike essential oils, which do not contain fat and evaporate without leaving any residue. Many plant-based oils are foodstuffs, while some, such as castor oil, are of pharmaceutical importance. Chemically, oils are composed of triglycerides. According to the proportion of unsaturated fatty acids, a distinction is made between non-drying (olive oil), semi-drying (soy or rapeseed oil) and drying oils (linseed or poppy seed oil). The term "drying" does not refer to evaporation, but to the solidification of the oil as a result of oxidation and polymerization of unsaturated fatty acids. Plant-based oils provide numerous health benefits such as reduced risk of cardiovascular disease, better metabolism and digestion, reduced likelihood of breast cancer, and supplying the body with omega-3 fatty acids. In this chapter, the aim is to give information about plant-based oil production technology and the importance of the homogenization process. A good understanding of the chemistry and processing of plant-based oils is very important for both

* Corresponding Author's Email: gokcen.yildiz@btu.edu.tr.

In: A Closer Look at Homogenization
Editor: Gulcin Yildiz
ISBN: 979-8-89113-656-4

the industry and consumers to define product quality. It is thought that the study will be useful for academicians, students, field experts, and future studies.

Keywords: plant-based oils, homogenization, fat

Introduction

Plant-based oil is a triglyceride obtained from a plant. These types of oils have been part of human culture for thousands of years. The term "plant-based oil" can be defined narrowly to refer to substances that are liquid only at room temperature, or it can be defined broadly to refer to substances that are liquid only at room temperature. For this reason, plant-based oils that are solid at room temperature are sometimes called plant-based fats. Plant-based oils consist of triglycerides, unlike waxes, which do not contain glycerin in their structure. Although oil can be obtained from many plant parts, in commercial practice the oil is primarily obtained from seeds [1].

Plant-based oils are obtained from oil-containing seeds, fruits or nuts by different pressing methods, solvent extraction or a combination of these. The obtained crude oils are subjected to a series of refining processes, both physical and chemical. There are many plant-based oils available from various sources. These include popular plant-based oils: oilseed oils, soybean, cottonseed, peanut and sunflower oils, as well as palm oil, palm kernel oil, coconut oil, castor oil, rapeseed oil and others. Their yields, different compositions and, accordingly, physical and chemical properties determine their usefulness in various applications other than edible uses [2].

To produce plant-based oils, oil must first be extracted from oil-bearing plant components, typically seeds. This can be done through mechanical extraction using an oil mill or chemical extraction using a solvent. The extracted oil can then be purified and refined or chemically modified if needed.

In this study, information about plant-based oil production technology will be given. It is thought that the study will be useful for academicians, students, field experts, and future studies.

Fundamental Chemistry of Oils

People have been using plant-based oils for centuries. Fat-containing nuts and animal fats were consumed as energy sources long before nutritional concepts were considered. Oils were also used early on for illumination, as medicine, as cosmetics in religious ceremonies, and were applied to weapons and utensils. Sesame and olive, ancient oils of the Middle East, were valued for their longevity. Sunflowers were cultivated in the Arizona-New Mexico region in B.C., and seeds from the Missouri-Mississippi river basins were among the first plants transferred to Europe by explorers. The invention of the cotton gin in the late 1700s led to a major cotton export trade in the United States in the early 1800s and the development of cottonseed oil as the first new oil of the Industrial Age in the mid-1800s. The first methods of caustic refining, bleaching, deodorization, winterization, and hydrogenation were among the innovations developed, including the continuous screw press and the development of the first herbal shortening "Crisco" (short for crystallized cottonseed oil). Processing of soybeans, a product first developed in China, led to further oil industry innovations in the mid-1900s, including the development of continuous solvent extractors and steam distillation technologies to reduce or eliminate the original crude flavor in crude oil. As flavor and stability have improved, mankind has expanded the uses of oils: "baking, frying, baking shortenings, salad dressings, food lubricants (such as release agents in baking and confectionery processes), flavor carriers, and dust control agents." Each of the applications requires oils with specific physical and chemical properties. Other oils such as palm oil, regular canola oil, high oleic and low linolenic canola oil, high oleic sunflower oil, high oleic safflower oil were all commercialized much later than animal oil and cottonseed oil [3].

Oil Composition

Fat is a mixture of 96-98% fatty acid triacylglycerols (often called "triglycerides"), with the remainder consisting of other fat-dispersible or fat-soluble compounds. Triglycerides consist of three fatty acids substituted at the hydroxyl (alcoholic) sites of a glycerin (glycerol) backbone. The structure of a simple triglyceride is shown in Figure 1, where each fatty acid is represented by a different "R". Depending on the extent to which the three former hydroxyl groups of glycerol are replaced by fatty acids, the resulting compounds are known as follows. Monoglycerides form when one of the three hydroxyl groups of glycerol is replaced by a fatty acid. Diglycerides are formed by replacing two of the three hydroxyl groups of glycerol with the same or different fatty acids. Triglycerides are formed by replacing all three of the hydroxyl groups of glycerol with fatty acids also called neutral fat [4].

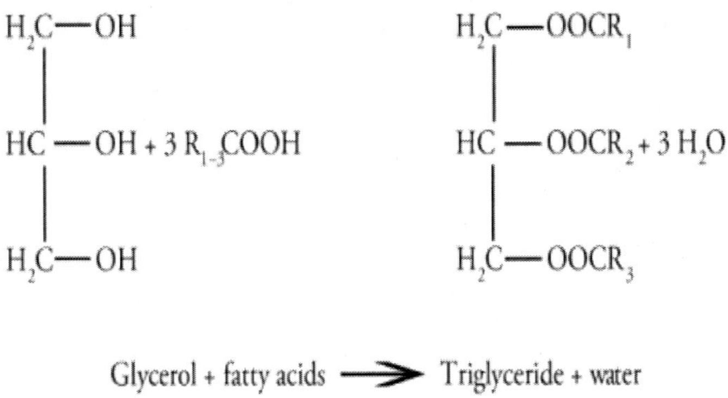

Figure 1. Formation of triglycerides [4].

Differences between Oils and Fats

Fats and oils can be considered the essential macronutrients of the animal body. Both fats and oils consist of carbon (C), hydrogen (H), and oxygen (O). The combination of these elements forms molecular chains called fatty acids. Individual fatty acids can be saturated or unsaturated molecules. The combination of fatty acids determines whether the final chain is saturated or unsaturated. The main difference between fats and oils is that fats consist of high amounts of saturated fatty acids, which will take a solid form at room temperature, whereas oils consist mainly of unsaturated fatty acids, which will take a liquid form at room temperature. Fats and oils do not mix with water. Fats and oils have many domestic and industrial uses. For example, soap can be made from tallow or coconut oil and a substance called a strong alkali, such as sodium hydroxide. Soap molecules have water-loving heads and water-repelling tails. In dirty water, tails gather around dirt particles and thus the tail dissolves in the water and is carried with it [5].

Fatty Acids in Common Plant-Based Oils

Fatty acids are the building blocks of triglycerides. They generally contain four to 22 carbon atoms and have a linear structure. Sometimes fatty acids are called "short-chain" (4-8 carbon atoms), "medium-chain" (10-12 carbon atoms), and "long-chain" (14 or more carbon atoms). The following fatty acids are most common in plant-based oils:

- Saturated fatty acids: Lauric (C12), Palmitic (C16), Stearic (C18), Arachidic (C20), Behinic (C22)
- Unsaturated fatty acids: Oleic (C18:1) Linoleic (C18:2) Linolenic (C18:3)

Oleic acid, which has one double bond, is called a "monounsaturated fatty acid," while linoleic and linolenic acids are called "polyunsaturated fatty acids" because they contain more than one double bond (2 and 3, respectively) [4].

Saturated and Unsaturated Fatty Acids

Nowadays, the annual demand for meat is increasing due to population growth and increasing income, education level and nutritional adequacy awareness. Meat is enriched with proteins, amino acids, fats, minerals and other organic components that are important for the development of the body. However, some aspects of human health must be considered, especially the composition of fatty acids in meat. Previous studies have reported that red meat causes some physical ailments such as heart problems, hypertension and cholesterol, and this is associated with the meat's higher content of saturated fatty acids compared to unsaturated fatty acids. Therefore, it is important to regulate meat consumption in appropriate portions, as inappropriate proportions lead to a negative impact on human health [6]. High saturated fatty acid (SFA) consumption is closely associated with serum low-density lipoprotein (LDL)-cholesterol concentration, which is a risk factor for coronary heart disease [7]. Saturated fatty acid has a stable polyhydrocarbon chain, it cannot be easily oxidized and can turn into unsaturated fatty acid or other harmful chemicals. Saturated fatty acid is found in animal products (e.g., meat, cheese and milk) or plant products (e.g., palm oil and palm kernel oil). Saturated fatty acid can also increase LDL cholesterol, which directly increases total cholesterol, which consists of LDL and HDL [8].

Unsaturated fatty acid is essential for human health, for example, it lowers blood pressure, has better nutritional content and is an antioxidant. Monounsaturated fatty acid (MUFA) and some polyunsaturated fatty acids (PUFA) are antithrombonenic. Ruminant fats contain higher SFA than PUFA compared to SFA in non-ruminant fats due to hydrogenation of the rumen unsaturated fatty acid load. Therefore, the strategy of increasing the ratio of PUFA to SFA in intramuscular fat will ultimately increase consumers' perspective on healthy meat. Cholesterol is a vital substance in the regulation of body life, so it is found in all cells of the body. In humans, cholesterol plays an important role both as a precursor of hormones and other molecules and as a structural element of cell membranes. According to Tabas [9] cholesterol, cholesterol metabolites and biosynthetic precursors of cholesterol play important roles in cell membrane physiology, absorption of dietary nutrients, reproductive biology, responses to stress, salt and water balance and calcium metabolism. However, cholesterol levels in human blood greater than 250 mg/dL can have a negative impact on human health. Meat, especially red meat, is widely identified as the main source of dietary cholesterol [6].

Typical Behavior of Fatty Acids

Typical behavior of saturated and unsaturated fatty acids is described below.

Unsaturated Fatty Acids

Unsaturated fatty acids are unstable and very susceptible to oxidation even at ambient temperatures. They tend to [4].

- oxidize easily when exposed to air or oxygen.
- creates Aldehydes, ketones etc. creates.
- primarily forms oxidative polymers.
- forms cyclic compounds.

Saturated Fatty Acids

Saturated fatty acids are stable in dry and cold atmospheres. However, the presence of additional functions or unsaturated bonds in fatty acid chains can make them chemically reactive towards oxygen and moisture in ambient air.

The general rule is that fatty acid samples should be kept at temperatures between -20°C and -70°C and the samples should be protected from direct exposure to light. Samples should be stored in a closed container under N_2, He or Ar to protect against oxygen and moisture [10].

Goals of Proper Oil Processing

The aim of proper oil processing is to obtain finished oil with the following properties [4]:

- Long oxidative stability,
- Long thermal stability,
- Long flavor stability,
- Long storage stability and
- Long shelf life of food products formulated with oil.

It is critical that processors understand the basic components of the oil, its properties, and how to maintain process conditions that provide oil with the quality standards previously listed.

Non-Triglyceride Components of Oils

As previously mentioned, crude plant-based oils generally contain 96-98% triglycerides. Although these components are present in small amounts, they can be very effective in determining the overall stability and performance of the oil. These can be grouped as follows [4]:

- Non-major triglyceride components
- Minor nontriglyceride components

Major Nontriglycerides

The following components are generally found at high levels in crude oil and can be measured as a percentage [4]:

- Phospholipids
- Free fatty acids (FFA)
- Diglycerides
- Monoglycerides

Hydrable and Non-Hydrable Phospholipids

There are two types of phospholipids in crude oils, depending on their proximity to water [4]:

- Hydrable phospholipids
- Non-hydratable phospholipids

Treatment with water at 140-158°F (60-70°C) hydrates some of the phospholipids in crude oils, which precipitate or can be separated by centrifugation. For example, 600-800 ppm phosphorus in crude soybean oil can be reduced to 200 ppm or less with a simple water degumming process. Phospholipids that cannot be removed by water alone are considered "non-dewaterable." The purpose of acid pretreatment of crude oil is to convert non-hydratable phospholipids into hydratable forms by retaining (removing) absorbed divalent cations (such as calcium and magnesium metals) that inhibit their hydratability (Gupta, 2017).

Free Fatty Acids

Fatty acids separated from triglyceride molecules are called "free fatty acids," "FFA" and are divided into two parts: the link-accepting hydrogen ion and the link-giving residue. FFA formation in the oil of stored oilseeds is a natural event initiated by "lipase" enzymes. Crushing of the seed and subsequent small amounts of FFA are also formed during processing and storage of crude oil. Fatty acids attached to triglycerides are still reactive in oxidation and hydrogenation processes. FFA amounts in crude oil vary depending on the type of oil and the history of the sample. Typical FFA values in selected crude oils are shown in Table 1 [4].

Table 1. Typical Free Fatty Acid (FFA) Content of Common Crude Plant-bases Oils [4]

Oils	FFA content (%)
Most seed oils	0.5–1.5
Crude palm oil	1–4
Raw cottonseed oil	0.5–3
Virgin olive oil	<0.8
Extra virgin olive oil	<2

Monoglycerides and Diglycerides

Mono- and diglycerides are generally produced by the interesterification (glycerolysis) of triglycerides with glycerol. Monoglycerides can be separated from di- and triglycerides and glycerol by high-molecular thin-film molecular distillation. The appearance of monoglycerides varies from a pale straw to brown oily liquid to a white or slightly off-white waxy solid. Antioxidants generally have a phenolic structure and are good free radical scavengers, inhibiting or interfering with the free radical mechanism of auto-oxidation. Bread is by far the largest application for monoglycerides in food technology. Any function of monoglycerides and other emulsifiers in baking depends on the dispersibility properties of the emulsifiers during dough mixing. Table margarines and low-fat spreads are widely used in the food industry. Industrial applications such as special cake, cream and puff pastry margarines are also known [11].

Minor Nontriglycerides

Minor nontriglyceride components of crude oil at parts per million include [4]:

- Tocopherols and tocotrienols
- Sterols and sterol esters
- Volatile and non-volatile compounds formed from the decomposition of triglycerides
- Color compounds
- Trace metals

Tocopherols

Tocopherols are antioxidants naturally found in plant-based oils and are one of nature's protection against fat oxidation. There are four types of tocopherols: alpha, beta, gamma and delta. Sometimes these forms are identified by the Greek letters α, β, γ and δ, respectively. Alpha (α) tocopherol protects the oil against photooxidation (oxidation under visible light). The functions of beta (β) tocopherol, which is found in very low concentrations in oils, are not fully known. Gamma (γ) and delta (δ) tocopherols protect the oil against autoxidation. Autoxidation is the primary pathway for fat oxidation, and fat degradation occurs even in the absence of light. This type of oxidation process occurs during processing, storage, distribution of fat and fat-containing food ingredients, as well as the production and storage of food products. The reaction is initiated by the formation of a free radical from the unsaturated oil by a metal initiator. The reaction continues and propagates until the oil runs out of oxygen or unsaturated fatty acids [4].

Sterols and Sterol Esters

Sterols are ubiquitous in eukaryotic cells but not in prokaryotes [12]. Sterols are an essential component of cell membranes, and both animals and plants produce them. They play a key role in cell membrane function. The sterol ring is common to all sterols; the differences are in the side chain. Plant sterols include a wide variety of molecules that are structurally similar to cholesterol; prime examples are 4-desmethyl sterols. Plant sterols are C-28 or C-29 sterols and differ from cholesterol (C-27) by the presence of an extra methyl (campesterol) or ethyl (sitosterol) group on the cholesterol side chain [13]. Although over 40 plant sterols have been identified from seven different plant classes [14] campesterol (C-28), stigmasterol (C-29), and especially β-sitosterol (C-29) are the most abundant [15].

The terms plant sterol and phytosterol are sometimes used as generic terms to include both unsaturated sterols and saturated stanols but are used here to specifically refer to unsaturated compounds. Saturated plant sterols, called plant stanols, do not have double bonds in the ring structure. Plant stanols are produced by the hydrogenation of sterols and are not abundant in nature. Sitostanol and campestanol are saturated plant sterols that are found in nature in much lower amounts than plant sterols [16].

Cholesterol in foods is found either as free alcoholic sterol or cholesteryl esters, whereas plant sterols are found as free plant sterols, esterified plant sterols, plant sterile glycosides and acylated plant sterile glycosides. Clearly, the chemical, physical, and nutritional properties of these classes of phytosterols can be very different. Different fractions are assumed to be located in different parts of the plant cell. Free plant sterols are a structural part of the cell wall. Plant sterile esters are generally believed to be storage products. They can be found in the cytosol of plant cells, in droplets or vesicles. The largest amount of plant sterile glycosides has been found in the microsomal fraction of the plant cell, and acylated sterile glycosides are believed to exist in the mitochondria [17].

Volatile and Non-Volatile Compounds

Autoxidation produces numerous oil degradation products, including [4]:

- Primary oxidation products, such as peroxide value (PV).
- Secondary oxidation products, e.g., aldehydes, ketenes, etc.
- Tertiary oxidation products, such as alcohols, acids, oxidation polymers, epoxides, cyclic fatty acids, etc.

Most of these compounds have low molecular weight and volatilize when the oil is heated. However, some fatty acid derivatives are too large and do not evaporate. These compounds have different effects on oil and product aromas and their stability.

Color Compounds

The main color compounds in plant-based oils are carotenes and chlorophylls, but other chromophoric compounds are also present. Among plant-based oils, palm oil contains the highest amount of carotene. However, soybeans and canola contain the highest amount of chlorophyll. Most of the carotenes are removed from the oil by heat bleaching in deodorization, which will be described later. Most of the chlorophyll is removed from the oil during the bleaching process using bleaching clay. Most of the carotenes are retained in deodorized palm oil, called "red palm oil," using a very special process. This oil is sold as a naturally carotene-rich oil. The carotene content of this oil is 500-600 ppm compared to 600-800 ppm in crude palm oil (CPO). The benefits of carotenes on human vision have been demonstrated in human studies conducted in India and the Far East, and red palm oil is promoted for this nutritional property. This oil also has a higher tocopherol and tocotrienol content than conventionally processed palm oil or palm olein [18].

Trace Metals

Trace metals are undesirable in processed oils because they initiate the autoxidation reaction and shorten the storage stability of oils and food products formulated with them. The most common metals found in crude oil are iron, calcium, magnesium, and sometimes very low levels of copper. Toxic "heavy metals" can also be found in very low concentrations in crude oils. Trace metals are removed from the crude oil by bleaching clay and bound by citric acid after deodorization [4].

Oil Analysis Used in Plant-Based Oil Industry and Their Importance

Quality control and quality assurance testing are becoming increasingly important at every point in the food supply chain, from production and packaging to distribution and retail. The focus on effective and reliable food analysis has increased over the last few years as cases of food fraud and adulteration have come to light. Plant-based oil is analyzed at various stages of processing. Each analysis provides specific information to both the processor and the users [19].

Plant-Based Oil Production Technology

To produce plant-based oils, oil must first be extracted from oil-bearing plant components, typically seeds. This can be done through mechanical extraction using an oil mill or chemical extraction using a solvent. The extracted oil can then be purified and refined or chemically modified if needed.

Plant-Based Oil Production

The plant-based oil processing industry involves the extraction and processing of oils and fats from vegetable sources. Plant-based oils and fats are used primarily for human consumption, but are also used for animal feed, medicinal purposes and some technical applications. Oils and fats are obtained from various fruits, seeds and nuts. Preparation of raw materials includes peeling, cleaning, crushing and conditioning. Extraction processes are usually mechanical (boiling for fruits, pressing for seeds and nuts) or involve the use of solvents like hexane. After boiling, the oil is skimmed off; after pressing, the oil is filtered; and after solvent extraction, the crude oil is separated, and the solvent is evaporated and recovered. Residues are conditioned (e.g., dried) and reprocessed to obtain by-products such as animal feed. Crude oil refining includes resin removal, neutralization, bleaching, deodorization, and advanced refining [20, 21].

Plant-Based Crude Oil Refining

Chemical refining is a traditional method used since ancient times. It can be used for all fats and oils, even if they are slightly spoiled. Each step of the refining process has specific functions to remove some undesirable compounds. Chemical refining generally follows six processes [22].

- Degumming to remove phospholipids and mucilaginous gums.
- Neutralization that eliminates free fatty acids (FFA), phospholipids, metals and chlorophylls
- Washing and drying to remove soap and water residues
- Bleaching aimed at removing pigments, peroxides and residues of both fatty acids and salts
- The main purpose is to remove wax in wax-rich oils.
- The final stage of chemical refining is the deodorization process, which removes volatiles, carotenoids and free fatty acids.

However, chemical refining has several disadvantages, as each processing step also participates in the removal of certain bioactive molecules. They mainly consist of tocopherols and polyphenols, which can act as antioxidants. Likewise, chemical refining requires higher costs and may result in the release of polluting waste [23].

Degumming

Degumming is a very important step in the refining process of plant-based oils. It enables the elimination of "gums" or "mucilage," which consists of compounds such as carbohydrates, proteins and trace metals, as well as phospholipids in crude oil. Phospholipids or phosphatides occur naturally in oils. These compounds are important biochemical intermediates in the growth and functioning of plant cells. Phosphatidylcholine (PC), phosphatedylethanolamine (PE), phosphatidylserine (PS) and phosphatidylinositol (PI) are the main types of phospholipids. In general, plant-based oils contain two types of phospholipids: hydratable and non-hydratable. These compounds can trap metallic ions (copper + iron) and prevent their catalytic activity related to free radical production in crude oils [24].

Moreover, the presence of these compounds in crude oils poses many problems for storage and processing. Phospholipids are often associated with heavy metals that are catalysts in oxidation reactions and can sometimes act as pro-oxidants in plant-based oils. Failure to completely remove phosphorus-rich components during alkaline neutralization creates a number of subsequent refining difficulties that result in a dark color during storage. Therefore, it is imperative that they are eliminated from crude fats. In fact, the degumming stage consists in eliminating all compounds (phospholipids, glycolipids, proteins, etc.) that are prone to becoming insoluble through hydration. There are four types of degumming processes: "water degumming, acid degumming, dry degumming and enzymatic degumming." Chemical refining usually begins with the degumming step. Pre-degumming conditioning consists of separating anhydrous phospholipids by mixing the oil with a small amount of acid, such as phosphoric or citric acid [25].

Some oils may be degummed beforehand with water to remove hydratable phospholipids. Recovered gums represent crude lecithin. Dry degumming process is recommended for oils with low phospholipid content. This technique uses a concentrated acid (phosphoric or citric) along with bleaching earth (1 to 3 g/100 g). The acid (0.05 to 1.2 g/100 g) is dispersed in oil at 353°K (80°C). This acid breaks down unhydrated phosphatides into phosphatidic acid and is eliminated by centrifugation. The remaining amount is further adsorbed through the bleaching earth. Dry degumming has been developed for palm, palm kernel and coconut type oils that contain small amounts of phospholipids. The degumming process combines the acid degumming step with the bleaching process, eliminating the addition of water and centrifugation of the gums. This technique is performed under low pressure at a temperature of 393 to 413°K (120-140°C). The main advantage is that no aqueous waste is generated except for the use of water in the vacuum system. The latest process in degumming is the enzymatic degumming process, which offers many benefits. Enzymatic purification is a type of biotechnological process in which a phospholipase, specifically phospholipase

C, converts anhydrous phospholipids into lysophospholipids. These components are insoluble in oil and must be removed by centrifugation [26].

Enzymatic degumming is a relatively new technology that has emerged in the last 20 years. The process was first reported in the 1990s by Roehm and Lurgi, citing the "EnzyMax Process" project. The enzyme has been used to hydrolyze non-hydratable phospholipids into their hydratable forms. Since then, recent studies have focused on the use and development of new enzymes to optimize the reduction of phosphorus levels. During the process, the crude oil is pretreated with a combination of citric acid and caustic soda. It is then mixed with water and an enzyme (Lecitase Ultra) using a high shear mixer. This forms a very stable emulsion that can be broken by centrifugation and subsequently separated from the oil into phospholipids and mucilaginous substances. Enzymatic degumming is a unique process that is very different from acid degumming. In fact, both hydratable and non-hydratable phospholipids present in oil are hydrolyzed to the corresponding lysophospholipids [26].

Alkali Refining (Neutralization)

The free fatty acid content is expressed as g/100 g of oleic acid, except for some oils such as palm oil, where it is calculated as g/100 g of palmitic acid, and coconut and palm kernel oils, where it is calculated as g/100 g of oleic acid. Acidity depends on the nature of the oil, which depends on its geographical origin, harvest, seed breaking conditions and storage time. It varies between 0.7 and 10 g/100 g, especially for some spoiled oils. Plant-based oils that contain high amounts of free fatty acids (via hydrolysis and/or oxidation) must be refined to be edible. The presence of these compounds in crude oils causes many problems in terms of storage and causes undesirable color and odor in the final product. Free fatty acids affect the chemical quality and organoleptic instability of the oil. Many methods have been developed to eliminate fatty acids to increase the value of degraded plant-based oils. Some include chemical treatment with caustic soda neutralization and physical treatment based on steam distillation. In chemical refining, the oil is treated with an alkaline solution (caustic soda), which reacts with the free fatty acids (FFA) present and converts them into soap stock according to the following equation [27].

$$R - COOH \text{ (acid)} + NaOH \text{ (base)} \longrightarrow R - COONa \text{ (soap)} + H_2O \text{ (water)}$$

Additionally, when neutralization is not done properly, caustic soda not only neutralizes fatty acids, which is the desired goal, but it also attacks the neutral fat in a manner called "parasitic" saponification. Therefore, the concentration and dosing rate of caustic soda need to be calculated based on the FFA of the degummed oil for optimal neutralization.

Triglyceride + Caustic Soda ⟶ Soap + Glycerol

From an economic point of view, caustic soda neutralization cannot be applied to oils with a percentage of free fatty acids greater than 15 g/100 g, because the loss of neutral oil in soaps becomes very significant. Additionally, other compounds can be removed in this step. These are excess phosphoric acid, residual proteins, residual gums, carbohydrates, oxidation products from FFA, trace amounts of metals and pigments, which facilitate bleaching and deodorization. Additionally, some of the bioactive molecules in the oil that can act as antioxidants, such as tocopherols and polyphenols, are also removed. The resulting soap is generally insoluble in oil. Therefore, it can be easily separated mechanically from oil based on the specific gravity difference between soap and neutral oil. The separated oil is then washed with water to remove soap, alkaline solution and other impurities, making it ready for color removal or deodorization. Soap stocks are highly alkaline (pH = 10–11). In addition to sodium soap and caustic soda, they contain water, salt, sodium phosphates, gums, carried neutral oil, coloring agents, oxidation byproducts and various contaminants. This residue is referred to as residual oleins, which are processed and valued in a special unit. Soap stocks have multiple-end uses. These include applications in lipochemistry as an alternative feedstock for biodiesel production and as an ingredient in animal feed [28].

Bleaching

Bleaching is a critical step in the refining process of oils and is usually preceded by degumming, neutralization and drying processes. Bleaching is a complex physical and chemical process used in refining plant-based oils. The purpose of bleaching (or decolorization) is to reduce the levels of colored pigments (carotenoids and chlorophylls). It also removes residual traces of phosphatide, soap, phospholipid contaminants, lipid peroxidation products and other impurities. Finally, it indirectly affects the color of the deodorized oil. Adsorption bleaching clays, activated carbon, special silica or a combination of these are used to achieve bleaching. Bleaching earth is the most popular absorbent for oil decolorization and the most widely used adsorbent material by the oil industry. Bleaching clay is preferred over other adsorbents such as silica-based and activated carbon products due to its low cost and relatively high adsorption capacity [29].

In fact, bentonite is the most preferred bleaching clay used in the oil industry. In general, activated earth does not have bleaching properties in its natural state. Their chemical composition does not indicate that they can be activated. In fact, activation is the transformation of silicates into colloidal silica, which has a significant adsorbing power. Activation is the chemical reaction of strong inorganic acid (sulfuric or hydrochloric acids) at temperatures between 353°K (80°C) and 403°K (130°C). Chemical treatment significantly changes its textural properties. Strong acids act by replacing cations with protons, especially when increasing the adsorbing surface. The quality of the earth depends on the amount and nature of the acid used, contact time and temperature. The degree of bleaching depends on the level of cation substitution by hydrogen ions of the acid in the clay structure, according to the following equation [30].

$$\text{Cation- Clay} + 2H^+ \longrightarrow \text{H- Clay} + \text{Cation}$$

For cost control reasons, sulfuric acid is preferred over hydrochloric acid. New evidence has shown that other acids (mainly phosphoric, acetic and oxalic acid) can be used to activate bentonite clay for the removal of organic dyes (Mordant Red 73). Also, acid-activated bleaching earth is sometimes called bentonite. In the final stage, the active soil is washed to remove as much as possible acid and metallic salts resulting from acid action. They are then dried and crushed. Another important adsorbent (powder or granule) in the bleaching process is activated carbon; It is an excellent adsorbent to remove unwanted color or other organic contaminants such as polycyclic aromatic hydrocarbons, benzo(a)pyrene, benzo(a)anthracene, benzo(b)fluoranthene and chrysene. Polycyclic aromatic hydrocarbons (PAHs) are a class of universal chemical contaminants in plant-based oils [31].

On the other hand, activated carbon removes residues of some polyaromatic hydrocarbons of mineral oils, among other products. Because they have long been used in the sugar industry, research into their adsorbent properties dates back to World War I, when they were used in gas masks. Activated carbon is produced from various carbon-containing substances. Activated carbon contains 95 to 98 g/100 g of carbon and is characterized by its porosity. It is accepted that the specific surface of 1 g of activated carbon reaches 600 to 1000 m^2. Activation improves capillary structure and opens tar-clogged pores. Preparation of activated carbon consists of two main steps. The first is carbonization, which is the thermal decomposition of base materials in an inert atmosphere (temperature 1073°K (800°C)), and the second is the activation of carbonized products. This activation can be achieved by physical and/or chemical treatment to expand the diameter of small pores and create new pores. In order to obtain high adsorption capacity in the bleaching of some oils, a mixture of activated carbon and bleaching earth is used in the refinery industry. In general, the amount of activated carbon should be in the range of 5–10 g/100 g, depending on the amount of bleaching earth. This technique is well documented in the literature. The general bleaching method occurs through the adsorption of pigments on an adsorbent material. In general, when an absorbent comes into contact with oil, the absorbent attracts colored pigments and other compounds to its surface that need to be eliminated. Attraction condenses the molecules, and they form a sheath in which the concentration of a substance adsorbed in the oil differs from the initial concentration. Langmuir's and Freundlich's equations below give the theoretical adsorption capacity [32].

$$\text{Langmuir}\ \frac{\text{Xe}}{(x/m)} = \frac{1}{a} + \frac{b}{a} \cdot \text{Xe},$$

$$\text{Freundlich}\ \text{Log}\left(\frac{x}{m}\right) = N \cdot \log(\text{Xe}) + \text{Log}(K)$$

Here m is the amount of adsorbent, x is the amount of adsorbed substance, when equilibrium is reached, the absorbent no longer acts on the oil, but only changes its color. The bleaching process is carried out under vacuum for 20–40 minutes in the contact range of 353–393°K (80–120°C). Generally, the process is carried out under a slight vacuum to prevent oxidation, which is enhanced by the distribution of oil on soil particles. The amount of adsorbent used varies between 0.1 and 3 g/100 g, depending on the crude oil quality. However, other higher percentage bleaching materials can be used to meet final color requirements. The pretreated oil is heated to 363–383°K (90–110°C) under vacuum and then vigorously mixed with the adsorbent (bleaching earth or/or activated carbon) in bleach. After a retention time of 20-40 minutes, the oil adsorbent mixture is filtered. Only filters were used to separate the oil adsorbent. Because the centrifuge process is not suitable for this separation. Therefore, short filtration times, efficient filtration and minimization of oil retention on the adsorbent material are required. After this filtration, the oil is now ready for deodorization [30].

Odor Elimination

The final stage of refining involves high temperatures and requires great care. Deodorized oil is devoid of any taste, including pleasant ones. Deodorization is a simple distillation process. This process allows the elimination of free fatty acids and removes odors, different unpleasant components, pollutants (pesticides, light polycyclic aromatic hydrocarbons and other volatile components). On the other hand, odor removal also has other negative effects. Among these, important bioactive molecules such as tocopherols, squalene, sterols and polyphenols can be extracted in this step. The most important are undesirable side reactions such as destruction of some essential nutrients, cis-trans-isomerization, conjugation and polymerization. It also sometimes produces other undesirable compounds such as 3-MCPD-esters. The latter was identified as a new class of oil refiner contaminants. Careful execution of this process improves the oil's stability and color while preserving its nutritional value. Deodorization is a vacuum steam distillation. It does not contain any technological aids and proceeds with a simple injection of water vapor into the oil. The process involves passing steam through layers of oil held in trays and heating it to high temperatures of 453–513°K (180–240°C) using a high-pressure steam boiler. Using a very high vacuum of 2 to 8 mmHg, the process eliminates unpleasant odors caused by aldehydes, ketones, alcohols, short-chain fatty acids and heat-resistant pigments. It is the stripping of substances that carry taste and odor, which are more volatile than oil, with steam. The thermodynamic equilibrium of fat and solute (flavor-releasing substance) is given by Raoult's law [33].

$$\frac{P_{vo}}{P_v} = \frac{V}{H}$$

P_{vo} is the partial pressure of the volatile component dissolved at a given temperature, P_v is the partial pressure this component will have at the same temperature, V is the number of moles of volatile components, and H is the number of moles of oil. The resulting oil is then conditioned under nitrogen to protect it from oxidation. Careful execution of these processing steps ensures that fully refined oils have good organoleptic and physicochemical qualities [12].

Wax Removal or Winterization (Cooling)

Waxes are esters of long-chain primary alcohols and long fatty acids. These acids have low solubility in oils, high melting properties, and crystallize at low and room temperatures, usually in winter. Wax generally does not adversely affect the functionality of the plant-based oils. The presence of balwax affects the quality of the oil and gives a cloudy appearance, especially in winter. Its hazy appearance is due to the precipitation of dissolved waxes. The wax removal process is also called winterization. The term "winterization" occurs when the waxes in the oil crystallize during the winter months when the temperature is low, thus giving the oil a hazy appearance. This process is relevant to several types of wax-rich oils, such as rice bran, canola, corn oil, sunflower and olive pomace. The waxing process involves three main process steps. In the first, the bleached oil must be heated to 328°K (55°C) to ensure that the oil is completely liquid. Second, the oil is slowly cooled to 283–288°K (10–15°C). Ideally, the cooled oil is kept at this temperature for several hours. In the third step, after crystallization is complete, the cooled oil is pumped into a filter machine to separate the wax from the plant-based oils. Filtration yields a clear oil and by-product waxes [34].

Hydrogenation

Hydrogenation is complex in nature and a heterogeneous reaction process that occurs between hydrogen (gas phase) and unsaturated fatty acids (liquid phase), converting some or all of the unsaturated fatty acids into saturated fatty acid (stearic acid). This includes the following physical and chemical reactions:

- Mixing and dispersing the catalyst with a mechanical mixer.
- Diffusion of hydrogen gas from the oil mass to the catalyst surface.

- Adsorption of reactants (hydrogen and unsaturated fatty acids) on the catalyst surface.
- Partial or complete saturation of unsaturated fatty acids on the catalyst surface.
- Desorption of reaction products and unreacted fatty acids and oil molecules from the catalyst surface.
- Heat release due to the exothermic nature of the hydrogenation reaction.

Hydrogen gas and unsaturated fatty acids disperse throughout the bulk of the oil and reach the catalyst surface. These reactants pass through most of the oil by mechanical agitation. Two reactants, hydrogen and unsaturated fatty acids, diffuse through a stagnant oil microfilm on the surface of the catalyst to reach the active sites on the catalyst. The reaction between unsaturated fatty acids and hydrogen gas takes place on the catalyst surface. This is sometimes called the chemisorption process, which refers to a chemical reaction supported by the adsorption process. The reaction products, which are typically saturated fatty acids, unsaturated fatty acids (cis and trans isomers), and triglyceride molecules, leave the catalyst surface through the desorption process. More fresh reactants are absorbed on the surface of the catalyst and the hydrogenation reaction continues. Heat is produced in the reaction. The average heat of reaction in hydrogenation is 1.65 Btu/lb oil for one unit decrease in iodine value (IV). This is a valuable source of process heat and is recovered through a heat recovery system in many facilities [4].

Physical Refining

The process consists of the same steps described in chemical treatment, except for the alkaline neutralization process. Chemical refining involves removing free fatty acids by adding caustic soda and separating the soap by centrifugation (mechanical separation), while physical refining removes free fatty acids and other compounds through steam distillation in the final step. This process is also known as steam refining. Therefore, physical refining of crude oils overcomes the disadvantages of neutralization with sodium hydroxide. As a matter of fact, this process, which is considered environmentally friendly, minimizes the formation of liquid waste [35].

Another advantage of this process over chemical treatment is that it is more economical (e.g., less chemicals used, lower investment cost, less energy input and better efficiency). However, this process is not suitable for all oil types as it is extremely sensitive to crude oil quality. Physical refining is used for oils with high acidity. Considering the phospholipid content, Dumont and Narine [28] proposed two physical refining processes depending on the phospholipid content in the crude oil. In general, physical refining includes the following three main processing steps:

- Degumming to remove phosphatides
- Bleaching and filtration to remove color pigments
- Deodorization ensures the elimination of free fatty acids and other volatile compounds.

Sometimes dewaxing is added for wax-rich types of oils (e.g., high-melting-point esters of fatty acids and fatty alcohols that have low solubility in oils), such as corn, rice bran, canola, and sunflower oils. The principle of physical purification includes the following three important steps. The first step consists of subjecting the oil to phosphoric acid in a short-mix chemical refining process. It is the most important step for degumming and should be done carefully. An analysis test called "Degummimg Efficiency" was developed to evaluate the degummimg effectiveness of a particular refined oil sample. This test is determined by the following formula

$$\text{Degumming Efficiency"}\left(\frac{\text{g}}{100\,\text{g}}\right) = \frac{(P_0 - P_d)}{P_0} * 100$$

Here P_0 is phospholipids (ppm) crude oil and P_d is phospholipids (ppm) degummed oil. Bleaching (or decolorization) is the second step. The purpose is the same as chemical treatment. The main purpose of bleaching is to reduce the levels of some colored pigments (carotenoids and chlorophylls), but it also removes phosphatide residues, phospholipid residues, pollutants, lipid peroxidation products and other impurities. The oil is then mixed with acid-activated bleaching earth or another adsorbent. The standard bleaching process temperature is 368–378 °K (95–108 °C). The spent adsorbent along with some precipitated carotenoids and other impurities are then removed by filtration. After this step, the oil is ready for deodorization. The final step in the physical refining of oils is deacidification and deodorization. Oils are deodorized under the conditions described above for chemical purification. Odor removal has three main purposes:

- Removal of volatile components such as free fatty acids, different flavors, pollutants (pesticides, light polycyclic aromatic hydrocarbons, etc.). However, some bioactive components (tocopherols, sterols, etc.)
- Thermal bleaching of colored pigments and peroxides
- To achieve better quality, more acceptable appearance and greater flavor stability throughout shelf life.

The deodorization process is completely determined by four process parameters: amount of stripping steam, time, pressure and temperature. Odor removal is usually performed at high temperatures (>473°K) (>200°C) with low vacuum pressure. The use of high temperatures and vacuum often results in the formation of negative byproducts. The effects of process conditions on standard quality parameters and nutritional quality of oil are well documented in the literature. In conclusion, physical refining offers numerous advantages over chemical refining for plant-based oils [36].

Conclusion

Plant-based oils and fats are important components of foods and are essential components of our daily diet. Plant-based oils are obtained by mechanical extraction or solvent extraction of oilseeds (soybean, rapeseed, sunflower, etc.) or oily fruits such as palm and olive. Plant-based oils generally contain triglycerides (about 98 g/100 g) (3), triesters formed by the reaction between glycerol and fatty acids, and small amounts of other substances. Some of these, such as diglycerides, vitamins, phytosterols, tocopherols and polyphenols, have important health benefits in humans 6) and therefore should not be removed during processing. Other compounds known to have adverse effects on the quality and stability of oils include free fatty acids, unsaponifiables, waxes, pigments, solid impurities (mostly fibres), oxidation products (peroxides, aldehydes, ketones, alcohols and oxidized fatty acids). These compounds are non-toxic. However, their presence in oils is undesirable as they affect stability and sensory acceptability by consumers. Indeed, these molecules can impart bad taste and odor and affect the functional properties of the oil. Plant-based oils may also contain some contaminants: "pesticides, trace metals, mineral oil aromatic hydrocarbons (MOAH), aflatoxins, dioxins, polycyclic aromatic hydrocarbons (PAH) and residues of organic solvents." The origin of these pollutants can be attributed to the environment in which oil crops are grown, seeds are transported and stored, and crude oils are processed and stored. Today, there is global pressure from both consumers and food industries for the quality of refined oil to meet the specifications set for food safety. The oil should be odorless, fairly neutral in taste, clear and colorless, and free of contaminants. The process that removes unwanted and toxic components from crude oils is known as "refining". Refining is practically essential for crude oils that cannot be consumed as virgin oil to provide a product with an attractive appearance, a neutral taste, and greater resistance to oxidation. Likewise, it allows obtaining oils more suitable for various industrial uses and minimizing oil loss during processing and getting rid of undesirable substances such as pesticide residues, metal residues, polycyclic aromatic hydrocarbons, dioxins and alteration products.

Although refining extends the shelf life of oil, it has several disadvantages. One of the main disadvantages is the loss of substances responsible for healthy, pharmaceutical properties and technological interest in oils, such as tocopherols, phospholipids, squalene, polyphenols and phytosterols. Another important disadvantage of refining is the formation of undesirable compounds such as glycidyl esters, 3-MCPD-esters, harmful trans-fatty acids and polymeric triacylglycerols. These can directly affect the safety level of refined oils. Various studies have been conducted to determine the effects of refining on minor bioactive compounds such as sterols and tocopherols. Therefore, tocopherols are significantly reduced in plant-based oils and directly affect the reduction of the shelf life and nutritional quality of the oils. In this context, such a refining process for crude oils should be done to remove undesirable compounds and cause the least possible damage to the desired components. It is also important to minimize fat and produce lower levels of harmful compounds such as 3-MCPD-esters and unhealthy trans-fatty acids.

A good understanding of the chemistry and processing of plant-based oils is crucial to defining quality products for both industry and consumers. The aim of both types of refining is to achieve better quality, a more acceptable appearance, a lighter odor or color, a longer stability and good safety by the elimination of contaminants and FFA oxidation products. The problem is that many of these processes also remove substances that contribute to the healthy properties of oils. This directly affects the stability of refined plant-based oils.

Effective refining processes must be developed to purify crude oils from undesirable compounds while causing the least possible damage to interesting components and not producing higher levels of undesirable compounds.

References

[1] Raji, A. (2018). *Extraction and characterization of vegetable oil using bread fruit seed.* Afribary. Retrieved from https://afribary.com/works/extraction-and-characterization-of-vegetable-oil-using-bread-fruit-seed-3138.

[2] Uluata, S. (2010). *Characterization of some vegetable oils.* IU Institute of Science and Technology, Malatya, Turkey.

[3] Gupta, M. K. (2017). *Practical Guide to Vegetable Oil Processing.* Academic Press and AOCS Press: New York, NY, USA.

[4] Kaya, C. (2006). Biodiesel Production from Vegetable Oils. DU Institute of Science and Technology, Diyarbakir, Turkey.

[5] Byjus (2022). Difference Between Fats And Oils. Retrieved from https://byjus.com/chemistry/difference-between-fats-and-oils/.

[6] Baltušnikienė, A., & Jukna, V. (2011). Impact of genotype on cattle growth, beef chemical composition and cholesterol level. *Veterinarija ir zootechnika, 54(76),* 14-19.

[7] French, P., Stanton, C., Lawless, F., O'Riordan, E. G., Monahan, F. J., Caffrey, P. J., & Moloney, A. P. (2000). Fatty acid composition, including conjugated linoleic acid, of intramuscular fat from steers offered grazed grass, grass silage, or concentrate-based diets. *Journal of Animal Science, 78,* 2849-2855.

[8] Tabas, I. (2002). Cholesterol in health and disease. *The Journal of Clinical Investigation, 110,* 583-590.

[9] Brondz, I. (2016). *Lipids, Fatty Acids. In: Reedijk, J., Ed., Reference Module in Chemistry, Molecular Sciences and Chemical Engineering.* Elsevier, Waltham, 2nd Edition, 76-88.

[10] Moonen, H., & Bas, H. (2014). *Mono-and diglycerides. In Emulsifiers in Food Technology,* V. Norn (Ed.). 73-92.

[11] Cheng, W. W., Liu, G. Q., Wang, L. Q., & Liu, Z. S. (2017). Glycidyl fatty acid esters in refined edible oils: A review on formation, occurrence, analysis, and elimination methods. *Comprehensive Reviews in Food Science and Food Safety, 16(2)*, 263-281.

[12] Haslenda, H., & Jamaludin, M. Z. (2011). Industry to industry by-products exchange network towards zero waste in palm oil refining processes. *Resources, Conservation and Recycling, 55(7),* 713-718.

[13] Tahuk, P. K., Dethan, A. A., & Sio, S. (2018). The composition of saturated and unsaturated fatty acids and cholesterol in meat of Bali bull fattened using greenlot system in small holder farming. *Jurnal Kedokteran Hewan, 12(3),* 66-70.

[14] Law, M. (2000). Plant sterol and stanol margarines and health. *British Medical Journal, 320,* 861-864.

[15] Jong, A., Plat, P., & Mensink, R. P. (2003). Metabolic effects of plant sterols and stanols. *The Journal of Nutritional Biochemistry, 14,* 362-369.

[16] Taşan, B., Bilgin, Ü., & Demirci, A. Ş. (2006). Phytosterols as functional food ingredients. *Journal of Tekirdag Agricultural Faculty, 3,* 153-159.

[17] Alptekin, E. (2007). Enzymatic hydrolysis of waste activated bleaching earth oil. ITU Institute of Science and Technology.

[18] Dunkle, M. N., David, F., Sandra, P., and Vollmer, M. (2013). *Vegetable Oil Analysis.*

[19] World Bank (2007). *Vegetable Oil Processing.*

[20] Vaisali, C., Charanyaa, S., Belur, P. D., & Regupathi, I. (2015). Refining of edible oils: a critical appraisal of current and potential technologies. *International Journal of Food Science & Technology, 50(1),* 13-23.

[21] Gharby, S. (2022). Refining Vegetable Oils: Chemical and Physical Refining. *The Scientific World Journal, 2022,* 1-10.

[22] Delgado, A., Al-Hamimi, S., Ramadan, M. F., Wit, M. D., Durazzo, A., Nyam, K. L., & Issaoui, M. (2020). Contribution of tocols to food sensorial properties, stability, and overall quality. *Journal of Food Quality, 2020,* 1-8.

[23] Zufarov, O., Schmidt, Š. & Sekretár, S. (2008). Degumming of rapeseed and sunflower oils. *Acta Chimica Slovaca, 1(1),* 321-328.

[24] Sharma, H., Giriprasad, R., & Goswami, M. (2013). Animal fat-processing and its quality control. *Journal of Food Processing & Technology, 4(8),* 1000252.

[25] Dijkstra, A. J. (2010). Enzymatic degumming. *European Journal of Lipid Science and Technology, 112(11),* 1178-1189.

[26] Ruiz-Méndez, M. V., Márquez-Ruiz, G., & Dobarganes, M. C. (1997). Relationships between quality of crude and refined edible oils based on quantitation of minor glyceridic compounds. *Food Chemistry, 60(4),* 549-554.

[27] Dumont, M. J., & Narine, S. S. (2007). Soapstock and deodorizer distillates from North American vegetable oils: Review on their characterization, extraction and utilization. *Food Research International, 40(8),* 957-974.

[28] Sabah, E., Cinar, M., & Çelik, M. S. (2007). Decolorization of vegetable oils: Adsorption mechanism of β-carotene on acid-activated sepiolite. *Food Chemistry, 100(4),* 1661-1668.

[29] Usman, M. A., Ekwueme, V. I., Alaje, T. O., & Mohammed, A. O. (2012). Characterization, acid activation, and bleaching performance of Ibeshe clay, Lagos, Nigeria. *International Scholarly Research Notices, 2012,* 1-5.

[30] Ma, Y., Shi, L., Liu, Y., & Lu, Q. (2017). Effects of neutralization, decoloration, and deodorization on polycyclic aromatic hydrocarbons during laboratory-scale oil refining process. *Journal of Chemistry, 2017,* 1-9.

[31] Liu, Y., Huang, J., & Wang, X. (2008). Adsorption isotherms for bleaching soybean oil with activated attapulgite. *Journal of the American Oil Chemists' Society, 85(10),* 979-984.

[32] Sherazi, S. T. H., & Mahesar, S. A. (2016). Vegetable oil deodorizer distillate: a rich source of the natural bioactive components. *Journal of Oleo Science, 65(12),* 957-966.

[33] Omar, S., Girgis, B., & Taha, F. (2003). Carbonaceous materials from seed hulls for bleaching of vegetable oils. *Food Research International, 36(1),* 11-17.

[34] Tasan, M., & Demirci, M. (2005). Total and individual tocopherol contents of sunflower oil at different steps of refining. *European Food Research and Technology, 220(3),* 251-254.

[35] Siragakis, G., Antonopoulos, K., Valet, N., & Spiratos, D. (2006). Olive oil and pomace olive oil processing. *Grasas y Aceites, 57(1),* 56-67.

Index

E

F

H

I

L

M

N

O

P

Q

R

S

T

U

V

W

Y